AF616827

Wastewater Treatment

Advanced Processes and Technologies

Wastewater Treatment

Advanced Processes and Technologies

edited by

D. G. Rao

R. Senthilkumar

J. Anthony Byrne

S. Feroz

CRC Press is an imprint of the
Taylor & Francis Group, an **informa** business

Co-published by IWA Publishing, Alliance House, 12 Caxton Street, London SW1H 0QS, UK

Tel. +44 (0)20 7654 5500, Fax +44 (0)20 7654 5555

publications@iwap.co.uk

www.iwapublishing.com

ISBN13: 978-178040-034-1

CRC Press
Taylor & Francis Group
6000 Broken Sound Parkway NW, Suite 300
Boca Raton, FL 33487-2742

Printed in the United States of America on acid-free paper
Version Date: 20120501

International Standard Book Number: 978-1-4398-6044-1 (Hardback)

Library of Congress Cataloging-in-Publication Data

Wastewater treatment : advanced processes and technologies / editors, D.G. Rao ... [et al.].
p. cm.
Summary: "Emphasizing new technologies that produce clean water and energy from the wastewater treatment process, this book presents recent advancements in wastewater treatment by various technologies such as chemical methods, biochemical methods, membrane separation techniques, and nanotechnology. It addresses sustainable water reclamation, biomembrane treatment processes, advanced oxidation processes, and applications of nanotechnology for wastewater treatment. It also includes integrated cost-based design methodologies.. Equations, figures, photographs and tables are included within the chapters to aid reader comprehension. Case studies and examples are included as well"-- Provided by publisher.
Includes bibliographical references and index.
ISBN 978-1-4398-6044-1 (hardback)
1. Sewage--Purification. I. Rao, D. G. (Dubasi Govardhana)

TD745.W36184 2012
628.1'62--dc23 2012014607

Visit the Taylor & Francis Web site at
http://www.taylorandfrancis.com

and the CRC Press Web site at
http://www.crcpress.com

Contents

Preface

The importance of wastewater treatment in the modern industrial world is very high in view of the fact that more than 97%, dormant in polar regions, of the available water is saline (in seas and oceans) and 2% of the freshwater is unavailable for human consumption. Thus, very little quantity of water is available for human consumption. The world population is increasing, and the per capita water consumption is also increasing day by day, which lays a heavy burden on science, technology, and engineering to meet the challenges of water treatment and supply in the future. Economic and social growth cannot be ensured without industrialization, which is in turn a culprit in spoiling the available water resources due to the generation of large quantities of wastewater. It is paradoxical but true. To add another dimension to the existing problem is the increased day-by-day legislative restrictions that are being imposed by various governments all over the world in view of the safety and health concerns of the citizens. Urbanization with overconcern for hygiene also generates huge quantities of wastewater that is known as *graywater*. It comes from household kitchens, toilets, and restaurants. The graywater from kitchens and restaurants is not toxic but is not suitable for human consumption. In the present complex scenario, the only alternative is to treat the available wastewater to make it as clean as possible. The treated water may not be exactly suitable for potable purpose, but can at least be used for various other purposes, viz., recycling partly for industrial purposes, steam generation, or gardening and agriculture.

The treatment of wastewater is complicated because of the heterogeneous nature of the water streams coming from the various domestic and industrial sources. The industrial sources are as diverse as drugs and pharmaceutics, pesticides, food processing, fermentation, vaccines manufacturing nuclear processing, and metallurgical and animal processing industries. The pollutants generated can be physical, chemical, and biological in nature, and they can be toxic or nontoxic. Hence, the treatment methods are also varied in nature in order to process the diverse effluent wastewaters coming from various sources.

This book is an honest attempt to present important concepts, technologies, and issues in this direction by various experts in the field of wastewater treatment. The treatment methods cover various process industries and utilize various technologies for the purpose. Chapters 2–4 deal with advanced oxidation processes including processes based on Fenton and photo-Fenton, ozonolysis, photocatalysis, and sonolysis. Various types of reactors used in wastewater treatment are dealt with in Chapters 5, 9, and 13. Microbial treatment methods, in general, for wastewater treatment are described in Chapter 6, whereas those used in various process industries are covered in Chapter 8.

Effluent treatment methods, usually practiced in food processing industries, are comprehensively dealt with in Chapter 10. Removal of low-molecular-weight substances from wastewater is a challenging task, and hence special methods for their removal are needed, which are all described in Chapter 11. Seaweeds are good adsorbents and may be applied in wastewater treatment for the removal of toxic substances (Chapter 7). The treatment of graywater needs a special attention in view of its increasing magnitude. Chapter 12 describes such treatment methods with a case study of the Muscat municipality. A special concept of central effluent treatment plants (CETPs) is gaining prominence in the treatment and release of wastewater from small-scale processing units into municipal water lines, after meeting the stringent legislative requirements. It is dealt with in the introductory chapter (Chapter 1).

All efforts have been made by the editors and authors to judiciously blend most of the treatment processes and technologies in one single book in order to make the diverse subject matter as comprehensible as possible. It is, indeed, difficult to make it concise with the whole gamut of advanced processes and technologies in a single book of this nature; hence, enthusiastic readers are advised to consult the original references for complete understanding of any process or technology. This book is ideally suited for researchers and professionals working in the area of wastewater treatment. Each chapter is specific in its own way and, hence, may cater to the requirements of professionals interested in that area. The bibliography given at the end of each chapter would act as a guide for comprehensive information in that particular area. Hence, most of the chapters end with a comprehensive list of literature references.

At the very outset, we would like to thank all our contributing authors, who have done an excellent job in drafting and delivering the chapters. The success of this publication is largely due to them. We would also like to extend our sincere thanks to the staff of the editorial and publication department of CRC Press, who have been very helpful and cooperative throughout the preparation of this material and have been largely responsible for the book in its present form. We thank all the authors, publishers, and industries whose works have been referred to and who have extended the copyright permissions to utilize their published information in this book in some form or the other. We would like to extend our sincere thanks to the executives and management of Caledonian College of Engineering, Muscat (Sultanate of Oman), and to the staff of the University of Ulster (United Kingdom), for their encouragement and support for this work. We also thank our families, who had largely extended their moral support during the last 2 years while preparing (editing) this book.

This publication is a sincere effort made by us to put in a nutshell the vast subject matter of wastewater treatment, which is so vital in the twenty-first century. We are aware of the fact that this book may not be holistic in its approach; but still we feel we are richly rewarded if the publication meets at least partly the requirements of researchers, professionals, and young students working in the area of wastewater treatment. Since this book is an edited version of the works of so many authors in the field, we are afraid that there may be some mistakes or omissions. We request the readers to kindly bring them to the notice of the editors (e-mail addresses enclosed) by contacting us with their views and positive criticisms for the overall improvement of the book.

D. G. Rao
R. Senthilkumar
J. Anthony Byrne
S. Feroz

Contributors

Mushtaque Ahmed
College of Agricultural and Marine Sciences
Sultan Qaboos University
Al-Khod, Muscat, Sultanate of Oman

Abdullah Al-Buloshi
College of Agricultural and Marine Sciences
Sultan Qaboos University
Al-Khod, Muscat, Sultanate of Oman

Ahmed Al-Maskary
College of Agricultural and Marine Sciences
Sultan Qaboos University
Al-Khod, Muscat, Sultanate of Oman

Isabel Oller Alberola
Plataforma Solar de Almería
Carretera Senés
Tabernas, Spain

J. Allen
School of Engineering
Deakin University
Geelong, Australia

Rakshit Ameta
Department of Pure and Applied Chemistry
University of Kota
Kota, India

Suresh C. Ameta
Department of Chemistry
M.L. Sukhadia University
Udaipur, India

Mahad S. Baawain
Department of Civil and Architectural Engineering
Sultan Qaboos University
Al-Khod, Muscat, Sultanate of Oman

J. A. Byrne
Nanotechnology and Integrated BioEngineering Centre
University of Ulster
Northern Ireland, UK

P. Fernández-Ibáñez
Plataforma Solar de Almería
Carretera Senés
Tabernas, Spain

S. Feroz
Caledonian College of Engineering
Muscat, Sultanate of Oman

V. Jegatheesan
School of Engineering
Deakin University
Geelong, Australia
and
School of Engineering and Physical Sciences
James Cook University
Townsville, Australia

Sourish Karmakar
School of Biochemical Engineering
Banaras Hindu University
Varanasi, India

Nikolaus Klamerth
Plataforma Solar de Almería
Carretera Senés
Tabernas, Spain

Anil Kumar
Department of Chemistry
M.P. Government P.G. College
Chittorgarh, India

Arka Pravo Kundu
Department of Mining Engineering
Banaras Hindu University
Varanasi, India

Kanika Kundu
Chemistry Section
Banaras Hindu University
Varanasi, India

Subir Kundu
School of Biochemical Engineering
Banaras Hindu University
Varanasi, India

N. Meyyappan
Sri Venkateswara College of Engineering
Sriperumbudur, Chennai, India

D. Navaratna
School of Engineering
Deakin University
Geelong, Australia
and
School of Engineering and Physical Sciences
James Cook University
Townsville, Australia

J. Naylor
School of Engineering
Deakin University
Geelong, Australia

R. Parthiban
Sri Venkateswara College of Engineering
Sriperumbudur, Chennai, India

S. Pinchon
School of Engineering
Deakin University
Geelong, Australia

P. B. Punjabi
Department of Chemistry
M.L. Sukhadia University
Udaipur, India

D. G. Rao
Caledonian College of Engineering
Muscat, Sultanate of Oman

Sixto Malato Rodríguez
Plataforma Solar de Almería
Carretera Senés
Tabernas, Spain

E. Searston
School of Engineering
Deakin University
Geelong, Australia

R. Senthilkumar
Caledonian College of Engineering
Muscat, Sultanate of Oman

H. K. Shon
Faculty of Engineering
University of Technology Sydney
Broadway, Australia

L. Shu
School of Engineering
Deakin University
Geelong, Australia

Ana Zapata Sierra
Plataforma Solar de Almería
Carretera Senés
Tabernas, Spain

C. Teil
School of Engineering
Deakin University
Geelong, Australia

Joseph V. Thanikal
Head, Built and Natural Environment department
Caledonian College of Engineering
Muscat, Sultanate of Oman

M. Velan
Department of Chemical Engineering
Anna University
Chennai, India

K. Vijayaraghavan
Institute for Water and River Basin Management
Department of Aquatic Environmental Engineering
Karlsruhe Institute of Technology
Karlsruhe, Germany
and
Singapore-Delft Water Alliance
National University of Singapore
Singapore

J. Virkutyte
Pegasus Technical Services Inc.
Cincinnati, Ohio, USA

Y. Wang
School of Engineering
Deakin University
Geelong, Australia

Z. P. Xu
ARC Centre of Excellence for Functional Nanomaterials
Australian Institute for BioEngineering and Nanotechnology
The University of Queensland
Brisbane, Australia

1

Introduction

D. G. Rao, R. Senthilkumar, J. A. Byrne, and S. Feroz

One of the greatest challenges of the twenty-first century would be to have an incessant supply of safe drinking water and clean air to breathe for the millions of living things all over the world. The major concern in this is not the depletion of air and water but the indiscriminate damage that is being done to them under the guise of industrial development. The day is not far off when they will become rare commodities. The problem being addressed in this book is concerned with the wastewater treatment.

The worldwide concern for the depletion of global water sources is rising day by day. It is more than just the depletion of sources; with the ever-increasing population and growing economy, demands for water are also continuously growing. Water sources, however, are not as abundant as they seem at first, since only in a very limited number of situations can available water be used without any treatment. A casual observation of the world map would suggest that the supply of water is endless since it covers over 80% of the earth's surface. Unfortunately, however, we cannot use it directly since 97% is in the salty seas and oceans, 2% is tied up in the polar ice caps, and most of the remainder is beneath the earth's surface. When a huge amount of water is required for different industrial processes, only a small fraction of the same is incorporated into their products and lost by evaporation; the rest finds its way into the water courses as wastewater. Wastewaters are those waters that emanate from (i) domestic sources, (ii) restaurants and establishments, and (iii) factories and industries. Of them, industries are the main polluters of natural bodies of water. Newer technologies lead to newer and more toxic wastes; these wastes take longer periods of time for decomposition, and most of the time, toxic wastes are deeply buried in the ocean or land. But this is far from a permanent solution as it degrades the earth. Newer technologies are being researched every day, but much less development has occurred in the field of waste treatment. The world depends on earth for disposal, but what will happen to earth. Little thought has been given to this. Recently, the world saw a major disaster in the Mexican Gulf, where BP (M/s British Petroleum) lost an oil well, creating an oil slick of millions of gallons and deeply endangering marine and human life nearby.

Anthropogenic activities include rapidly growing industrialization, series of new constructions, manyfold increases in transportation, aerospace movements, development and enhancement in technologies, that is, nuclear power, pharmaceutical, pesticides, herbicides, agriculture, etc. All of these are most desirable activities for human development and welfare but they also lead to the generation and release of objectionable materials into the environment. Thus, they pollute the whole environment, making our life on this beautiful earth quite miserable. The situation, if not controlled in a timely manner, could become a malignant problem for the survival of mankind on the earth. To have a neat, clean, healthy, and green environment, there is an urgent need to search for such an approach, which may be applicable at room temperature, safe to handle, economic, eco-friendly, and above all, the main requirement is that it should not be harmful to the environment in any manner.

There are many sources of water pollution, but two general categories exist: direct and indirect contaminant sources. Direct sources include effluent outfalls from industries, refineries, waste treatment plants, etc. Indirect sources include contaminants that enter the water supply from soils/groundwater systems and from atmosphere via rain water. Soils and groundwater contain residues of human agricultural practices (fertilizers, pesticides, etc.) and atmospheric contaminants that come from various human practices (such as gaseous emissions from automobiles, factories, etc.). Pollutants in water include a wide spectrum of chemicals and pathogens, with different physical chemistries or sensory changes. There are a number of ways to treat wastewaters based on the type of contaminants. These various treatment methods can be conveniently classified into the following:

1. Physical methods
2. Chemical methods
3. Combination of physical and chemical methods
4. Biological methods

In general, contaminants are categorized into two broad classes, namely organic and inorganic. Some organic water pollutants include industrial solvents, volatile organic compounds (VOCs), insecticides, pesticides, dyes, and food processing wastes. Inorganic water pollutants include metals, fertilizers, acidity caused by industrial discharges, etc. There are three alternatives for the disposal of liquid wastes:

1. Direct disposal of wastes into streams without any treatment
2. Discharge of wastes into municipal sewers for combined treatment
3. Separate treatment of industrial wastes before discharging into water bodies

The selection of a particular process depends on the self-purification capacity of streams, permissible levels of pollutants in water bodies, and the economic interests of both the municipalities and the industries. Depending on the mode of discharge of the waste and the nature of the constituents present in it, most of the treatments are based on conventional technologies, for example, equalization, neutralization, physical treatment, chemical treatment, and biological treatment.

A number of water treatment technologies are desired to at least partially cleanse the water to serve the following purposes even though it is certain that the treated water cannot be as safe and pure as freshwater for potable purposes:

1. The treated water may be used for some other beneficial purposes.
2. The effluents do not mix directly with streams, lakes, and beaches and cause them to become polluted.
3. The treated water may be used for agricultural purposes.
4. In small quantities, the treated water may be used for raising kitchen gardens, horticultural crops, etc.

Most wastewater treatment processes cannot effectively respond to diurnal, seasonal, or long-term variations in the composition of wastewater. A treatment process that may be effective in treating wastewater during one time of the year may not be as effective at treating wastewater during another time of the year. Some of the major concerns of treated water for reuse are as follows:

1. How reliable are the treatment methods so that the treated water may be reused for the intended purpose, if not directly for the potable purpose for human consumption?
2. How safe is the water for protecting public health?
3. To what extent does the treated water gain public acceptance?

Nowadays, much attention is given to the treatment of industrial wastes, due to their growing pollution potential arising out of the rapid industrialization. Streams can assimilate certain amounts of waste before they are polluted, and a municipal sewage treatment plant can be designed to handle any kind of industrial waste.

In addition to the treatment by municipalities, there is also an approach known as *common effluent treatment plants* (cetps), which is mostly in vogue in most of the industrial estates in India to treat the industrial wastewaters. These treatment plants are established in industrial areas. Effluents from some of the small-scale processing plants are transported to the CETP where they are treated to safe limits based on the following:

1. Composition of effluents
2. Type of processing plant
3. Time of delivery from processing plant

The CETPs are a wonderful concept in wastewater treatment and are especially helpful to small and medium industries that cannot afford to have a treatment plant of their own (Rao 2010, p. 410). However, such treatment plants can meet the requirements of a particular group of processing industries, namely pharmaceutical industries, textile industries, food processing industries, etc. They obviously may not be able to treat a wide range of effluent waters. But the effluents can be classified in terms of their pollutant constituents on the basis of some physicochemical parameters such as flow rate, pH, TSS (total soluble solids), COD (chemical oxygen demand), BOD (biological oxygen demand), etc. One such treatment plant in operation in India is in Vatva Industrial Estate (in Gujarat state), where the processing industries include dyes, dye intermediates, bulk drugs and pharmaceuticals, fine chemicals, and textiles. The characteristics of the effluents were consolidated by M/s Sudarshan Chemicals, Pune, based on which the design for CETP was made by M/s Advent Corporation, USA (Figure 1.1). The extended aeration technique in biological treatment process is the main criterion for treatment in this unit.

A similar kind of effluent treatment plant, operating in the Industrial estate in Pattancheru (Hyderabad, India) and catering to the needs of local bulk drug and pharmaceutical manufacturing units, is Enviro-Tech Ltd. The unit works on the dissolved air floatation principle and was supplied by M/s Krofta Engineering (Krofta Technologies Corporation, USA). A coagulant (alum) is used along with a small dosage of a polyelectrolyte to coagulate the suspended solids (Rao 2010, p. 410). A special decanter is used to scoop the floated material (sludge) with the help of a patented "Krofta spiral scooper" and push it to the stationary central section from where it is discharged (Figure 1.2).

FIGURE 1.1 (See color insert)
Common Effluent Treatment Plant in Vatva Industrial Estate in Gujarat (India).

FIGURE 1.2 **(See color insert)**
Krofta spiral scooper.

Major industries have their own wastewater and effluent treatment plants. Most of the chemical processing units release wastewaters in some form or the other. Some of the major categories of processing industries releasing effluents are summarized in Table 1.1. The contaminants in wastewaters released from any of the above-mentioned processing industries can be broadly classified as follows:

- Particulates
- Suspended solids
- Soluble solids
- VOCs
- Organic materials
- Inorganic materials
- BOD components
- COD components
- Oils and fats
- Greases
- Proteins and proteinaceous materials
- Soluble vitamins and micronutrients
- Toxins and vaccines
- Microorganisms, bacteria, virus, etc

Hence, the treatments for them are also varied. The various treatment levels are as follows:

TABLE 1.1
Various Processing Industries with Possible Contaminants

S. No.	Processing Industry	Possible Contaminants in Wastewaters	Remarks/Treatment Methods
1	Bioprocess industries	Organic matter, soluble and suspended particles, toxins	Physical, chemical, and biological methods, BOD reduction
2	Drugs and pharmaceutical industries	Organic matter, soluble and suspended particles, chemical contaminants	Physical, chemical, and biological methods, neutralization
3	Dyes and dye intermediates	Chemical contaminants	Physical and chemical methods
4	Fermentation processes	Organic matter, soluble and suspended particles	Physical, chemical, and biological methods, BOD reduction
5	Fine chemicals	Chemical contaminants	Physical and chemical methods
6	Flavor and aromatic industries	Oils, fats, and greases	Physical and chemical methods
7	Food processing industries	Organic matter, soluble and suspended particles	Physical, chemical, and biological methods, BOD reduction (see Chapters 8, 9, and 11)
8	Leather processing industries	Organic matter, soluble and suspended particles	Physical, chemical, and biological methods, BOD reduction
9	Man-made fiber industries	Chemical contaminants	Physical and chemical methods
10	Natural food colors and flavors	Organic matter, soluble and suspended particles, chemical contaminants	Physical, chemical, and biological methods, BOD reduction
11	Nuclear thermal units	Nuclear wastes, suspended solids	Physical and chemical methods (see Chapter 8)
12	Paints and varnishes	Chemical contaminants, oils, fats, and greases	Physical and chemical methods
13	Paper and pulp industries	Alkalis, chemicals, dyes	Physical and chemical methods
14	Petroleum refining and petrochemicals	Chemical contaminants, oils, fats, and greases	Physical and chemical methods (see Chapter 8)
15	Sugar- and molasses-based industries	Organic matter, soluble and suspended particles	Physical, chemical, and biological methods, BOD reduction
16	Textile industries	Chemical contaminants, soluble and suspended particles	Physical and chemical methods
17	Vaccine manufacturing units	Organic matter, soluble and suspended particles	Physical, chemical, and biological methods, BOD reduction
18	Vegetable oil mills	Chemical contaminants, oils, fats, and greases	Physical and chemical methods

- Primary
- Secondary
- Tertiary
- Advanced tertiary processes

All these treatment methods utilize a number of separation processes that are classically known as *unit operations* (UOs). UOs are a set of physical separation processes that all can be broken down into a number of simple mathematical expressions that, on integration, will unify all of the processing operations. Various UOs are shown in the Table 1.2 for various combinations of phases (Rao 2010, p. 362). The information in the table is more general in nature and is particularly applicable to bioprocessing.

Heavy metal ions present in the wastewaters of various chemical industries (listed in Table 1.1) have been noticed to have adverse effects on the performance of treatment methods, and hence their impact on the receiving environment needs careful consideration. Reckless and uncontrolled discharge of wastewaters containing heavy metals into the environment will pose detrimental effects to humans, animals, and plants. As a result, removal and recovery of heavy metals from industrial wastewaters before subjecting them to biological treatment have gained significant attention in recent years to protect the environment. Lead, mercury, chromium, cadmium, copper, zinc, nickel, and cobalt are the most frequently found heavy

TABLE 1.2

Various Unit Operations for the Treatment of Suspended and Soluble Particulates in Wastewater Treatment

System	Type	With Phase Change	Without Phase Change
Solid–liquid	Soluble	Adsorption	Flocculation Ultrafiltration[a] Reverse osmosis
	Insoluble	Drying[a]	Filtration Sedimentation Decanting
Liquid–liquid	Soluble	Distillation[a], evaporation[a], extraction[a]	Chromatography[a]
	Insoluble	Air floatation, Foaming	Flocculation Sedimentation Centrifugation
Liquid–liquid–solid	Miscible	Adsorption	
	Immiscible	Air floatation, foaming	Centrifugation Sedimentation Decanting

Source: Rao, D.G., *Introduction to Biochemical Engineering*, 2nd edn, Tata McGraw Hill Education Pvt. Ltd, New Delhi, 2010. With permission.

[a] Rarely used methods.

metals in industrial wastewaters. Methods used for removing heavy metals from wastewaters are also based on physical, chemical, and biological methods. Physicochemical methods such as precipitation, adsorption, ion exchange, and solvent extraction require high capital and operating costs and may produce large volumes of solid wastes, so these methods are often restricted because of technical and/or economic constraints. Among the

TABLE 1.3

Treatment Methods/Equipment/Reactors Dealt with in the Book

S. No.	Treatment Methods/ Equipment/Reactor	Remarks	Chapter No.
1	Treatment of greywaters by aeration, filtration, and disinfection	The treated greywaters were used for irrigation purposes by pot trials in greenhouses with barley plantation.	13
2	Advanced oxidation processes (AOPs)	Photo-Fenton AOPs for wastewater reclamation; overview of photo-Fenton processes	2 and 4
3	Solar photocatalytic processes	Solar collectors and concentrators were described. Semiconductor photocatalysis was described to produce reactive O_2 species for the destruction of organic contaminants and inactivation of microorganisms.	2 and 3
4	Removal of low-molecular-weight substances from wastewaters	Adsorption and absorption and use of nanoparticles to remove low-molecular-weight substances that normally escape other methods. It is a general review.	12
5	Biological treatment of wastewaters	A general review covering biodegradation, bioaccumulation, biosorption, and phytoremediation	6
6	Biological treatment method using anaerobic fixed bed reactor	Used for the treatment of highly concentrated industrial wastewaters coming from a winery effluent	10
7	Anaerobic reactor for the treatment of wastewaters	A tapered fluidized bed anaerobic reactor was used for treating wastewaters coming from a synthetic sago industry effluent using mesoporic activated carbon.	9
8	Impinging-jet ozone bubble column reactors	Chemical reaction engineering aspects of bubble columns and use of neural network analysis for modeling bubble column reactors	5
9	Microbial treatment of wastewaters from various process industries	A general review was given on wastewater treatment in mining industry, oil industries, and nuclear power plants.	8
10	Wastewater treatment in food processing industries	A general review on wastewater treatment in food processing industries	11
11	Removal of heavy metals	Continuous flow sorption studies in a glass column	7

various biological methods, biosorption has emerged as a cost-effective and efficient alternative treatment technology for heavy metals. Biosorption is the process of uptake of heavy metal ions and radio nuclides from aqueous solutions by biological materials. Different types of biomass in nonliving form are found to be suitable for the uptake of heavy metals. Bacteria, fungi, algae, plant leaves, and root tissues are used as biosorbents for the recovery of metals from industrial discharges (Chapters 6 and 7). Among these different types of biomass, seaweeds are extensively used for metal biosorption due to their high uptake capacities.

In addition to the above classical physical and biological processes, we may also use membrane separation processes, reverse osmosis (RO), and ultrafiltration processes. However, their application in wastewater treatment is usually discouraged in view of their prohibitive costs and large quantities of wastewater to be handled. These processes are time-consuming and can be used at a small-scale level as in the case of downstream processing steps in chemical or bioprocessing industries. There are some advanced techniques, such as photocatalytic and *photo-Fenton* processes, which are being increasingly tried upon for wastewater treatments. The application of solar energy either in the form of photovoltaic effect or in a concentrated form is another emerging area used for wastewater treatments (see Chapters 2–4). The application of nanotechnology and nanoparticles for wastewater treatment is another fascinating area and has been attracting the attention of researchers in the recent years in wastewater treatment.

Thus, wastewater treatment has many facets that need to be attended to in order to cleanse the wastewaters and make them as pure as possible. The benchmark is to make them fully potable. If not, they at least should be used for agricultural purposes and various other non-potable purposes. The approach (cleaning process) protects the environment from the contaminants of wastewaters. This book addresses some of these issues covering a wide range of wastewaters produced from different processing industries by utilizing a variety of treatment methods. Some are traditional methods, while others are advanced processes. The treatment methods also use a wide variety of equipment for various UOs, solar panels, solar heaters, and photo-Fenton processes, while others use a wide variety of biochemical reactors for the biological treatment of wastewaters. They are all summarized in a nutshell in Table 1.3.

Reference

Rao, D.G. 2010. *Introduction to Biochemical Engineering*, 2nd edn. New Delhi: Tata McGraw Hill Education Pvt. Ltd., pp. 362, 410.

2

Solar Photo-Fenton as Advanced Oxidation Technology for Water Reclamation

Sixto Malato Rodríguez, Nikolaus Klamerth, Isabel Oller Alberola, and Ana Zapata Sierra

CONTENTS

2.1 Introduction

During the last 30 years, environmental chemistry has concentrated almost exclusively on "conventional pollutants," mainly pesticides and a large number of industrial chemicals. Nevertheless, in terms of the large number of commercial chemicals, these are only a small percentage of the pollutants found in the environment (Daughton and Ternes 1999). Over the recent decades, biologically active synthetic substances for use in agriculture, industry, and medicine have been dumped into the environment without any consideration for possible negative consequences. Recently, at the end of 2008, the European Commission approved a new Directive (2008/105/EC) on environmental quality standards in the field of water policy. The new directive considers the identification of the causes of chemical pollution of surface waters and the dealing with emissions at the sources, in the most economically and environmentally effective manner as a matter of priority. Concerning the 33 priority substances and priority hazardous substances (alachlor, anthracene, atrazine, benzene, and so on), Directive 2008/105/EC expresses the environmental quality regulations in terms of annual averages, providing protection against long-term exposure and maximum permissible concentrations for short-term exposure.

Apart from the priority pollutants with established risk, there are hardly any studies on most of the organic compounds, and there are no environmental quality criteria for them yet. Current developments in analytical techniques, such as gas chromatography–mass spectrometry (GC-MS or GC-MS/MS) and liquid chromatography–mass spectrometry (LC-MS or LC-MS/MS), have made the detection and analysis of many of these new organic compounds in the environment, the analysis of which was hitherto difficult, possible (Petrovic and Barceló 2006; Hogenboom et al. 2009; Pietrogrande and Basaglia 2007; Gómez et al. 2009). "Emerging contaminants" (ECs) are defined as a group of unregulated substances that could be candidates for future regulation, depending on the findings of research on their effects on human health and aquatic biota and surveillance data on the frequency of their presence in the environment. A wide range of compounds, for example, detergents, pharmaceuticals, personal hygiene products, flame retardants, antiseptics, industrial additives, steroids, and hormones, have recently been found to be particularly relevant. The main characteristic of these pollutants is that they do not need to be persistent to cause negative effects, because their rates of removal are compensated by their constant introduction into the environment. The increasing use of these substances directly increases their concentration in treated and other waters (Fono et al. 2006; Jackson and Sutton 2008; Nakada et al. 2008), as conventional wastewater treatment plants are not able to remove them entirely (Göbel et al. 2007; Teske and Arnold 2008). As most of these ECs have xenobiotic, endocrine-disrupting, nonbiodegradable, toxic, or persistent properties, they must be degraded and removed prior to their release into the environment. This is even more important if the water is reused for irrigation, as these contaminants could accumulate in soil and crops (Radjenović et al. 2007; Muñoz et al. 2009; Snow et al. 2007).

Traditionally, wastewater treatment has focused on pollution abatement, public health protection, and environmental protection by removing biodegradable materials, nutrients, and pathogens (Levine and Asano 2004). At present, wastewater reclamation is one of the tools available to better manage the water resources diverted from the natural water cycle to the anthropic cycle. The way water is reused should always be linked to health protection, public acceptance, and its perceived value in the community. The main objective of wastewater reclamation and reuse projects is to produce water of sufficient quality for all nonpotable uses (uses that do not require the standards of drinking water). Using reclaimed water for these applications would save significant volumes of freshwater that would otherwise be wasted (Sala and

Serra 2004). Reusable water should be free of these persistent, toxic, endocrine-disrupting, or nonbiodegradable substances (Radjenović et al. 2007; Teske and Arnold 2008); hence, an effective tertiary treatment method is required to remove these substances completely. Conventional municipal wastewater treatment plants (MWTPs), typically based on biological processes, are capable of removing some substances, but nonbiodegradable compounds may escape the treatment and be released into the environment (Carballa et al. 2004; Ternes et al. 2007). Antibiotic drugs have been identified as a particular category of trace chemical contaminants (Le-Minh et al. 2010). Much of the concern regarding the presence of antibiotics in wastewater and their persistence through wastewater treatment processes is related to the concern that they may contribute to the prevalence of resistance to antibiotics in bacterial species in wastewater effluents and surface water near wastewater treatment plants (Auerbach et al. 2007; Jury et al. in press). ECs and priority substances have been found in the MWTP effluents at mean concentrations ranging from 0.1 to 20 µg/L (Martínez Bueno et al. 2007; Richardson 2007; Zhao et al. 2009). Concern about the growing problem of the continuously rising concentrations of these compounds must be emphasized, and therefore, the application of more thorough wastewater treatment protocols, including the use of new and improved technologies, is a necessary task. Conventional secondary wastewater treatment processes appear to be highly variable in their ability to remove most of these compounds, with their performance apparently dependent upon specific operational conditions, such as the specific retention time (SRT). Accordingly, tertiary and advanced treatment processes may be necessary to provide a further reduction of these compounds, in order to minimize environmental and human exposure.

Among the advanced processes that can degrade these ECs, advanced oxidation processes (AOPs) are a particularly attractive option (Westerhoff et al. 2009; Klavarioti et al. 2009). Although there are different reacting systems (see http://www.jaots.net/), all of them are characterized by the same chemical feature: production of hydroxyl radicals (•OH), which can oxidize and mineralize almost any organic molecule, yielding CO_2 and inorganic ions. Rate constants (k_{OH}) for the formation of •OH radicals by the rate expression ($r = k_{OH}$ [•OH] C) for most reactions involving hydroxyl radicals in aqueous solution are usually of the order of 10^6–10^9 M^{-1}/s. They are also characterized by their nonselective attack, which is a useful attribute for wastewater treatment to solve pollution problems. The versatility of AOPs is also enhanced by the fact that there are different ways of producing hydroxyl radicals, facilitating compliance with the specific treatment requirements. Methods based on UV, H_2O_2/UV, O_3/UV, and H_2O_2/O_3/UV combinations use the photolysis of H_2O_2 and ozone to produce the hydroxyl radicals. Heterogeneous photocatalysis and homogeneous photo-Fenton are based on the use of a wide-bandgap semiconductor and the addition of H_2O_2 to dissolved iron salts, respectively, and the irradiation with UVA-visible light (Pignatello et al. 2006; Comninellis et al. 2008; Shannon et al. 2008).

Some of the disadvantages associated with AOPs are their high operating costs depending on the specific process: (i) high electricity demand (e.g., ozone and UV-based AOPs), (ii) relatively large amounts of oxidants and/or catalysts consumed (e.g., ozone, hydrogen peroxide, and iron-based AOPs) and slow kinetics (photocatalysis with TiO_2), and (iii) the pH required (e.g., Fenton and photo-Fenton). By using solar energy as a light source, optimizing the pH, and optimizing the amounts of oxidant/catalyst, processes such as photo-Fenton may be used for commercial applications.

AOP efficiency in the removal of ECs has typically been studied in demineralized water at bench scale in the initial concentration range of few milligrams to grams. This may not be realistic compared with the concentrations detected in real water and wastewaters. Hence, this work focused on solar photo-Fenton degradation of ECs typically found in the effluents of MWTPs, leaving the treated wastewater suitable for reuse. Moreover, to make the process suitable for practical applications, high iron concentrations (mM range) and excessive amounts of H_2O_2 were avoided. The results presented were obtained in a pilot-scale solar photo-Fenton treatment plant run with starting concentrations of 5 mg/L Fe and 50 mg/L H_2O_2. Real effluent wastewaters (REs) to which a mixture of 15 ECs at low concentrations, consisting of pharmaceuticals, pesticides, and personal-care products, selected from a list of 80 compounds found in MWTP effluents in previous studies (Martínez Bueno et al. 2007), was added (100 μg/L or 5 μg/L each were tested in this study). RE without spiking with any EC was also tested and evaluated by LC-MS. Water reuse is required to deal not only with ECs but also with the potential problems of pathogens. Therefore, the preliminary results of the removal of pathogens are also presented.

2.2 Solar Photo-Fenton

2.2.1 Fenton and Photo-Fenton

For the treatment of industrial wastewaters, Fenton and Fenton-like processes are probably among the most applied AOPs (Legrini et al. 1993; Suty et al. 2004). This is not the case for studies related to ECs degradation as tertiary treatment in MWTPs, but this information will be discussed in detail later. The first proposals for wastewater treatment applications were reported in the 1960s. Yet, it was not until the early 1990s when first works of the application of the photo-Fenton process for the treatment of wastewater were published by the groups of Pignatello, Lipcznska-Kochany, Kiwi, Pulgarín, and Bauer (Pignatello et al. 2006). Much of the literature that deals with the photo-Fenton process takes into account the possibility of driving the process with solar radiation. This is due to the fact that a priori the photo-Fenton process seems to be the most apt of all AOPs to be driven by sunlight, because soluble iron hydroxide (and especially iron–organic acid complexes) absorbs even part of the visible light spectrum (Malato et al. 2009). Though several excellent and comprehensive reviews on the process exist (Neyens and Baeyens 2003; Pignatello et al. 2006), we will give a short summary of the principles of reactions that occur in the photo-Fenton system for the sake of completeness and clarity of the following discussion.

Hydrogen peroxide is decomposed into water and oxygen in the presence of iron ions in an aqueous solution in the Fenton reaction, Equation 2.1, which was first reported by Fenton (1894). A mixture of ferrous iron and hydrogen peroxide is called *Fenton's reagent*. If ferrous is replaced by ferric iron, then the mixture is called Fenton-like reagent. Equations 2.1 through 2.7 show the reactions of ferrous iron, ferric iron, and hydrogen peroxide in the absence of other interfering ions and organic substances. The regeneration of ferrous iron from ferric iron, shown in Equations 2.4 through 2.6, is the rate limiting step in the catalytic iron cycle, if iron is added in small amounts.

$$Fe^{2+} + H_2O_2 \rightarrow Fe^{3+} + OH^- + OH^\bullet, \tag{2.1}$$

$$Fe^{2+} + OH^\bullet \rightarrow Fe^{3+} + OH^-, \tag{2.2}$$

$$Fe^{2+} + HO_2^\bullet \rightarrow Fe^{3+} + HO_2^-, \tag{2.3}$$

$$Fe^{3+} + H_2O_2 \rightarrow Fe^{2+} + HO_2^\bullet + H^+, \tag{2.4}$$

$$Fe^{3+} + HO_2^\bullet \rightarrow Fe^{2+} + O_2 + H^+, \tag{2.5}$$

$$Fe^{3+} + O_2^{\bullet -} \rightarrow Fe^{2+} + O_2, \tag{2.6}$$

$$OH^\bullet + H_2O_2 \rightarrow H_2O + HO_2^\bullet. \tag{2.7}$$

Furthermore, radical–radical reactions (Equations 2.8 through 2.10) have to be taken into account:

$$2OH^\bullet \rightarrow H_2O_2, \tag{2.8}$$

$$2HO_2^\bullet \rightarrow H_2O_2 + O_2, \tag{2.9}$$

$$HO_2^\bullet + OH^\bullet \rightarrow H_2O + O_2. \tag{2.10}$$

If organic substances (such as quenchers, scavengers, and pollutants in the case of wastewater treatment) are present in the system $Fe^{2+}/Fe^{3+}/H_2O_2$, they react in many ways with the generated hydroxyl radicals. Yet, in all cases, the oxidative attack is electrophilic and the rate constants are close to the diffusion-controlled limit. The following reactions with organic substrates have been reported (Legrini et al. 1993): hydrogen abstraction from aliphatic carbon atoms (Equation 2.11), electrophilic addition to double bonds or aromatic rings (Equation 2.12), and electron transfer reactions (Equation 2.13).

$$OH^{\bullet} + RH \rightarrow R^{\bullet} + H_2O, \tag{2.11}$$

$$R - CH = CH_2 + OH^{\bullet} \rightarrow R - C^{\bullet}H - CH_2OH, \tag{2.12}$$

$$OH^{\bullet} + RX \rightarrow RX^{\bullet +} + OH^{-}. \tag{2.13}$$

The generated organic radicals continue reacting, prolonging the chain reaction, and thereby contribute to reduce the consumption of oxidants in wastewater treatment by Fenton and photo-Fenton methods. In the case of aromatic pollutants, the ring system is usually hydroxylated before it is broken up during the oxidation process. Substances containing quinone and hydroquinone structures are typical intermediate degradation products. Anyway, sooner or later, ring-opening reactions occur, which further carry on the mineralization of the molecules (Chen and Pignatello 1997). But there is one major setback of the Fenton method: especially when the treatment goal is the total mineralization of organic pollutants, the carboxylic intermediates cannot be further degraded. Carboxylic and dicarboxylic (L: monocarboxylic and dicarboxylic acids) acids are known to form stable iron complexes, which inhibit the reaction with peroxide (Kavitha and Palanivelu 2004). Hence, the catalytic iron cycle reaches a standstill before total mineralization is accomplished, as shown in Equation 2.14.

$$Fe^{3+} + nL \rightarrow [FeL_n]^{x+} \xrightarrow{H_2O_2,\ dark} \text{no further reaction.} \tag{2.14}$$

In the photo-Fenton system, the primary step in the photoreduction of dissolved ferric iron is a ligand-to-metal charge-transfer (LMCT) reaction. Subsequently, the intermediate complexes dissociate as shown in reaction 2.15. The ligand can be any Lewis base that is able to form a complex with ferric iron (OH^-, H_2O, HO_2^-, Cl^-, $R\text{–}COO^-$, R–OH, $R\text{–}NH_2$, etc.). Depending on the reacting ligand, the product may be a hydroxyl radical, such as the ones shown in reactions 2.16 and 2.17, or another radical derived from the ligand. The direct oxidation of an organic ligand as well is possible, as shown in reaction 2.18, for carboxylic acids.

$$[Fe^{3+}L] + h\nu \rightarrow [Fe^{3+}L]^{*} \rightarrow Fe^{2+} + L^{\bullet}, \tag{2.15}$$

$$[Fe(H_2O)]^{3+} + h\nu \rightarrow Fe^{2+} + OH^{\bullet} + H^{+}, \tag{2.16}$$

$$[Fe(OH)]^{2+} + h\nu \rightarrow Fe^{2+} + OH^{\bullet}, \tag{2.17}$$

$$[Fe(OOC-R)]^{2+} + h\nu \rightarrow Fe^{2+} + CO_2 + R^{\bullet}. \tag{2.18}$$

Depending on the ligand, the ferric iron complexes will have different light absorption properties. Hence, reaction 2.15 takes place with different quantum yields and also at different wavelengths. Consequently, the pH plays a crucial role in the efficiency of the photo-Fenton reaction, because it strongly influences the complexes that are formed. Thus, a pH of 2.8 was frequently postulated as an optimum pH for photo-Fenton treatment (e.g., Pignatello 1992), because at this pH, precipitation does not take place and the dominant iron species in solution is $[Fe(OH)]^{2+}$, which is the most photoactive ferric iron–water complex. In fact, as shown in its general form in reaction 2.15, ferric iron can form complexes with many substances and undergo photoreduction. Of special importance are carboxylic acids, because they are the intermediate products frequently produced in an oxidative treatment. Such ferric iron–carboxylate complexes can have much higher quantum yields than ferric iron–water complexes. It is, therefore, a typical observation that a reaction shows an initial lag phase, until intermediates are formed, which can regenerate ferrous iron from ferric iron more efficiently by accelerating the process. This behavior is observed in most of the degradation results shown in the following sections.

Fe(III) complexes present in mildly acidic solutions, such as $Fe(OH)^{2+}$, absorb light appreciably in the UV and visible regions. The quantum yield for Fe^{2+} formation in reaction 2.17 is dependent on the wavelength. It is 0.14–0.19 at 313 nm and 0.017 at 360 nm (Faust and Hoigne 1990). Fe(III) may also complex with certain contaminants or their organic by-products. These organic complexes typically have higher molar absorption coefficients in the near-UV and visible regions than the aquo complexes. Polychromatic quantum efficiencies in the UV/visible range from 0.05 to 0.95 are common (Pignatello et al. 2006). This is why the photo-Fenton process is apt to be driven by sunlight.

2.2.2 Solar Photocatalysis Hardware

For many of the solar detoxification system components (Blanco and Malato 2003), the equipment used is identical to that used for other types of water treatment, and the construction materials for such treatments are commercially available. Most piping may be made of polyethylene or polypropylene, avoiding the use of metallic or composite materials that could be degraded by the oxidation conditions of the photocatalytic process. The materials must not be reactive and must not interfere with the photocatalytic process. All materials used must be inert to degradation by UV solar light, in order to be compatible with the minimum required lifetime of the system. Photocatalytic reactors must transmit UV light efficiently because of the process requirements. With regard to the reflecting/concentrating materials, aluminum is the best option because of its low cost and high reflectivity in the solar UV spectrum on the earth's surface. The reflectivity (reflected radiation/incident radiation) of traditional silver-coated mirrors is very low (between 300 and 400 nm) and, therefore, aluminum-coated mirrors are the best option in this case. Aluminum-coated surface is the only metal surface that is highly reflective throughout the ultraviolet spectrum. For aluminum, the reflectivity ranges from 92.3% at 280 nm to 92.5% at 385 nm, while the reflectivity values for silver are 25.2% and 92.8%, respectively.

The photocatalytic reactor must be transparent to UV radiation. The choice of materials that are both transmissive to UV light and resistant to its destructive effects is limited. Common materials that meet these requirements are fluoropolymers, acrylic polymers, and several types of glass. Quartz has excellent UV transmission as well as good temperature and chemical resistance, but high costs make it completely unfeasible for photocatalytic applications. Fluoropolymers are a good choice of plastic for photoreactors because of their good UV transmittance, excellent ultraviolet stability, and chemical inertness. But one of their greatest disadvantages is that, in order to achieve a desired minimum pressure resistance, the wall thickness of a fluoropolymer tube has to be increased, which in turn will lower its UV transmittance. Acrylics could also be potentially used, but they are very brittle. Other low-cost polymeric materials are significantly more susceptible to attack by •OH radicals. Standard glass, used as a protective surface, is not satisfactory because it absorbs part of the incident UV radiation due to its iron content. Borosilicate glass has good transmissive properties in the solar range with a cutoff of about 285 nm (Blanco et al. 2000). Therefore, such a low-iron-content glass would seem to be the most adequate one. Therefore, both fluoropolymers and glasses are valid photoreactive materials.

The original solar photoreactor designs (Goswami 1995) for photochemical applications were based on line-focus parabolic-trough concentrators (PTCs). In part, this was a logical extension of the historical emphasis on trough units for solar thermal applications. Furthermore, PTC technology was relatively mature, and the existing hardware could be easily modified for photochemical processes. The main disadvantages are that these collectors (i) use only direct radiation, (ii) are expensive, and (iii) have low optical efficiencies. On the other hand, one-sun (nonconcentrating) collectors have no moving parts or solar tracking devices. They do not concentrate the radiation. So, efficiency is not reduced by the factors associated with concentration and solar tracking. As there is no concentrating system (with its inherent reflectivity), the optical efficiency of these collectors is higher as compared with PTCs. They are able to utilize both the diffuse and direct portions of the solar UV-A. An extensive effort in the design of small nontracking collectors has resulted in the testing of several different nonconcentrating solar reactors (Blanco-Galvez et al. 2007). Although one-sun collector designs possess important advantages, the design of a robust one-sun photoreactor is not trivial, due to the need for weather-resistant and chemically inert UV transmitting reactors. In addition, nonconcentrating systems require significantly more photoreactor area than concentrating photoreactors. Hence, as a consequence, full-scale systems must be designed to withstand the operating pressures of fluid circulation.

2.2.2.1 Compound Parabolic Concentrators

To design a solar collector for photocatalytic purposes, there is a set of main constraints for performing the optimization: (1) collection of UV radiation, (2) working temperatures as close as possible to ambient temperature, and (3) quantum efficiency. Finally, its construction must be economical and should be efficient, with a low pressure drop. As a consequence, the use of tubular photoreactors has a decisive advantage because of the inherent structural efficiency of the tubing. The tubing is also available in a large variety of materials and sizes and is a natural choice for a pressurized fluid system. There is a category of low-concentration collectors called compound parabolic concentrators that are used in thermal applications. They are an interesting option between parabolic concentrators and static flat systems. Thus, they also constitute a good option for solar photochemical applications (Ajona and Vidal 2000). Compound parabolic collectors (CPCs) are static collectors with reflective surfaces designed to be ideal in the sense of nonimaging optics and can be designed for any given reactor shape. They do so, illuminating the complete perimeter of the receiver, rather than just the "front" of it, as in conventional flat plates. These concentrating devices have ideal optics, thus maintaining the advantages of both the PTCs and the static systems (Colina-Márquez et al. 2009). The concentration factor (R_C) of a two-dimensional CPC collector is given by Equation 2.19 and is defined in Figure 2.1, where A is the aperture of the solar collector.

$$R_{C,CPC} = \frac{1}{\sin\theta_a} = \frac{A}{2\pi r}. \tag{2.19}$$

The normal values for the semiangle of acceptance (θ_a), for photocatalytic applications, are between 60° and 90°. A special case is the one in which $\theta_a = 90°$, whereby $R_C = 1$ (nonconcentrating solar system). When this occurs, all the UV radiation that reaches the aperture area of the CPC (direct and diffuse) can be collected and redirected to the reactor. If the CPC is designed for an acceptance angle of +90° to −90°, all incident solar diffuse radiation can be collected (Figure 2.1). The light reflected by the CPC is distributed all

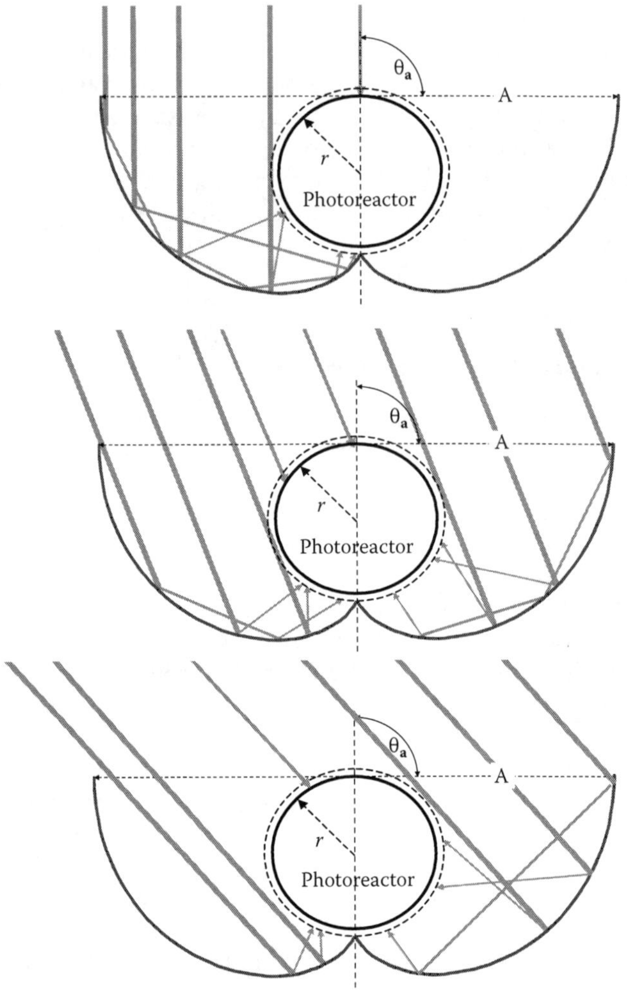

FIGURE 2.1
Schematic drawing of CPC with a semiangle of acceptance of 90° under different solar angles.

around the tubular receiver so that almost the entire circumference of the receiver tube is illuminated. CPCs have the advantages of both technologies (PTCs and nonconcentrating collectors) and none of the disadvantages, so they seem to be the best option for photocatalytic processes based on the use of solar radiation. They can make highly efficient use of both direct and diffuse solar radiations, without the need for solar tracking.

One important factor related to the photoreactor design is its diameter. It seems obvious that a uniform flow must be maintained at all times in the reactor, since a nonuniform flow causes nonuniform residence times, which can lower the efficiency when compared with the ideal. As already commented, the Fenton reactant consists of an aqueous solution of hydrogen peroxide and ferrous ions providing hydroxyl radicals. When the process is complemented with UV/visible radiation, it is called *photo-Fenton*. In this case, the process becomes catalytic. Fe^{3+} (related species and organic complexes) absorbs solar photons as a function of its absorptivity. This effect must be considered when determining the optimum load as a function of light-path length in the photoreactor. The optimum concentrations of 0.2–0.5 mM Fe as a function of the photoreactor diameter have been proposed after many experiments with different photoreactors under sunlight at the Plataforma Solar de Almeria (PSA) installation (Malato et al. 2009). In the results shown in this chapter, attending to the diameter of the photoreactor used, 0.35 mM mg/L of Fe has been used.

2.3 Experimental Setup

2.3.1 Solar Pilot Plant

Experiments were performed in a pilot CPC solar plant designed at the Plataforma Solar de Almería for solar photocatalytic applications (Figure 2.2). This reactor is composed of two modules (11 L each) with 12 Pyrex glass tubes (30 mm O.D.) mounted on a fixed platform tilted to 37° (local latitude). The water flows (20 L/min) directly from one module to the other and finally to a reservoir tank (10 L). The material chosen for the piping and the valves (3 L) between the reactor and the tank is black high-density polyethylene (HDPE) because it is highly resistant to chemicals, weather-proof, and opaque to avoid any photochemical effect outside of the collectors. The total illuminated area is 3 m^2. Polished aluminum is used as the reflective material because of its high UV reflectivity in the concerned UV range of 300–400 nm. The total volume (two modules + reservoir tank + piping and valves) is 35 L (V_T), and the irradiated volume is 22 L (V_i). The incident solar ultraviolet radiation (UV) was measured by a global UV radiometer (KIPP&ZONEN, model CUV 3) mounted on a platform tilted to 37° (the same as CPCs). The temperature inside the reactor was continuously recorded by a PT-100 inserted in the piping. With Equation 2.20, a combination of the data obtained from several days' experiments and their comparison with those obtained from other photocatalytic experiments are possible, where t_n is the experimental time for each sample, UV is the average solar ultraviolet radiation ($\lambda < 400$ nm) measured between t_{n-1} and t_n, and t_{30W} is the "normalized illumination time." In this case, time refers to a constant solar UV power of 30 W/m^2 (typical solar UV power on a perfectly sunny day around noon).

$$t_{30W,n} = t_{30W,n-1} + \Delta t_n \frac{UV}{30} \frac{V_i}{V_T}; \quad \Delta t_n = t_n - t_{n-1}; \quad t_0 = 0\,(n = 1). \tag{2.20}$$

(a)

(b)
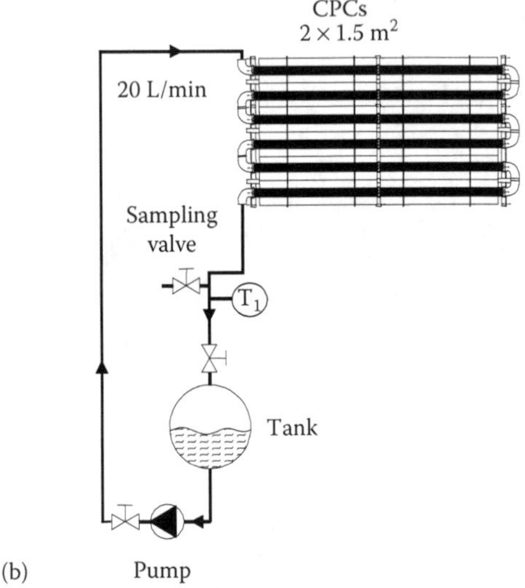

FIGURE 2.2 (See color insert)
(a,b) Solar pilot plant: scheme and view of CPCs.

2.3.2 Reagents

All reagents used for chromatographic analyses, namely acetonitrile, methanol, and ultrapure (Milli-Q) water, were of HPLC grade. The analytical standards for chromatography analyses were purchased from Sigma-Aldrich. Table 2.1 lists the 15 compounds (pharmaceuticals, pesticides, and personal-care products) used. Photo-Fenton experiments were performed using iron sulfate ($FeSO_4 \cdot 7H_2O$), reagent-grade hydrogen peroxide (30% w/v), and sulfuric acid, all provided by Panreac. The filters used were syringe-driven 0.2 μm Millex nylon membrane filters from Millipore. RE were taken downstream of the Almería and El Ejido MWTP

TABLE 2.1

Name and Structure of the 15 Selected Compounds

Name	Structure	Name	Structure
Acetaminophen analgesic/ antipyretic	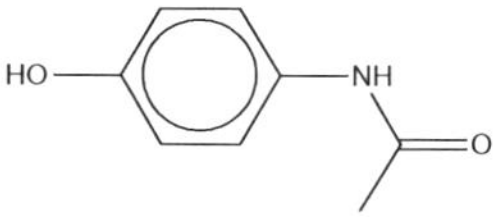	Ibuprofen nonsteroidal anti-inflammatory	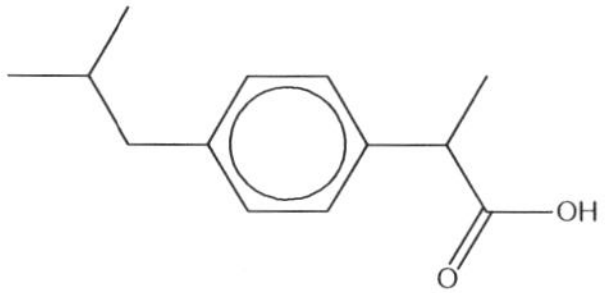
Antipyrine analgesic	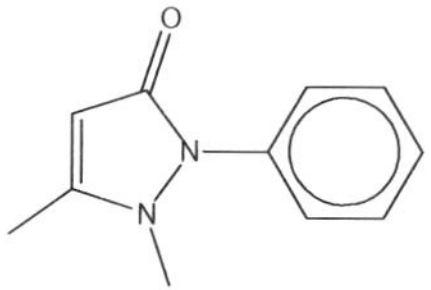	Isoproturon phenylurea herbicide	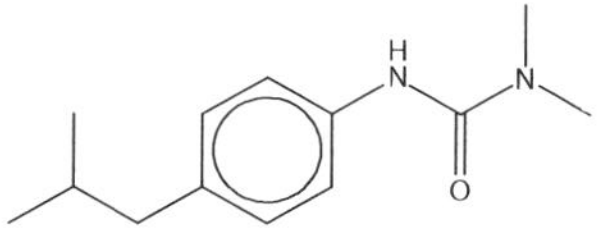
Atrazine herbicide	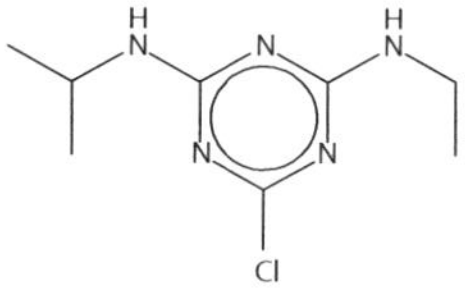	Ketorolac anti-inflammatory	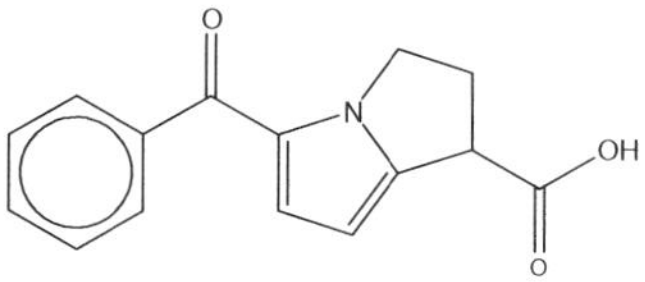
Caffeine stimulant	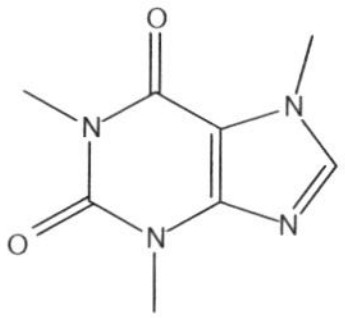	Ofloxacin Gram-negative antibiotic	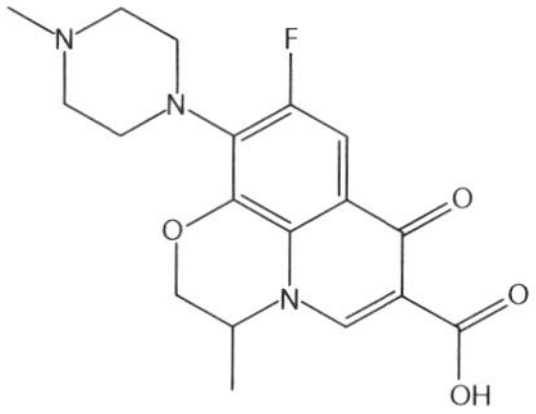

(continued)

TABLE 2.1 (Continued)

Name and Structure of the 15 Selected Compounds

Name	Structure	Name	Structure
Carbamazepine anticonvulsant	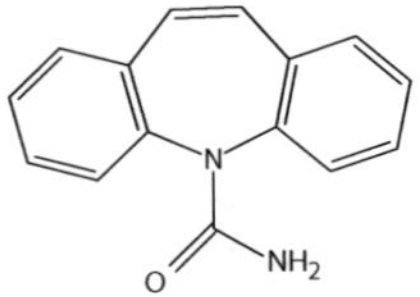	Progesterone steroid hormone	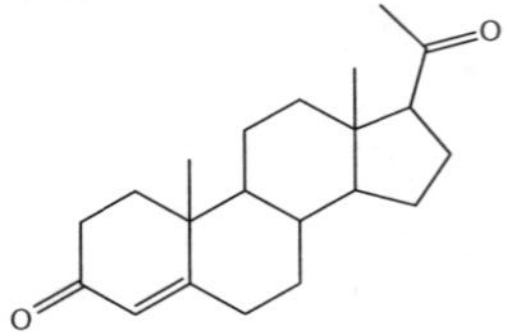
Diclofenac anti-inflammatory	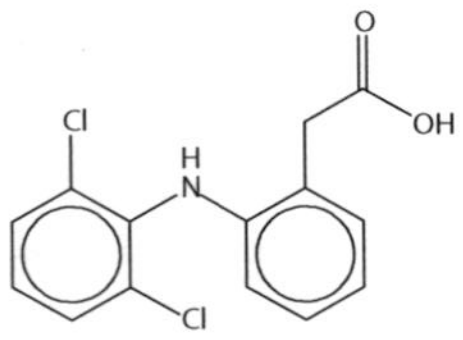	Sulfamethoxazole bacteriostatic antibiotic	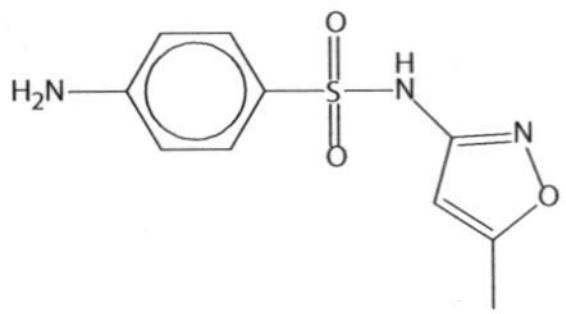
Flumequine broad-spectrum antibiotic	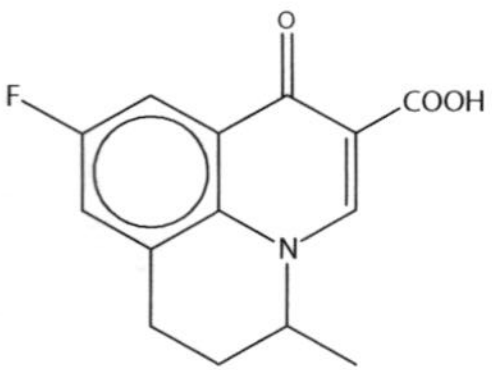	Triclosan antibacterial/antifungal agent	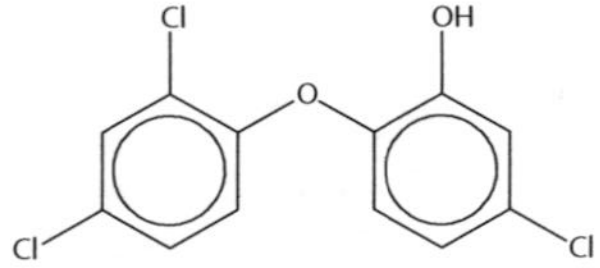
Hydroxybiphenyl biocide	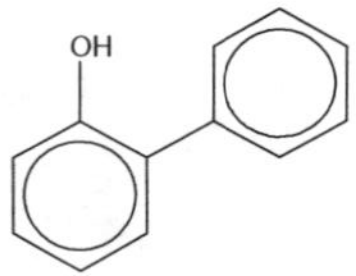		

secondary biological treatment (in the province of Almería, Spain) and used as received within the next 2 days. Initial COD, DOC, and TIC were between 57–86 mg/L, 20–34 mg/L, and 100–126 mg/L, respectively.

2.3.3 Analytical Measurements

The dissolved organic carbon (DOC) and total inorganic carbon (TIC) were measured by the direct injection of samples filtered with 0.2 μm syringe-driven filters into a Shimadzu 5050A TOC analyzer. Spectrophotometry for the determination of the iron and hydrogen peroxide concentrations was performed with a UNICAM 2 spectrophotometer. The total iron concentration was determined with 1,10-phenantroline according to ISO 6332. The hydrogen peroxide concentration was analyzed using titanium (IV) oxysulfate (DIN 38 402 H15 method), which allows the H_2O_2 concentration to be determined immediately based on a yellow complex with maximum absorption at 410 nm formed during the reaction of H_2O_2. The peroxide and iron concentrations are calculated using calibration curves. The concentration profile of each compound during degradation was determined by UPLC-UV (series 1200, Agilent Technologies, Palo Alto, CA). The analytes were separated using a reversed-phase C-18 analytical column (Agilent XDB-C18 1.8 μm, 4.6 × 50 mm) using acetonitrile (mobile phase A) and ultrapure water (25 mM formic acid, mobile phase B) at a flow rate of 1 mL/min. A linear gradient progressed from 10% A (original conditions) to 82% A in 12 min. The reequilibration time was 3 min. The limit of quantification was between 1.5 and 10.0 μg/L, depending on the EC.

A 25 mL sample of RE spiked with 100 μg/L was filtered through a 0.2 μm syringe-driven filter, the filter was washed with 3 mL ACN, the two solutions were mixed, and an aliquot was injected into the UPLC-UV system. A 200 mL sample of RE spiked with 5 μg/L was extracted with solid-phase extraction (SPE) and recovered in 2 mL of acetonitrile/water (1/9), filtered through 0.2 μm syringe-driven filter, and injected into the UPLC-UV. Under these conditions, the recovery factor was >80%. This procedure was used also for RE (without any spiking) evaluation.

The method for the analysis of the target compounds with HPLC-QTRAP-MS (Martínez Bueno et al. 2007) was developed for the 3200 QTRAP-MS/MS system (Applied Biosystems, Concord, ON, Canada). The separation of the analytes was performed using an HPLC (series 1100, Agilent Technologies, Palo Alto, CA) equipped with a reversed-phase C-18 analytical column (Zorbax SB, Agilent Technologies) of 5 μm particle size, 250 mm length, and 3.0 mm-i.d. For the analysis in positive mode, the compounds were separated using acetonitrile (mobile phase A) and HPLC-grade water with 0.1% formic acid (mobile phase B) at a flow rate of 0.2 mL/min. A linear gradient progressed from 10% A (initial conditions) to 100% A in 40 min, after which the mobile-phase composition was maintained at 100% A for 10 min. The reequilibration time was 15 min. The compounds analyzed in the negative mode were separated using acetonitrile (mobile phase A) and HPLC-grade water (mobile phase B) at a flow rate of 0.3mL/min. An LC gradient started with 30% mobile phase A and was linearly increased to 100%, in 7min, after which the mobile-phase composition was maintained at 100% A for 8min. The reequilibration time was 10min. The injection volume was 20 μL in both modes.

2.3.4 Experimental Procedures

Three approaches were used: (i) spiking RE with 100 μg/L of each contaminant; (ii) spiking the RE with 5 μg/L of each contaminant as the typical EC concentrations in the effluent are in the 0.1–15.0 μg/L range, with an SPE follow-up (see below for details) in which the samples were preconcentrated 100-fold; and (iii) treating the RE and analyzing the EC devolvement with HPLC-QTRAP-MS after the same SPE preconcentration.

The mixture of the 15 compounds dissolved in methanol at 2.5g/L (mother solution) was added directly into the pilot plant and well homogenized by turbulent recirculation for 30min. The pH in the RE was between 7.1 and 8.5, depending on the day when the water was collected, and the recirculation time for this process was usually from 60 to 120min. After stabilizing the desired pH, H_2O_2 at a concentration of 50mg/L was added and homogenized by recirculating for 15min. Finally, $FeSO_4 \cdot 7H_2O$ was added (Fe^{2+} = 5mg/L). After recirculating for 15min, during which the Fenton reaction started, the collectors were uncovered and the photo-Fenton process began. The hydrogen peroxide and iron concentrations were measured in every sample taken. The experiments normally lasted 3–4h. The peroxide was sometimes consumed completely and 10mg/L more of it was added at a time during the tests.

2.4 Results and Discussion

Figure 2.3 shows the photo-Fenton treatment of RE. The initial DOC, TIC, and COD were 36mg/L, 106mg/L, and 60mg/L, respectively. In this case, 406mg/L H_2SO_4 was added to reach <20mg/L TIC, and the recirculation time in the pilot plant, necessary to remove the carbonate species, was 30min. The spiked amount of 100 μg/L of each EC is high compared with that normally found in real wastewaters (<10 μg/L), but it is low enough to simulate real conditions and gain insight into the behavior of the photo-Fenton process.

The residual concentrations of the contaminants at the end of the experiments (t_{30W} > 100min) can be seen in Figure 2.3. Only atrazine remained at around 20% of the initial concentration. The pH varied from the original 8.1 to 4.5 at the end, while the iron concentration varied from 5g/L to 3.2mg/L. The overall amount of peroxide consumed was 29mg/L. It should be mentioned, although not relevant to the purpose of the experiments, because

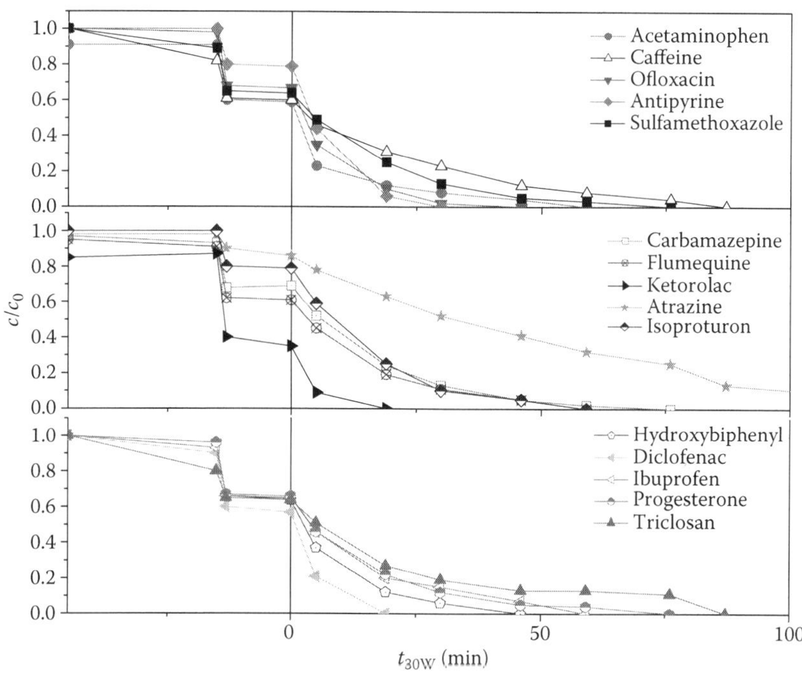

FIGURE 2.3
Degradation of the 15 ECs (0.1 mg/L each) by photo-Fenton (5 mg/L Fe, no pH adjustment) in RE acidified for carbonate release.

the main purpose was to degrade ECs, that DOC mineralization was rather low (around 25%). *Vibrio fischeri* toxicity was evaluated during the treatment of RE with ECs at 0.1 mg/L. Activation was observable in the RE experiment, which in some way favored the growth of *V. fischeri*. When the photo-Fenton process started, the degradation of parent pollutants resulted in more toxic intermediates, which increase the inhibition rate. In the last steps, the inhibition with RE was quite constant, probably because the toxic organic intermediates took longer than 100 min to mineralize. It should be remarked that 0.1 mg/L of each EC is quite high compared with the normal concentration found in effluents from MWTP. Therefore, toxicity assessment, as well as EC degradation, should be taken into account during AOP wastewater treatment.

RE was spiked with 5 μg/L of each EC and treated by photo-Fenton, and an iron concentration of 5 mg/L. 499 mg/L of H_2SO_4 was necessary for the pH adjustment, with a final TIC = 4.8 m/L. Samples were extracted with SPE as described in Section 2.3.2. The HPLC chromatograms indicating the degradation profiles of each of the 15 ECs are shown in Figure 2.4. All ECs were below their limit of detection (LOD) after t_{30W} = 64 min. DOC evolved from 21 to 14 mg/L, while the H_2O_2 consumption was 38 mg/L. This means that

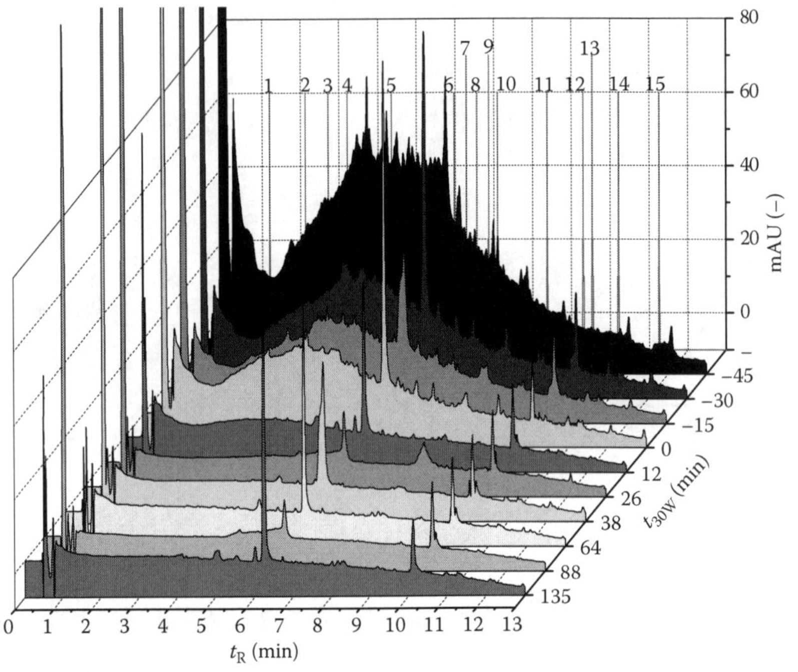

FIGURE 2.4 (See color insert)
Degradation profile (HPLC-UV 245 nm chromatograms) of the 15 ECs (5 μg/L each) by photo-Fenton (5 mg/L Fe). 1 Acetaminophen, 2 caffeine, 3 ofloxacin, 4 antipyrine, 5 sulfamethoxazole, 6 carbamazepine, 7 flumequine, 8 ketorolac, 9 atrazine, 10 isoproturon, 11 hydroxybiphenyl, 12 diclofenac, 13 ibuprofen, 14 progesterone, and 15 triclosan.

ECs are easily degraded by •OH and that the organic content of the RE does not significantly compete with them.

Since the optimum concentrations of 0.2–0.5 mM Fe have been proposed for this photoreactor, as it was stated in the solar photocatalysis hardware description, Figure 2.5 shows the degradation profile of the selected ECs, comparing 20 mg/L with 5 mg/L of Fe. Evolution of DOC also shows the mineralization of organic compounds. While it is obvious that the degradation of ECs was slower with 5 mg/L than with 20 mg/L of Fe, it was still efficient and rapid. The main differences in treating the RE with 5 mg/L and 20 mg/L of Fe are related to (i) less degradation by dark Fenton at Fe of 5 mg/L, (ii) similar photo-Fenton treatment times to reach the LOD of all ECs, (iii) less mineralization of the LOD of ECs with 5 mg/L Fe, and (iv) less H_2O_2 consumption for reaching the LOD at Fe of 5 mg/L. Figure 2.5 shows that by using a lower concentration of Fe, complete degradation of ECs is possible using lower doses of hydrogen peroxide and, therefore, with less mineralization of the overall organic content of the RE. This means that ECs are easily degraded by •OH and that the organic content of the RE does

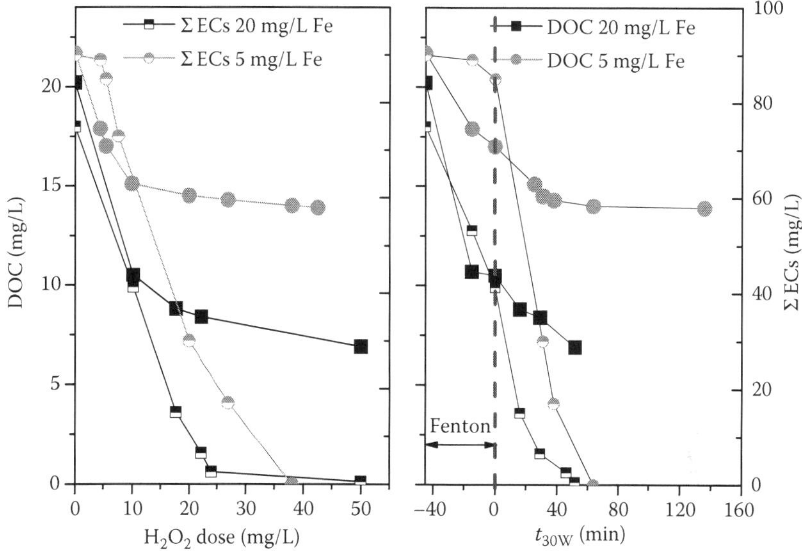

FIGURE 2.5
Mineralization of organic content and sum of all ECs present in RE as a function of the hydrogen peroxide consumption and illumination time during the experiments with 15 selected ECs at 5 μg/L each with 5 mg/L and 20 mg/L of Fe. The LOQ corresponds to the last of 15 compounds, which was possible to quantify. Dashed line indicates illumination start-up.

not significantly compete with them. This is important as it enables the rapid degradation of ECs with low iron dose and low hydrogen peroxide consumption.

In view of the aforementioned results, RE without spiking with other ECs was treated with photo-Fenton and 5 mg/L of Fe (see Figure 2.6). The original COD was 57 mg/L and DOC was 12.5 mg/L, containing different ECs as stated in Table 2.2. About 100 mg/L of H_2O_2 was consumed, and the DOC dropped to 7 mg/L at t_{30W} = 100 min and to 6.5 mg/L at the end of the experiment (t_{30W} = 140 min), with four compounds still present.

Table 2.2 shows the 52 ECs detected and analyzed with HPLC-QTRAP-MS after SPE and their degradation time. It should be remarked that the four substances that have a concentration above their limit of quantification (LOQ) after t_{30W} = 140 min of treatment are nicotine, caffeine, chlorfenvinphos, and cotinine. Figure 2.6 shows the H_2O_2 consumption, DOC mineralization, and sum of all ECs present in the water (around 33 μg/L among all 52 ECs). The significant changes in the EC concentration profile took place during dark Fenton, but almost complete degradation below their limit of quantification (LOD) took place mainly within the first 19 min under illumination. This means that over 98% of ECs can be successfully degraded, adding only 35 mg/L of H_2O_2 to the RE, almost the same quantity of peroxide

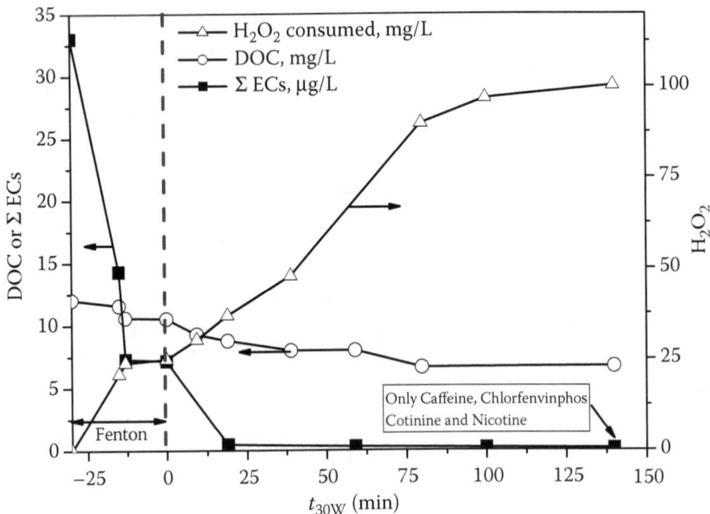

FIGURE 2.6
Mineralization of the organic content, hydrogen peroxide consumption, and sum of all ECs present in RE as a function of illumination time during experiment with 5 mg/L of Fe. Dashed line indicates illumination start-up.

that was necessary for the removal of 98%–99% of ECs (t_{30W} = 60 min). And to remove ECs until 99.5%, 100 mg/L of H_2O_2 and 140 min were necessary. Fenton degradation was very efficient when just 5 mg/L of Fe^{2+} was added but proceeded no further after 15 min in the dark until the reactor was uncovered (t_{30W} = 0 min), as Fe^{3+} reduction to Fe^{2+} is inefficient without illumination. It is unclear why the four aforementioned substances were more resistant to degradation than the rest of the EC and analyzed in this experiment. Different experiments performed with other RE (not the same ECs and not the same concentration) presented similar results (almost complete degradation of all ECs in less than 30 min under solar irradiation) without any specific behavior of nicotine, caffeine, chlorfenvinphos, and cotinine. These compounds were rapidly degraded in most of the cases.

The overarching goal for the reclamation and reuse of water in the future is to capture water directly from nontraditional sources such as industrial or municipal wastewaters and restore it. Municipal wastewaters are commonly treated by activated sludge systems that use suspended microbes to remove organics and nutrients and large sedimentation tanks to separate the solid and liquid fractions. This level of treatment produces wastewater effluent suitable for discharge to surface waters or for restricted irrigation and some industrial applications.

Concerning water disinfection, the most important issue is, of course, safe drinking water. In addition to the well-known task of drinking water

TABLE 2.2

Initial Concentration of 52 ECs Found in RE Using a Method Developed by Martínez Bueno et al. (2007) and Their Degradation Time with 5 mg/L of Fe

Compound	C_0 (ng/L)	Deg. Time[a] t_{30W} (min)	Compound	C_0 (ng/L)	Deg. Time t_{30W} (min)
4-AA	2257	F[b]	Gemfibrozil	2968	<19
4-AAA	4515	<59	Hydrochlorothiazide	314	<19
4-FAA	2236	<19	Ibuprofen	781	<19
4-MAA	5684	F	Indomethacin	98	F
Amitriptyline HCl	22	F	Isoproturon	20	F
Antipyrine	770	<59	Ketoprofen	261	<19
Atenolol	627	<19	Lincomycin	61	F
Atrazine	15	<19	Mefenamic acid	27	F
Azithromycin	75	F	Mepivacaine	25	<19
Bezafibrate	205	<19	Metoprolol	15	<19
Caffeine	3782	8 ng/L at 140 min	Metronidazole	16	F
Carbamazepine	89	19	Nadolol	13	F
Chlorfenvinphos	651	99 ng/L at 140 min	Nicotine	166	47 ng/L at 140 min
Ciprofloxacin	392	F	Norfloxacin	179	F
Citalopram HBr	219	F	Ofloxacin	1139	<19
Clarithromycin	109	F	Paraxanthine	2050	<19
Clofibric acid	16	F	Pravastatin	426	F
Cotinine	172	11 ng/L at 140 min	Primidone	82	<19
Diazepam	9	F	Propranolol	20	F
Diclofenac	793	<19	Ranitidine	162	F
Diuron	213	F	Salicylic acid	48	F
Epoxy-carbamazepine	11	F	Simazine	8	F
Erythromycin	80	F	Sulfamethoxazole	284	<19
Famotidine	20	F	Sulfapyridine	161	<19
Fenofibric acid	81	<19	Trimethoprim	26	F
Furosemide	429	F	Venlafaxine	188	F

Source: Martínez Bueno, M.J., Agüera, A., Gómez, M.J., Hernando, M.D., García-Reyes, J.F., and Fernández-Alba, A.R., *Anal. Chem.*, 79, 9372–9384, 2007. With permission.

[a] LOD was reached before this time, which corresponded with the sampling time.

[b] F means that the compound was degraded during Fenton process before illumination.

disinfection, the second most critical issue is the disinfection of water for reuse mainly in agriculture. Bearing in mind that the daily drinking-water requirement per person is only 2–4 L, it is often forgotten that it still takes 2000–5000 L of water to produce a person's daily food requirement (FAO, AQUASTAT 2008). Eighty percent of the land cultivated worldwide is still

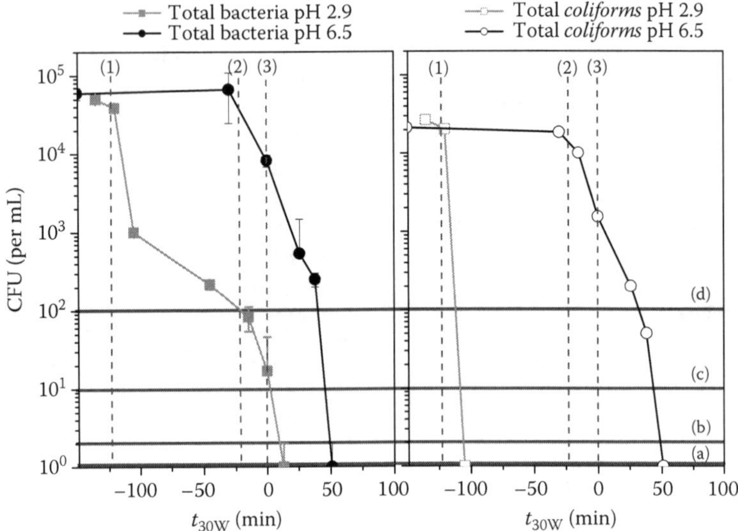

FIGURE 2.7
Disinfection of RE (see Table 2.2) as a function of illumination time during experiment with 5 mg/L of Fe. (1) Acid added, (2) H_2O_2 + Fe added, (3) illumination. Limits according to Spanish Directive RD 1620/2007 are also included. (a, 100 CFU per mL): irrigation of trees and cereal crops and industrial use (not food processing); (b, 10 CFU per mL): irrigation of pastures and aquifer recharge; (c, 2 CFU per mL): urban irrigation (gardens) and recreational use (as golf courses); (d, 1 CFU per mL): irrigation of fresh vegetables.

exclusively rainfed land. Water reuse requires effective management to deal not only with ECs but also with the potential problems of pathogens. Therefore, photo-Fenton may be also used to treat water to avoid the presence of pathogens and disinfect effluents in a wastewater treatment plant, for further water reuse. Figure 2.7 presents the preliminary results in this way, including the information about the pathogen limits in water according to Spanish Directive RD 1620/2007, which regulates different thresholds attending to the final use of the reclaimed water. Two different approaches were tested: photo-Fenton at approximately neutral pH and at optimum pH. By both procedures it was possible to attain the thresholds specified in the legislation for different purposes, being quicker at initial pH = 2.9. Water pH should be adjusted, increasing its salinity too. Photo-Fenton at neutral pH was also effective but slower. Otherwise, it should be remarked that under these conditions, water would be suitable for reuse after the treatment. These results open a new way to treat municipal wastewater effluents for avoiding both ECs and pathogens. The main advantage of this system is the low cost (low consumption of hydrogen peroxide and less use of solar energy). Moreover, it does not require the posttreatment removal of reagents or catalysts.

Acknowledgment

The authors wish to thank the Spanish Ministry of Science and Innovation for financial support (EDARSOL project http://www.psa.es/webesp/projects/edarsol/index.php, CTQ2009-13459-C05-01).

References

Ajona, J.A. and A. Vidal. 2000. The use of CPC collectors for detoxification of contaminated water: Design, construction and preliminary results. *Sol. Energy* 68: 109–120.

Auerbach, E.A., E.E. Seyfried, and K.D. McMahon. 2007. Tetracycline resistance genes in activated sludge wastewater treatment plants. *Water Res.* 41: 1143–1151.

Blanco, J. and S. Malato. 2003. *Solar Detoxification*. France: UNESCO Publishing.

Blanco, J., S. Malato, P. Fernández, et al. 2000. Compound parabolic concentrator technology development to commercial solar detoxification applications. *Sol. Energy* 67: 317–330.

Blanco-Galvez, J., P. Fernández-Ibáñez, and S. Malato-Rodríguez. 2007. Solar photocatalytic detoxification and disinfection of water: Recent overview. *J. Sol. Energy Eng.* 129: 4–15.

Carballa, M., F. Omil, J.M. Lema, et al. 2004. Behaviour of pharmaceuticals, cosmetics and hormones in a sewage treatment plant. *Water Res.* 38: 2918–2926.

Chen, R. and J.J. Pignatello. 1997. Role of quinone intermediates as electron shuttles in Fenton and photoassisted Fenton oxidations of aromatic compounds. *Environ. Sci. Technol.* 31: 2399–2406.

Colina-Márquez, J., F. Machuca-Martínez, and G. Li Puma. 2009. Photocatalytic mineralization of commercial herbicides in a pilot-scale solar CPC reactor: Photoreactor modeling and reaction kinetics constants independent of radiation field. *Environ. Sci. Technol.* 43: 8953–8960.

Comninellis, C., A. Kapalka, S. Malato, S.A. Parsons, I. Poulios, and D. Mantzavinos. 2008. Advanced oxidation processes for water treatment: Advances and trends for R&D. *J. Chem. Technol. Biotechnol.* 83: 769–776.

Daughton, C.G. and T.A. Ternes. 1999. Pharmaceuticals and personal care products in the environment: Agentes of subtle change? *Environ. Health Perspect.* 107: 907–938.

FAO's Information System on Water and Agriculture. 2008. NRL http://www.fao.org/nr/water/aquastat/main/indexstm (accessed November 5, 2010).

Faust, B.C. and J. Hoigne. 1990. Photolysis of FeIII-hydroxy complexes as sources of OH radicals in clouds, fog and rain. *Atmos. Environ.* 24A: 79–89.

Fenton, H.J.H. 1894. Oxidation of tartaric acid in presence of iron. *J. Chem. Soc.* 65: 899–910.

Fono, L.J., E.P. Kolodziej, and D.L. Sedlak. 2006. Attenuation of wastewater-derived contaminants in an effluent dominated river. *Environ. Sci. Technol.* 40: 7257–7262.

Göbel, A., C. McArdell, A. Joss, H. Siegrist, and W. Giger. 2007. Fate of sulfonamides, macrolides and trimethoprim in different wastewater treatment technologies. *Sci. Total Environ.* 372: 361–371.

Gómez, M.J., M.M. Gómez-Ramos, A. Agüera, M. Mezcua, S. Herrera, and A.R. Fernández-Alba. 2009. A new gas chromatography/mass spectrometry method for the simultaneous analysis of target and non-target organic contaminants in waters. *J. Chromatogr. A* 1216: 4071–4082.

Goswami, D.Y. 1995. Engineering of solar photocatalytic detoxification and disinfection processes. In *Advances in Solar Energy*, ed. K.W. Böer. American Solar Energy Society: Boulder, CO, pp. 165–209.

Hogenboom, A.C., J.A. van Leerdam, and P. de Voogt. 2009. Accurate mass screening and identification of emerging contaminants in environmental samples by liquid chromatography–hybrid linear ion trap orbitrap mass spectrometry. *J. Chromatogr. A* 1216: 510–519.

Jackson, J. and R. Sutton. 2008. Sources of endocrine disrupting chemicals in urban wastewater, Oakland, CA. *Sci. Total Environ.* 405: 153–160.

Jury, K.L., S.J. Khan, T. Vancov, R.M. Stuetz, and N.J. Ashbolt. 2011. Are sewage treatment plants promoting antibiotic resistance? *Crit. Rev. Environ. Sci. Technol.* 41: 243–270.

Kavitha, V. and K. Palanivelu. 2004. The role of ferrous ion in Fenton and photo-Fenton processes for the degradation of phenol. *Chemosphere* 55: 1235–1243.

Klavarioti, M., D. Mantzavinos, and D. Kassinos. 2009. Removal of residual pharmaceuticals from aqueous systems by advanced oxidation processes. *Environ. Int.* 35: 402–417.

Legrini, O., E. Oliveros, and A.M. Braun. 1993. Photochemical processes for water treatment. *Chem. Rev.* 93: 671–698.

Le-Minh, N., S.J. Khan, J.E. Drewes, and R.M. Stuetz. 2010. Fate of antibiotics during municipal water recycling treatment processes. *Water Res.* 44: 4295–4323.

Levine, A.D. and T. Asano. 2004. Recovering sustainable water from wastewater. *Environ. Sci. Technol.* 38: 201–208.

Malato, S., P. Fernández-Ibáñez, M.I. Maldonado, J. Blanco, and W. Gernjak. 2009. Decontamination and disinfection of water by solar photocatalysis: Recent overview and trends. *Catal. Today* 147: 1–59.

Martínez Bueno, M.J., A. Agüera, M.J. Gómez, M.D. Hernando, J.F. García-Reyes, and A.R. Fernández-Alba. 2007. Application of liquid chromatography/quadrupole-linear ion trap mass spectrometry and time-of-flight mass spectrometry to the determination of pharmaceuticals and related contaminants in wastewater. *Anal. Chem.* 79: 9372–9384.

Muñoz, I., M.J. Gómez-Ramos, A. Agüera, A.R. Fernández-Alba, J.F. García-Reyes, and A. Molina-Díaz. 2009. Chemical evaluation of contaminants in wastewater effluents and the environmental risk of reusing effluents in agriculture. *TrAC* 28: 676–694.

Nakada, N., K. Kiri, H. Shinohara, A. Harada, K. Kuroda, S. Takizawa, and H. Takada. 2008. Evaluation of pharmaceuticals and personal care products as water-soluble molecular markers of sewage. *Environ. Sci. Technol.* 42: 6347–6353.

Neyens, E. and J. Baeyens. 2003. A review of classic Fenton's peroxidation as an advanced oxidation technique. *J. Hazard. Mater.* B 98: 33–50.

Petrovic, M. and D. Barceló. 2006. Liquid chromatography-mass spectrometry in the analysis of emerging environmental contaminants. *Anal. Bioanal. Chem.* 385: 422–424.

Pietrogrande, M.C. and G. Basaglia. 2007. GC-MS analytical methods for the determination of personal-care products in water matrices. *TrAC* 26: 1086–1094.

Pignatello, J.J. 1992. Dark and photoassisted Fe^{3+}-catalyzed degradation of chlorophenoxy herbicides by hydrogen peroxide. *Environ. Sci. Technol.* 26: 944–951.

Pignatello, J.J., E. Oliveros, and A. MacKay. 2006. Advanced oxidation processes for organic contaminant destruction based on the Fenton reaction and related chemistry. *Crit. Rev. Environ. Sci. Technol.* 36: 1–84.

Radjenović, J., M. Petrović, D. Barceló, and D. Petrović. 2007. Advanced mass spectrometric methods applied to the study of fate and removal of pharmaceuticals in wastewater treatment. *TrAC* 26: 1132–1144.

Richardson, S. 2007. Water analysis: Emerging contaminants and current issues, review. *Anal. Chem.* 79: 4295–4324.

Sala, L. and M. Serra. 2004. Towards sustainability in water recycling. *Water Sci. Technol.* 50: 1–7.

Shannon, M.A., P.W. Bohn, M. Elimelech, J.G. Georgiadis, B.J. Mariñas, and A.M. Mayes. 2008. Science and technology for water purification in the coming decades. *Nature* 452: 301–310.

Snow, D.D., S.L. Bartelt-Hunt, S.E. Saunders, and D.A. Cassada. 2007. Detection, occurrence, and fate of emerging contaminants in agricultural environments. *Water Environ. Res.* 79: 1061–1084.

Suty, H., C. De Traversay, and M. Cost. 2004. Applications of advanced oxidation processes: Present and future. *Water Sci. Technol.* 49: 227–233.

Ternes, T., M. Bonerz, N. Herrmann, B. Teiser, and R. Andersen. 2007. Irrigation of treated wastewaters in Braunschweig, Germany: An option to remove pharmaceuticals and musk fragrances. *Chemosphere* 66: 894–904.

Teske, S.S. and R.G. Arnold. 2008. Removal of natural and xeno-estrogens during conventional wastewater treatment. *Rev. Environ. Sci. Biotechnol.* 7: 107–124.

Westerhoff, P., H. Moon, D. Minakata, and J. Crittenden. 2009. Oxidation of organics in retentates from reverse osmosis wastewater reuse facilities. *Water Res.* 43: 3992–3998.

Zhao, J.-L., G.-G. Ying, L. Wang, J.-F. Yang, X.-B. Yang, L.-H. Yang, and X. Li. 2009. Determination of phenolic endocrine disrupting chemicals and acidic pharmaceuticals in surface water of the Pearl River in South China by GC-MS. *Sci. Total Environ.* 407: 962–974.

3

Solar Photocatalytic Treatment of Wastewater

J. A. Byrne and P. Fernández-Ibáñez

CONTENTS

3.1 Introduction

Advanced oxidation processes (AOPs) are a group of related processes or technologies that lead to the generation of radical oxygen species, which result in the oxidative degradation of pollutants in water. Heterogeneous or semiconductor photocatalysis is an AOP wherein the photoexcitation of the semiconductor material results in the production of radical oxygen species that can degrade organic pollutants. Inorganic substances present in water can undergo oxidation or reduction to less harmful analogs. AOPs present alternative and complementary approaches to the treatment of water containing problematic pollutants that may not be effectively removed by conventional treatment methods. For example, pesticides, by their very nature, are designed to be stable, persistent, toxic, and active at low concentrations and, because of their widespread dispersion in the environment, they will inevitably find their way into water sources. AOPs may be employed as a specific measure to deal with persistent organic chemicals such as pesticides. Furthermore, there is evidence that conventional disinfection methods are not effective against certain species of pathogenic microoganisms and AOPs may present an alternative or complementary approach to dealing with disinfection-resistant strains. Semiconductor photocatalysis is an advanced oxidation technology, which has been widely studied for use in water and wastewater treatment.

3.2 Semiconductor Photocatalysis

Irradiation of a semiconductor particle with light of energy greater than or equal to the band gap energy results in the promotion of an electron from the valence band to the conduction band (Figure 3.1). The photoexcited electron leaves behind a positive hole in the valence band. Together, the valence band hole and the conduction band electron are referred to as an electron–hole pair and are responsible for the charge conduction in the semiconductor. Following formation, the charge carriers may recombine in the bulk dissipating the energy as heat or light, or they may move (via diffusion or migration) to the solid–liquid interface. For titanium dioxide (TiO_2), the valence band hole has a very positive electrochemical reduction potential (+2.4 V vs the standard hydrogen electrode (SHE) at pH 7) and the conduction band has a negative electrochemical reduction potential (−0.8 V (SHE) at pH 7). The positive hole is able to oxidize water or hydroxyl groups to yield hydroxyl radicals (•OH). The hydroxyl radical is an indiscriminate oxidant, which attacks organic species at or near the particle solution interface. In order to maintain electrical neutrality within the semiconductor, photogenerated electrons must be removed from the conduction band.

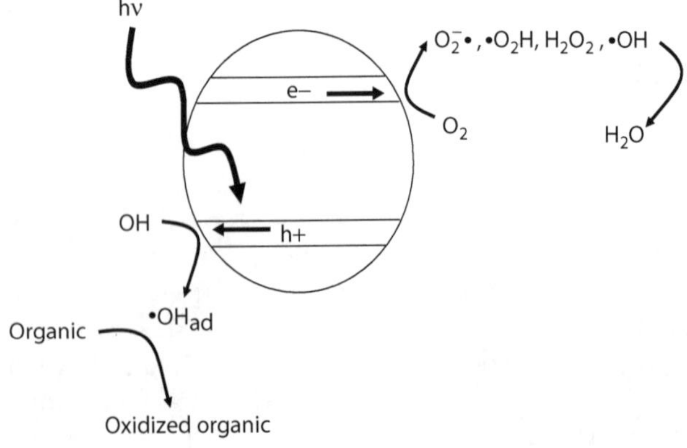

FIGURE 3.1
Schematic showing mechanism of semiconductor photocatalysis.

Any species with an electrochemical reduction potential more positive than that of the conduction band of the semiconductor can, in theory, accept electrons from the conduction band. For the oxidative destruction of organic pollutants in water, the most commonly employed oxidant (electron acceptor) is molecular oxygen, which is freely available from the air. The one-electron reduction of molecular oxygen yields a superoxide radical anion ($O_2^{-\bullet}$, or at low pH, a hydroperoxyl radical, $HO_2^{\bullet}$) and a two-electron reduction results in a hydrogen peroxide (H_2O_2). A further one-electron reduction will yield a hydroxyl ion OH^- and a hydroxyl radical ($\bullet OH$). With dissolved oxygen (DO) as the electron acceptor, the overall process of organic pollutant degradation can be described by a simplified Equation 3.1, and the mechanism is described in Figure 3.1.

$$\text{Organic pollutant} + O_2 \xrightarrow{\text{semiconductor} + h\nu} CO_2 + H_2O + \text{acids or salts.} \quad (3.1)$$

For more detailed information, the reader should refer to the excellent reviews in the literature (Herrmann et al. 1993; Pelizzetti and Minero 1993; Serpone 1994; Mills and Le Hunte 1997; Goswami 1997; Alfano et al. 2000; Tryk et al. 2000; Bahnemann 2004; Augugliaro et al. 2006; Gaya and Abdullah 2008; Fujishima et al. 2008; Malato et al. 2009; Herrmann 2010).

3.3 Selection of Photocatalyst

For wastewater treatment, the choice of semiconductor material is limited by boundary conditions, that is, the catalyst should be

- Photocatalytically active
- Chemically and photochemically stable (does not undergo photo-anodic corrosion)
- Nonsoluble in the pH range of the wastewater
- Nontoxic
- Appropriately priced

Of the semiconductor materials available, only a few meet the previously mentioned requirements. The most commonly used semiconductor material for water treatment applications is titanium dioxide (TiO_2). It is a wideband semiconductor (3.2 eV for anatase), insoluble under the normal pH range found in wastewater, photochemically stable, and also inexpensive in bulk form. The band edge potentials for TiO_2 are in appropriate positions to allow for the production of a hydroxyl radical and the reduction of molecular oxygen. Of the commercial forms of TiO_2 employed for research experiments, many have chosen to use Aeroxide P25 (formerly Degussa P25, supplied by Evonik Degussa Corporation), which has a recognized high photocatalytic activity. P25 TiO_2 is produced through the high temperature flame hydrolysis of $TiCl_4$ in the presence of hydrogen and oxygen and then treated with steam to remove the HCl. The product is 99.5% pure TiO_2 (70% anatase, 30% rutile), nonporous cubic particles with rounded edges. P25 TiO_2 powder has a surface area of ~50 m^2/g and an average particle diameter of ~21 nm (90% of the particles fall in the size range of 9–38 nm). In powder form, the particles do not exist in isolation, but rather as irreducible complex primary aggregates with a diameter of ~0.1 μm.

The band gap of anatase TiO_2 is 3.2 eV, which corresponds to wavelengths of light ≤387 nm, and therefore, TiO_2 can only be excited by UV light. The solar spectrum contains only a small proportion of UV (~5%) and this obviously limits the applications of TiO_2 photocatalysis in solar-driven water treatment. The apparent quantum efficiency for the degradation of organic compounds in water is usually reported to be around 1% (under good conditions) and under UV irradiation. Therefore, one can only reasonably expect an overall solar efficiency of around 0.05% for photocatalytic water treatment employing a UV band gap semiconductor. Current research efforts in the field are focused toward the development of stable visible-light active (VLA) photocatalytic materials, which can utilize the solar spectrum more effectively.

Of course, one can employ the use of UV light sources to drive the process, but this introduces the additional cost of electricity and the cost of the lamps themselves. The latter is probably the dominant cost factor. There are a wide range of UV sources that have been developed for water disinfection and other industrial processes. UVC (254 nm) sources are usually low-pressure mercury lamps that have reasonable lifetimes in the region of 5000 h. However, the downside of employing UVC sources is the requirement for quartz housing to allow light transmission and separation of the lamps from the water. Other sources may be employed including UVA lamps that employ a medium-pressure mercury lamp with a phosphor absorbing the 254 and 313 lines reemitting in the UVA along with the 365 nm line. The lifetime of these sources tend to be much less at around 2000 h.

The photocatalytic process is inherently a light-driven redox process. The oxidation path involves hole transfer to water and this must be balanced by the reduction pathway, which involves electron transfer from the conduction band to an electron acceptor. For the oxidative degradation of organic contaminants in water, molecular oxygen is normally employed as the electron acceptor species as it is a reasonably good electron acceptor, freely available from the air, and reasonably soluble in water. The one-electron electrochemical reduction of O_2 results in the production of a superoxide radical anion

($O_2^{-\bullet}$) and a hydroperoxyl radical ($HO_2^{\bullet}$), depending on the pH. Further reduction yields a hydrogen peroxide (H_2O_2) and a hydroxyl radical (•OH), all of which can feed into the oxidative attack on organic pollutants. Where the removal of electrons is slow, recombination will compete with hole transfer at the surface and quantum yields will be reduced. Gerischer and Heller (1991) reported that the reduction of O_2 may be rate-limiting in cases where the hole transfer is fast, the concentration of organic pollutant is high, or the concentration of DO is low (see also Gerischer 1993). Acceleration of the electron transfer rate to O_2 increases the quantum yield considerably and can be achieved by the deposition of metal clusters (Pd or Pt) on the surface of the semiconductor particle that act as electrocatalyst sites; however, the use of precious or semiprecious metals increases the cost of the catalyst material. At low light intensities, with air-sparging and agitation of the solution, the diffusion of O_2 to the semiconductor surface is fast enough to prevent O_2 depletion. Of course, DO is not the only species that can act as an electron acceptor. In theory, any species with an electrochemical reduction potential more positive than that of the conduction band can act as an electron acceptor. Indeed, the degradation of certain organic components may follow a reductive pathway. If the rate of reduction of O_2 is rate-limiting in the photocatalytic oxidation of organic pollutant, the addition of an electron acceptor that could withdraw electrons from the conduction band more rapidly should increase the rate of oxidation of the organic compound and the quantum yield of the process. Prairie et al. (1993a,b) reported increased rates for the photocatalytic oxidation of organic compounds when metal ions were used as electron acceptors in the absence of O_2. The organic compounds studied were salicylic acid, methanol, formic acid, ethylenediaminetetraacetic acid (EDTA), phenol, nitrobenzene, and other monosubstituted benzenes. The metal ions that gave rates of oxidation greater than with O_2 as the electron acceptor were Pt^{4+}, Hg^{2+}, Au^{3+}, Ag^{1+}, and Cr^{6+}. Brezova et al. (1995) reported that the addition of Fe^{3+} ions to TiO_2 suspensions, in the presence of O_2, increased the rate of degradation of phenol. Pelizetti et al. (1998) reported that the presence of metals such as Ag^+ and Fe^{3+} increased the activity of TiO_2 for the degradation of polychlorinated dioxins. However, Serpone et al. (1995) reported that the use of N_2O as a scavenger of conduction band electrons had no effect on the degradation of phenol in the absence of O_2 and suggested that the role of O_2 was more than just that of an electron acceptor. In the experiments investigating the degradation of phenol on TiO_2 photoanodes in a two-compartment cell, where the anode compartment was depleted of O_2, it was observed that the TiO_2 surface became discolored with a brown material that is presumed to be polyphenol in nature. Under UV irradiation in air, the TiO_2 surface recovered the bright white appearance. In the absence of DO, phenolic radicals generated by hole transfer reactions can combine to form polymeric species. Therefore, DO does not act merely as an electron scavenger in the photocatalytic oxidation of some organic pollutants (Byrne 1997).

3.4 Reaction Kinetics

It is widely reported that the intrinsic kinetics (where mass transfer is not limiting) of the photocatalytic degradation of organic substrates, using oxygen as the electron acceptor, fits a Langmuir–Hinshelwood kinetic scheme (Equation 3.2):

$$R_i = -d[S]_i/dt = k_S K_S [S]_i / (1 + K_S [S]_i), \quad (3.2)$$

where R_i is the initial rate of substrate removal, $[S]_i$ is the initial concentration of the organic substrate, and, traditionally, K_S is taken to be the Langmuir adsorption constant of species S on the surface of the semiconductor and k_S is a proportionality constant that provides a measure of the intrinsic reactivity of the photoactivated surface with S. Generally, the value of K_S derived from a kinetic study is not directly equivalent to the dark Langmuir adsorption isotherm for S on the semiconductor, with the latter values usually being much smaller.

The properties of oxide semiconductor materials will be significantly affected by the pH of an aqueous solution, including the surface charge on the semiconductor particles, the size of the aggregates formed, and the energies of the conductance and valence bands. Metal oxide particles suspended in water are amphoteric. For example, with TiO_2, the principal amphoteric functionality is the titanol moiety, TiOH. TiO_2 behaves as a simple diprotic acid in water. Hydroxyl groups on the surface of the TiO_2 undergo the following acid–base equilibria (Equations 3.3 and 3.4):

$$TiOH_2^+ \leftrightarrow TiOH + H^+ \rightarrow pK_{a1}, \quad (3.3)$$

$$TiOH \leftrightarrow TiO^- + H^+ \rightarrow pK_{a2}. \quad (3.4)$$

The point of zero charge (Equation 3.5) is given by one half of the sum of the two surface pK_as:

$$pH_{pzc} = 1/2(pK_{a1} + pK_{a2}). \quad (3.5)$$

For TiO_2 (P25), the surface acidity constants have been reported to be $pK_{a1} = 8$ and $pK_{a2} = 4.5$, resulting in a $pH_{pzc} = 6.25$. This implies that adsorption of anionic species will be favored at solution pH values <6.25 and adsorption of cationic species will be favored at pH values >6.25. The pH will also affect the particle size in solution, causing an increase or decrease in the agglomeration of the suspended particles, depending on the ζ-potential. As a rule of thumb, a ζ-potential of ±30 mV will result in dispersion.

Furthermore, the band edge potentials will also be shifted to become more negative by 59 mV with every increase of pH unit. Despite this, the rate of reaction for TiO_2 photocatalysis is not usually found to be strongly pH-dependent; typically, R_i varies by less than one order of magnitude from pH 2 to pH 12. Also, it has been found that the rate of photocatalytic mineralization is not strongly dependent upon pH between values ranging from 4 to 10 (Fernández-Ibáñez et al. 1999). The pH of water causes significant changes on the TiO_2 surface, including the electrical surface charge of the TiO_2 particles, the size of the aggregates it forms, and the positions of the conduction and valence bands (Fernández-Ibáñez et al. 2000). Furthermore, during photocatalytic reactions, intermediate by-products maybe be produced, which may behave differently depending on the solution pH. Therefore, a detailed analysis of the pH conditions should include not only the initial substrate but also the degradation intermediates produced during the process. Measurement of an overall parameter as dissolved organic carbon (DOC) or chemical oxygen demand (COD) should be used for determining the optimum pH, or at least for determining the effect of pH on the mineralization process.

Generally, it is reported that inorganic species such as perchlorate and nitrate anions have little effect on the reaction kinetics where TiO_2 is the photocatalyst. However, anions such as sulfate, chloride, and phosphate can significantly reduce the rate of reaction by 20%–70%, especially at concentrations greater than 10^{-3} mol/dm^3. This is mainly due to competitive adsorption between the inorganic ions and the organic species on the TiO_2 surface at the reaction sites.

3.5 Pollutants Degraded by Photocatalysis

Many processes utilized for water and wastewater treatment are removal processes involving phase transfer, for example, activated carbon. Photocatalysis is a destructive process and the pollutants undergo a systematic prolonged attack by radical species, leading eventually to mineralization. The fractional conversion achieved will be dependent on the residence time in contact with the photocatalyst. It is, therefore, necessary to ensure that the intermediate reaction products are either completely degraded in the process or that they do not pose a problem in the treated water. The mechanisms of photocatalysis are discussed in more detail in reviews by Mills et al. (1993), Fox and Dulay (1993), Hoffmann et al. (1995), and Mills and Le Hunte (1997). There is a large body of published literature relating to the application of photocatalysis for the degradation of a wide range of organic chemical pollutants, the treatment of inorganic pollutants, and the inactivation of a wide range of microoganisms. Table 3.1 summarizes some examples of pollutants that are

TABLE 3.1

Examples of Organic Pollutants Degraded by Semiconductor Photocatalysis

Class	Example
Alkanes	Methane, isobutane, pentane, heptane, cyclohexane, paraffin
Haloalkanes	Mono-, di-, tri-, and tetrachloromethane; tribromoethane; 1,1, 1-trifluoro-2,2,2-trichloroethane
Aliphatic alcohols	Methanol, ethanol, isopropyl alcohol, glucose, sucrose
Aliphatic carboxylic acids	Formic, ethanoic, dimethylethanoic, propanoic, oxalic acids
Alkenes	Propene, cyclohexene
Haloalkenes	Perchloroethene, 1,2-dichloroethene, 1,1,2-trichloroethene
Aromatics	Benzene, naphthalene
Haloaromatics	Chlorobenzene, 1,2-dichlorobenzene, bromobenzene
Nitrohaloaromatics	3,4-Dichloronitrobenzene, dichloronitrobenzene
Phenols	Phenol; hydroquinone; catechol; 4-methylcatechol; resorcinol; o-, m-, and p-cresol
Halophenols	2-, 3-, and 4-Chlorophenol; pentachlorophenol; 4-fluorophenol; 3,4-difluorophenol
Aromatic carboxylic acids	Benzoic, 4-aminobenzoic, phthalic, salicylic, m- and p-hydroxybenzoic, chlorohydroxybenzoic acids
Polymers	Polyethylene, poly(vinyl chloride) (PVC)
Surfactants	Sodium dodecyl sulfate (SDS), polyethylene glycol, sodium dodecyl benzene sulfonate, trimethyl phosphate, tetrabutylammonium phosphate
Herbicides	Methyl viologen, atrazine, simazine, prometron, propetryne, bentazon
Insecticides	DDT, parathion, lindane
Dyes	Methylene blue, rhodamine B, methyl orange, fluorescein
Pharmaceuticals	Trimethoprim, ofloxacin, enrofloxacin, clarithromycin and erythromycin, acetaminophen, diclofenac, carbamazepine

Source: Updated from Mills, A. and Le Hunte, S., *J. Photochem. Photobiol. A Chem.* 108, 1–35, 1997. With permission.

reported to have been effectively degraded by photocatalysis (at least under laboratory conditions).

Some of the more challenging pollutants to be degraded include persistent organic pollutants (POPs). These are becoming an increasing problem in the aquatic environment. Such substances can enter water from various sources and may not be effectively removed by conventional wastewater treatment processes. Pesticides have been classed as POPs due to their resistance to natural degradation processes and they can remain in the environment for long periods of time. By their very nature, they are designed to be toxic and kill unwanted organisms. They act by interfering with the biochemical and physiological processes that are common to a wide range of living systems, for example, parathion used for the control of insects in crops affects the central nervous system and the liver, whereas atrazine used for the control of broad leaf and grassy weeds inhibits photosynthesis. Although these compounds are designed to be organism-specific, they can attack nontarget organisms and, as a result, cause serious environmental damage. Atrazine is one of the most common pesticides found in groundwater sources and drinking water supplies. In some countries, the use of atrazine is banned while in others restrictions on its use have been implemented. Atrazine has been detected above the recommended levels (0.1 ppb or μg/L) throughout Europe and the United States and is considered as a priority pollutant in the European Union. It is a very stable compound due to the s-triazine ring, which inhibits natural degradation and it can persist in the environment for very long periods. It has also been reported to be an endocrine-disrupting chemical (EDC) (one that interferes with the hormone systems within the organisms). Photocatalysis has been shown to be effective in the degradation of a wide range of pesticides including the triazine herbicides, with TiO_2 being the most widely employed research photocatalyst for pesticide destruction in water. Out of all pesticides reported, the triazine herbicides are the only group resistant to total mineralization. So far, techniques such as ozonation and activated carbon adsorption have been considered to help eliminate atrazine from water. In 2006, the photocatalysis research group at the University of Ulster reported on the photocatalytic degradation of atrazine. In this work, we used a laboratory-scale stirred tank reactor (Figure 3.2), designed to study the intrinsic kinetics of degradation on immobilized photocatalyst films (McMurray et al. 2006). The photocatalytic degradation of the pesticide, atrazine, and the formation of intermediates, were followed using high performance liquid chromatography (HPLC), total organic carbon (TOC), and liquid chromatography mass spectrometry (LC–MS).

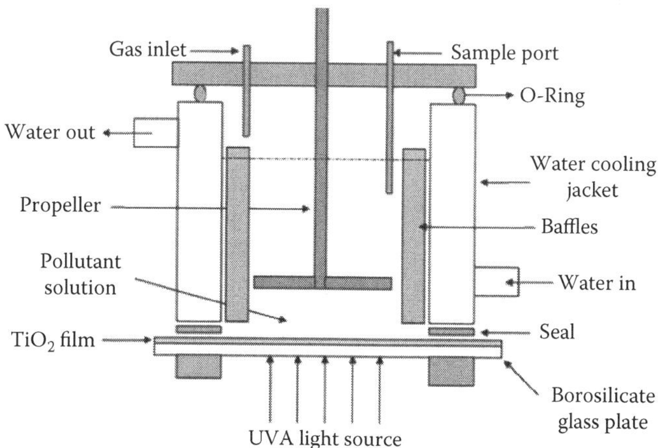

FIGURE 3.2
Custom-made stirred tank reactor for studying the intrinsic kinetics of photocatalytic degradation on photocatalytic films supported on glass or other UV transparent supporting substrates. (From McMurray, T.A., Dunlop, P.S.M., and Byrne, J.A., *J. Photochem. Photobiol. A Chem.*, 182, 43–51, 2006. With permission.)

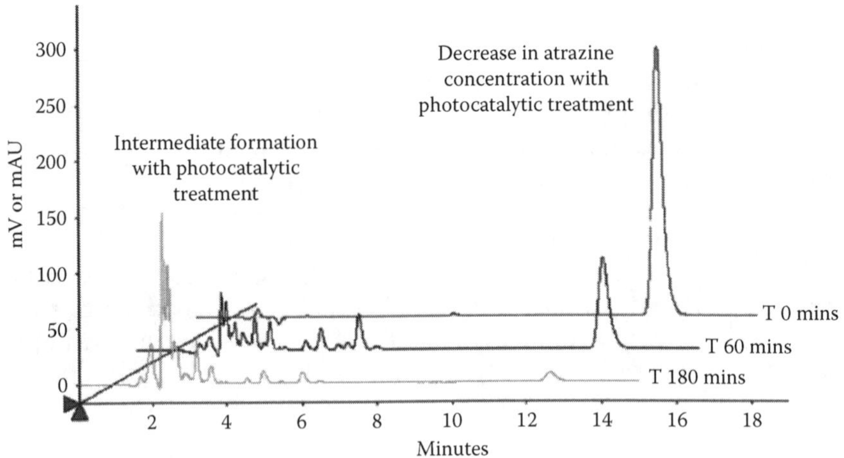

FIGURE 3.3
HPLC spectrum for time 0, 60, and 180min of photocatalytic treatment, showing decrease in atrazine and the increasing formation of the intermediates. (From McMurray, T.A., Dunlop, P.S.M., and Byrne, J.A., *J. Photochem. Photobiol. A Chem.*, 182, 43–51, 2006. With permission.)

The complete degradation of the parent compound, atrazine, was confirmed by HPLC (see Figure 3.3). However, the removal of TOC was much slower than the degradation of the parent compound and complete TOC removal was not observed. This was attributed to the removal of the carbon within the atrazine side chains.

A range of stable end products were determined, including cyanuric acid. Complete mineralization of atrazine to CO_2, water, and mineral acids was not observed within the timeframe of the experiment; however, the parent compound, which is both toxic and has suspected endocrine-disrupting properties, was degraded to a less problematic compound, cyanuric acid. Photocatalysis is therefore a possible treatment for water sources contaminated with atrazine.

In the degradative treatment methods using AOPs, it is not enough just to monitor the disappearance of the parent compound. In some cases, the toxicity of the intermediate breakdown products may be similar or greater than that of the parent compound. Therefore, it is desirable to monitor the toxicity of the intermediate products during the reaction to ensure that the treatment time is long enough to completely remove any toxic intermediates. For example, we studied the genotoxicity of p-nitrophenol (PNP) treated using photocatalysis (Shekler et al. 2004). Pesticide residues represent a major group of pollutants, particularly insecticides based on a parathion backbone. Parathion is a potent inhibitor of mammalian cholinesterase and is therefore highly toxic to humans. PNP is an analog of parathion and is a toxic intermediate in the photodegradation pathway of the parathion species. The photocatalytic degradation of organic pollutants proceeds via a

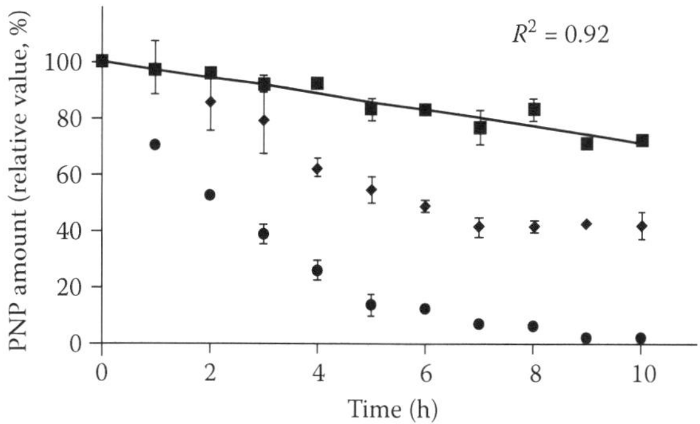

FIGURE 3.4
The relative values of the PNP amount (•), induction factor values (♦), and TOC values (▪) during the photocatalytic degradation of PNP. (From Sekler, M.S., Levi, Y., Polyak, B., Novoa, A., Dunlop, P.S.M., Byrne, J.A., and Marks, R.S., *J. Appl. Toxicol.*, 24, 395–400, 2004. With permission.)

series of intermediate steps, and it is important to assess the toxicity of the treated water as a function of the treatment time. In this work, we used a toxicity bioassay, the Vitotox* system, where the genes that trigger bioluminescence have been placed in *Salmonella typhimurium*. The bacteria emit light upon contact with the toxic substances, and the biotoxicity is easily quantified by measuring the extent of light production. The reactor used TiO_2 (99% anatase from Sigma Aldrich) immobilized on a borosilicate glass spiral tube. The light source was a UVA fluorescent lamp (TL 8W/05, main line emission 365 nm; Philips, The Netherlands), which was placed in the center of the glass spiral. Figure 3.4 shows the relative amount of PNP as a function of the photocatalytic treatment time as determined by TOC analysis, HPLC for the parent compound, and the induction factor as determined using the toxicity bioassay. It is clear that although the parent PNP is rapidly degraded, the induction factor is not degraded as rapidly and therefore intermediate products formed during the photocatalytic degradation have a measurable toxicity. Therefore, it is vital that one must be aware of the potential toxicity of the intermediate products formed in any degradative process used for wastewater treatment.

There is considerable concern about the environmental effects of estrogens and xenoestrogens in water supplies. These "estrogenic substances" act as EDCs and reported effects include the perturbation of sexual differentiation in embryos, cancers in reproductive organs, and altered glucose and fat metabolism. Some have proposed a link between EDCs in the environment and an observed decline in human male fertility. Other related effects in humans are increases in testicular cancer and breast cancer. The three main human estrogen compounds are 17-β-estradiol, estriol, and estrone, of which the former is the most potent. Its structure is shown in Figure 3.5.

The estrogens can exhibit biological effects at very low levels of concentration (~0.1 ng/L) and their detection at these concentrations is not possible by

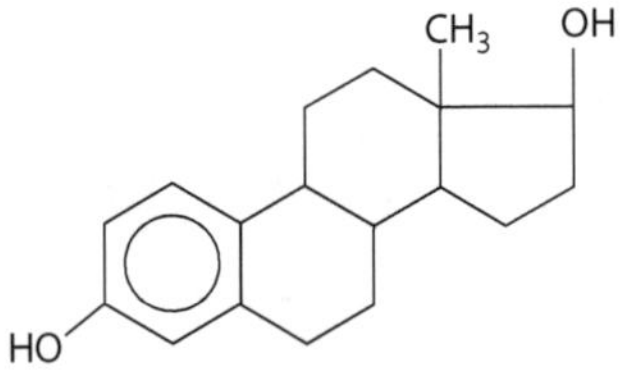

FIGURE 3.5
Structure of 17β-estradiol.

normal analytical methods. Estrogens may enter the aquatic environment from contraceptive pill residues, hormone replacement therapy residues, and diethylstilbestrol residues (used to promote livestock growth). Perhaps, the cause for greater concern are xenoestrogen substances or estrogen mimics. There is a wide range of diverse compounds that mimic the biological effects of estrogens. Substances classed as xenoestrogens include several pesticides, for example, DDT analogs and polychlorinated biphenyls (PBCs), and also alkylphenol polyethoxylates, which are surfactants used in detergents. Although not as potent as the estrogens, xenoestrogens are more abundant in the aquatic environment. Normal methods of water and sewage treatment are not completely effective in removing estrogenic substances. We demonstrated the potential for the use of photocatalysis to eliminate estrogenic substances (Coleman et al. 2000). However, it is not just enough to examine the degradation of the parent compound as the intermediate breakdown products may also show estrogenic activity. In another work, we examined the effect of photocatalysis on the estrogenic activity using a recombinant yeast bioassay (Coleman et al. 2004). Effectively, the yeast is genetically modified to respond to estrogenic substances. As can be seen in Figure 3.6, with

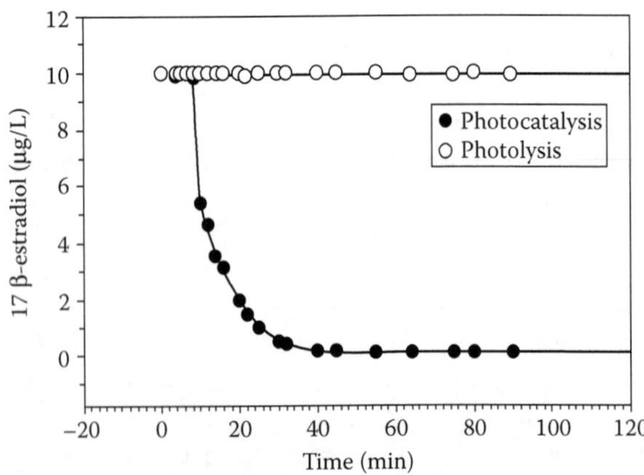

FIGURE 3.6
Photocatalysis and UVA photolysis of 17β-estradiol. (From Coleman, H.M., Routledge, E.J., Sumpter, J.P., Eggins, B.R., and Byrne, J.A., *Water Res.*, 38, 3233–3240, 2004. With permission.)

photocatalytic treatment (using Degussa P25 TiO_2 immobilized on titanium alloy), the estrogenic activity of 17β-estradiol was more rapidly destroyed as compared with UVA photolysis alone. Biological activity within the reactor is expressed as the amount of 17β-estradiol in a solution that is needed to produce the same magnitude response in the assay.

Thus, photocatalysis has been shown to be effective for the degradation of POPs and the emerging contaminants. However, photocatalysis should be seen as a complementary treatment technology for other biological and/or physicochemical processes. It may be applied for tertiary treatment of wastewaters that contain pollutants that cannot be effectively removed using conventional treatments. It may also be applicable for industrial wastewaters that contain refractory organics but should not be applied for secondary wastewater treatment where there are high levels of organic material that can be treated by biological processes, that is, domestic sewage or food-processing waste. There are a number of issues to be addressed for photocatalysis to be widely adopted as a robust treatment technology for the tertiary treatment of wastewater.

3.6 Reactor Engineering

In laboratory-scale experiments investigating photocatalytic destruction of chemical pollutants and the inactivation of microorganisms, many researchers employ a basic photoreactor system utilizing the photocatalyst as a powder suspension (slurry) in water. The optimum catalyst loading of a suspension reactor will be determined by the reactor configuration and the incident light intensity falling on the reactor. Below a certain catalyst loading, not all the incident photons may be absorbed and above a certain catalyst loading, the catalyst will begin to effectively mask itself with all incident light being absorbed by only a small fraction of the catalyst. In a suspension reactor where the depth of the suspension is greater than the light penetration, the degree of agitation of the suspension will be an important factor in determining the frequency at which each particle will be exposed to the incident radiation. The photocatalyst loading in suspension reactors has been investigated by various workers and optimum loadings reported range from 0.5 to >10 g/dm^3 (Matthews 1986; Okamoto et al. 1985). The optimum loading determined for a particular reactor configuration will be strongly dependent on the optical path length for the reactor.

As the surface area increases with decreasing particle size, one would expect that greater photocatalytic efficiency should be achieved with smaller particles. Size quantization in particles below 100 nm diameter can result in the blue shift of the band gap and an increase in the absorption coefficient. The use of small particles in water treatment obviously introduces the problem of posttreatment catalyst recovery. For example, Degussa P25 is a commercial form of TiO_2, which is commonly used in photocatalytic research. It has a small primary particle size (~30 nm) and is hydrophilic in nature, forming well-dispersed suspensions in water. The particles do tend to agglomerate to some extent, giving a particle size between 0.2 and 0.45 μm. In this agglomerated size range, the TiO_2 can remain in aqueous suspension for a period of a few days under quiescent conditions. The particle-settling velocity (v_s) is described by Stokes Law (Equation 3.6):

$$v_s = \frac{g(\rho_p - \rho)d_p^2}{18\mu}, \quad (3.6)$$

where ρ_p is the particle density, ρ is the density of water, d_p is the particle hydrodynamic diameter, and μ is the absolute viscosity of water. Taking the density of TiO_2 as 4.23 g/cm^3, it is calculated that it would take around 6 days for a 1.0 μm particle to settle 1 m. Posttreatment catalyst recovery would add to the overall capital and running costs of any system. Nevertheless, some workers have developed pilot-scale and even full-scale water treatment systems employing TiO_2 in suspensions (Zhang et al. 1994; Mehos and Turchi 1993).

Alternatively, the catalyst may be immobilized onto a suitable solid support matrix, which would negate the need for posttreatment removal. When immobilization of the catalyst is attempted, there is a concurrent large decrease in the surface area of the catalyst available for reaction. In addition, the catalyst must adhere to the solid supporting substrate in the reactor and, unless the supporting substrate is transparent to the band gap light, the reactor design is limited by optical constraints. Furthermore, mass transport in immobilized reactor systems would present a major problem.

Several methods for the immobilization of a photocatalyst onto a solid supporting substrate have been investigated. Where possible, the substrate used should have the following qualities:

- Be able to withstand environmental stress
- Have no deleterious effect on the catalyst
- Be nontoxic
- Inexpensive
- Transparent to wavelengths of light required for photocatalyst excitation

A wide range of reactor configurations has been reported in the literature. Some are simply photoreactors used for photolysis with immobilized or suspended photocatalyst added. Large-scale experiments have also been undertaken using parabolic trough concentrators adapted from solar thermal reactors. In the design of any photocatalytic reactor (including laboratory-scale reactors), the following key parameters must be addressed:

- Light source, intensity, wavelength, and radiation distribution
- Suspension or immobilized catalyst to be used
- Electron acceptor to be employed and introduced to the system
- Mass transfer efficiency within the system
- Concentration of the pollutant to be treated and fractional conversion required
- Interfering constituents in the water to be treated

The efficiency of solar photocatalysis, using wide band gap semiconductors such as TiO_2, will depend mainly on the UVA wavelengths present in sunlight. Solar UV at sea level is composed of roughly similar portions of both direct and diffuse electromagnetic radiations. Without cloud cover, the solar UVA spectrum is ~60% direct and ~40% diffuse. Therefore, the use of concentrating systems based on nonimaging optics with low concentrating factor has a clear potential compared with the imaging optics-based systems.

Work at the Plataforma Solar de Almeria (PSA) in Spain has focused on the use of nonconcentrating solar collectors for the enhancement of solar water treatment. Compound parabolic collectors (CPCs) are nonimaging concentrating systems with a diffuse focus. The concentrated rays are homogeneously distributed in the absorber. Their main advantage is that they concentrate diffuse radiation. Hence, they do not rely solely on direct solar radiation and are effective even on cloudy days. In addition, they concentrate radiation independent of the direction of sunlight and do not require sun tracking (in contrast to direction-dependent image forming systems). The major advantage with CPCs is that the concentration factor remains constant for all values of Sun zenith angle within the acceptance angle limit. Therefore, CPC enhancement can be utilized in the design of larger-scale water treatment systems. Figures 3.7 and 3.8 show photographs of CPC reactors at PSA, Spain.

FIGURE 3.7 (See color insert)
Photograph of a CPC reactor at PSA, Spain.

FIGURE 3.8 (See color insert)
Photograph showing CPC with borosilicate glass tubes (PSA, Spain).

Figure 3.9 shows a schematic of the overall photocatalytic water treatment unit. A photoreactor for use in solar photocatalytic wastewater treatment should have the following attributes:

- High illuminated volume/total volume ratio
- Operate under a low flow rate when utilizing immobilized photocatalyst to maximize the residence time in flow systems

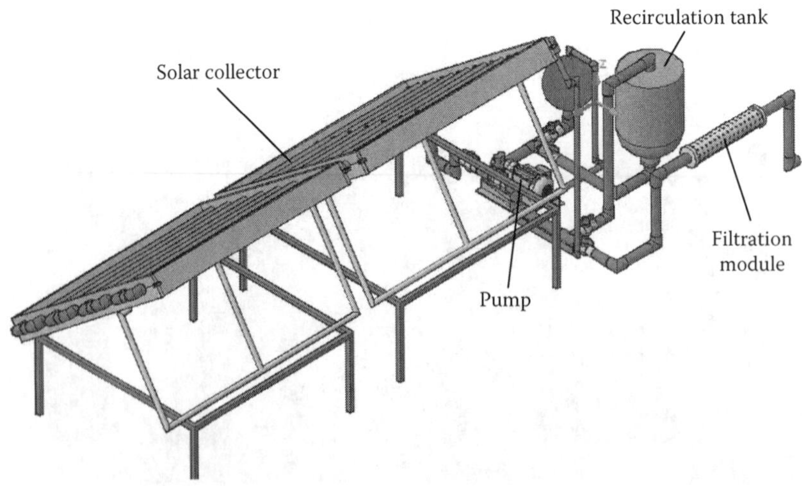

FIGURE 3.9 (See color insert)
Schematic of a CPC photocatalytic water treatment system.

- Include a CPC of good-quality, that is, high UVA reflectivity (for aluminum use to be 87%–90%) and that decreases minimally following environmental exposure
- A reaction vessel with high (90%) UVA transmission, for example, borosilicate glass, and robust under environmental conditions
- Low life cycle cost
- Low environmental impact
- Low maintenance requirements and easy access to replacement parts if necessary
- Low power requirements

In photocatalytic water treatment, the electron acceptor is normally DO, which is easily available in the air. In the presence of high concentrations of dissolved organics, the concentration of DO will be rapidly depleted and must be replenished to maintain photocatalytic activity. Furthermore, the solubility of oxygen in water decreases with increasing temperature, and the temperature within solar-irradiated reactors can reach 55°C. Novel reactor design must address the need for replenishment of DO in photocatalytic systems. Other oxidants, for example, H_2O_2, may be added to overcome the limitations of DO concentration; however, this would give rise to a dependence on consumable chemicals, which is undesirable.

3.7 VLA Photocatalytic Materials

As emphasized, the overall efficiency of TiO_2 under natural sunlight is limited by the need for UV excitation (for anatase, $\lambda \leq 387\,nm$). UV accounts for only around 5% of the incoming solar energy on the Earth's surface. Therefore, there has been a strong research effort in shifting the absorption spectrum of TiO_2 toward the visible region of the electromagnetic spectrum to make use of the greater availability of visible light photons. Various approaches have been attempted, including doping the TiO_2 with metal ions (Hamilton et al. 2008). One of the more promising approaches to achieve visible light activity is doping with nonmetal elements including N and S. For example, Asahi et al. (2001) reported the visible-light photoactivity of TiO_2 with nitrogen doping. Since then, many groups have demonstrated that nonmetal doping of TiO_2 extends the optical absorbance of TiO_2 into the visible-light region. The doping of TiO_2 may give rise to a color change in the material as a result of the absorption of visible light; however, an increase in visible absorption, in principle, does not guarantee visible light–induced activity.

Photocatalytic reactions proceed through redox reactions by photogenerated positive holes and photoexcited electrons. No activity may be observed if, for example, all of these species recombine. Various photocatalytic test systems with different model pollutants/substrates have been reported. Dyes are commonly used as model pollutants, partly because their concentration can be easily monitored; however, dyes also absorb visible light, and the influence of this photoabsorption by dyes should be excluded for the evaluation of the real photocatalytic activity of materials. Herrmann (2010) reported that a real photocatalytic activity test can be erroneously claimed if a noncatalytic side reaction or an artifact does occur. Dye decolorization tests can represent the most "subtle pseudo-photocatalytic" systems, hiding the actual noncatalytic nature of the reaction involved. An example of this dye-sensitized phenomenon was reported with the apparent photocatalytic "disappearance" of indigo carmine dye (Vautier et al. 2001). The indigo carmine was totally destroyed by UV-irradiated titania; however, its color also disappeared when using visible light, but the corresponding TOC remained intact. The loss of color actually corresponded to a limited transfer of electrons from the photoexcited indigo (absorbing in the visible) to the TiO_2 conduction band. Dye sensitization is well known and exploited in the "Gratzel" dye-sensitized photovoltaic cell (Oregan and Gratzel 1991). A dye that has been used widely as a test substrate for photocatalytic activity is methylene blue. Indeed, the degradation of methylene blue is a recommended test for photocatalytic activity in the ISO/CD10678. Yan et al. (2006) used methylene blue as a test substrate to evaluate the visible-light activity for S–TiO_2. Their results showed that a photoinduced reaction by methylene blue photoabsorption may produce results that could be mistaken to be evidence of visible-light photocatalytic activity. Therefore, Yan et al. (2006) recommend the use of model organic substrates that do not absorb in the spectral region being used for excitation.

The photoreactor to be used in the photocatalytic test reaction must be appropriate, and it is good practice to compare any novel material with a relatively well-established photocatalyst material, for example, Aeroxide (Degussa) P25 (Dunlop et al. 2010). Obviously, the catalyst should be compared in the same form, that is, suspension or immobilized. If one is employing a suspension system, the catalyst should be well dispersed and one should consider an analysis of the particle size distribution. Furthermore, the optimum loading for each catalyst should be determined if possible. Where an immobilized catalyst system is employed, the test reactor should be designed in such a way as to maximize mass transfer to the catalyst, otherwise the rate of degradation will simply be reflecting the mass transfer characteristics of the reactor. A high flow or a stirred tank system may be employed in an attempt to determine the intrinsic kinetics of the photocatalytic system.

There is a need to develop VLA photocatalytic materials for the destruction of organic pollutants in water; however, attention needs to be paid to the photocatalysis test protocols to ensure that the VLA materials are effective photocatalysts. Where possible,

- The light source should be appropriate with respect to the application and the emission spectrum should be known.
- More than one test substrate should be used, for example, the multi-activity assessment proposed by Ryu and Choi (2008).
- Substrates absorbing light within the emission spectrum of the light source are avoided.
- The reactor should be characterized, that is, for suspension systems the particle size distribution is determined. For immobilized systems, the photoreactor should be appropriate and well characterized in terms of mass transfer.
- The photonic efficiencies or formal quantum efficiencies (FQEs) are reported along with the emission spectrum of the illumination source. The FQE is given by

$$\text{FQE} = \text{rate of reaction/incident light intensity for polychromatic source}. \quad (3.7)$$

For multielectron photocatalytic degradation processes, the FQE will be much less than unity, unless a chain reaction is in operation. Therefore, it is most important that researchers specifically report their methods of quantum efficiency determination. Research and development for solar-driven water treatment should utilize experiments under simulated or real solar irradiation, not just visible-light sources. More research is required to determine if VLA materials can increase the efficiency of photocatalysis under solar irradiation.

3.8 Issues to be Addressed

For wastewater treatment, the photocatalytic reactor system should be robust, noncomplex, and require only low-level maintenance if possible. Therefore, photocatalyst regeneration stages are undesirable but will most likely be required. There needs to be more research into the longevity of photocatalyst materials under real working conditions. Reactors employing immobilized photocatalyst materials are desirable to reduce the complexity of the treatment system; however, catalyst stripping may be a problem due to erosion if the immobilization protocol does not produce a robust hard-wearing coating. Also, catalyst fouling by inorganic species present in the water can lead to a reduction in the photocatalytic efficiency over time.

Some researchers have undertaken pilot- or large-scale studies under real Sun conditions using synthetic and/or real wastewater. For example, Miranda-García et al. (2010), based at the PSA, investigated the degradation of 15 emerging contaminants in a photocatalytic pilot plant utilizing TiO_2 immobilized onto glass beads. The reactor utilized CPCs and consisted of two modules of 12 borosilicate glass tubes mounted on a fixed platform tilted 37° (local latitude). Two of the glass tubes were packed with the TiO_2-coated glass spheres. The total illuminated area was 0.30 m^2 and total volume was 10 L (0.96 L was the irradiated volume). The reactor was operated in a recirculating batch mode with a flow rate of 3.65 L/min. The degradation of the organic pollutants was achieved under solar irradiation, and, importantly, after five cycles, the photocatalyst activity was not decreased significantly. Bernabeu et al. (2011) investigated solar photocatalysis as a tertiary treatment for wastewater containing emerging pollutants (EPs). Pollutants examined included acetaminophen, diclofenac, caffeine, thiabendazole, acetamiprid, trimethoprim, ofloxacin, and carbamazepine. The experiments were carried out with real wastewater using two different pilot plants for wastewater detoxification (Solardetox Acadus-2006) based on compound parabolic collectors (CPCs) with a concentration factor of 1. One reactor (3.0 m^2 irradiated surface with 24 L irradiated volume) was located at the wastewater treatment plant and was employed to treat real wastewaters. It was loaded with 50 L of wastewater in each experiment. Experiments were undertaken under real Sun irradiation in Spain. Experiments with spiked samples (extra 5 mg/L of each pollutant, acetaminophen, diclofenac, caffeine, thiabendazole, acetamiprid, and trimethoprim) were carried out in a Solardetox Acadus-2001 (Ecosystem) plant, with a total surface of 0.26 m^2, and the irradiated volume was 1.83 L, which was located at the laboratories of Universidad Politécnica de Valencia. Where titanium dioxide was used as photocatalyst, they used Degussa P25 as a slurry with a loading between 0.2 and 0.5 g/L. Aqueous samples containing the EPs were taken from the effluent of the secondary settling tank of a wastewater treatment plant from the East of Spain. The plant consisted of a physical–chemical primary treatment followed by an aerobic activated sludge biological reactor. The treated water had the following characteristics: pH, 7.5; conductivity, 2.4–2.8 mS/cm; DOC, 15–50 mg/L; IC, 70–90 mg/L; and COD, 60–120 mg/L. They determined the concentration of pollutants at the inlet of the aeration tank of the biological reactor and at the effluent of the secondary settlement tank and found them in the range between 0.03 and 15 μg/L: five antibiotics (trimethoprim, ofloxacin, enrofloxacin, clarithromycin, and erythromycin), one analgesic (acetaminophen), one anti-inflammatory drug (diclofenac), one psychiatric drug (carbamazepine), one stimulating drug (caffeine), a fungicide (thiabendazole), and a pesticide (acetamiprid). Most of the EPs remained unaffected by the biological treatment. The solar photocatalytic treatment of the wastewater as a tertiary step proved to be effective for the destruction of almost all EPs studied (Figure 3.10). In addition to the destruction of the EPs

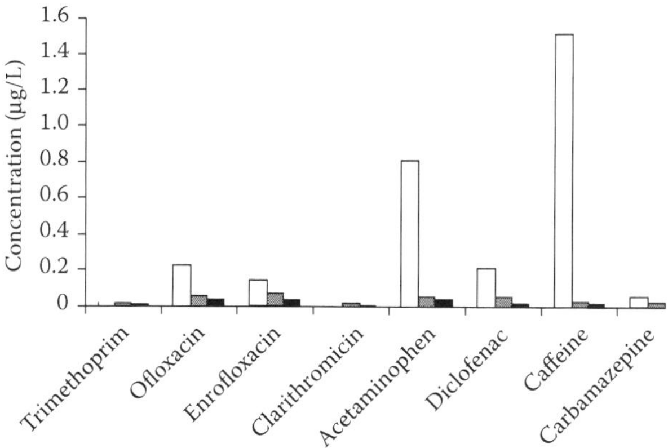

FIGURE 3.10
Concentration of emerging pollutants at the outlet of a wastewater treatment plant (white bars), after 3h (gray bars) and 6h (black bars) of solar photocatalysis with 0.2g/L of TiO_2. (From Bernabeu, A., Vercher, R.F., Santos-Juanes, L., et al., *Catal. Today*, 161, 235–240, 2011. With permission.)

studied, they found that photocatalysis was also effective for the inactivation of coliforms in the wastewater.

3.9 Conclusions

Semiconductor photocatalysis is an advanced oxidation technology that utilizes a semiconductor material along with light energy to produce reactive oxygen species, leading to the destruction of organic contaminants in water and the inactivation of microorganisms. Titanium dioxide is currently the most appropriate choice of photocatalyst for the treatment of water, although it is a wide band gap semiconductor and requires UV excitation. Only 5% of the solar spectrum is in the UV domain and this somewhat limits the solar efficiency of TiO_2 photocatalysis; however, it has been demonstrated, both on small scale and pilot scale, that solar TiO_2 photocatalysis is effective for the degradation of harmful pollutants, including emerging contaminants. Future research will focus on the development of VLA photocatalytic materials, which can utilize more of the solar spectrum. Solar photocatalysis is a promising clean technology that could be utilized as a tertiary treatment for wastewater to facilitate water recycling and reuse. Furthermore, photocatalysis may be utilized for the photocatalytic enhancement of the solar disinfection of water in developing countries (Byrne et al. 2011).

References

Alfano, O.M., D. Bahnemann, A.E. Cassano, R. Dillert, and R. Goslich. 2000. Photocatalysis in water environments using artificial and solar light. *Catal. Today* 58(2–3): 199–230.

Asahi, R., T. Morikawa, T. Ohwaki, K. Aoki, and K. Taga. 2001. Visible-light photocatalysis in nitrogen-doped titanium oxides. *Science* 293: 269–271.

Augugliaro, V., M. Litter, L. Palmisano, and J. Soria. 2006. The combination of heterogeneous photocatalysis with chemical and physical operations: A tool for improving the photoprocess performance. *J. Photochem. Photobiol. C Photochem. Rev.* 7(4): 127–144.

Bahnemann, D. 2004. Photocatalytic water treatment: Solar energy applications. *Sol. Energy* 77(5): 445–459.

Bernabeu, A., R.F. Vercher, L. Santos-Juanes, et al. 2011. Solar photocatalysis as a tertiary treatment to remove emerging pollutants from wastewater treatment plant effluents. *Catal. Today* 161: 235–240.

Brezova, V., E. Blazkova, M. Ceppan, and R. Fiala. 1995. The influence of dissolved metal ions on the photocatalytic degradation of phenol in aqueous TiO_2 suspensions. *J. Mol. Catal. A Chem.* 98: 109–116.

Byrne, J.A. 1997. Titanium dioxide photocatalysis for the treatment of polluted water. D.Phil. Thesis, University of Ulster.

Byrne, J.A., P.A. Fernandez-Ibanez, P.S.M. Dunlop, D.A. Alrousan, and J.W.J. Hamilton. 2011. Photocatalytic Enhancement for Solar Disinfection of Water: A Review. *Int. J. Photoenergy*, 2011: 1–12, Article ID 798051.

Coleman, H.M., B.R. Eggins, J.A. Byrne, F.L. Palmer, and E. King. 2000. Photocatalytic degradation of 17-β-oestradiol on immobilised TiO_2. *Appl. Catal. B Environ.* 24: 1–5.

Coleman, H.M., E.J. Routledge, J.P. Sumpter, B.R. Eggins, and J.A. Byrne. 2004. Rapid loss of estrogenicity of steroid estrogens by UVA photolysis and photocatalysis over an immobilised titanium dioxide catalyst. *Water Res.* 38: 3233–3240.

Dunlop, P.S.M., A. Galdi, T.A. McMurray, J.W.J. Hamilton, L. Rizzo, and J.A. Byrne. 2010. Comparison of photocatalytic activities of commercial titanium dioxide powders immobilised on glass substrates. *J. Adv. Oxid. Technol.* 13(1): 99–106.

Fernández-Ibáñez, P., S. Malato, and F.J. de las Nieves. 1999. Relationship between TiO_2 particle size and reactor diameter in solar photodegradation efficiency. *Catal. Today* 54: 195–204.

Fernández-Ibáñez, P., F.J. de las Nieves, and S. Malato. 2000. Titanium Dioxide/Electrolyte Solution Interface: Electron Transfer Phenomena. *J. Colloid Interface Sci.* 227: 510–516.

Fox, M.A. and M.T. Dulay. 1993. Heterogeneous photocatalysis. *Chem. Rev.* 93(1): 341–357.

Fujishima, A., X.T. Zhang, and D.A. Tryk. 2008. TiO_2 photocatalysis and related surface phenomena. *Surf. Sci. Rep.* 63(12): 515–582.

Gaya, U.I. and A.H. Abdullah. 2008. Heterogeneous photocatalytic degradation of organic contaminants over titanium dioxide: A review of fundamentals, progress and problems. *J. Photochem. Photobiol. C Photochem. Rev.* 9(1): 1–12.

Gerischer, H. 1993. Photocatalytic Purification and Treatment of Water and Air. In *Photocatalytic Purification and Treatment of Water and Air*, eds. Ollis, D.F. and H. Al-Ekabi. Amsterdam: Elsevier Science, pp. 1–18.

Gerischer, H. and A. Heller. 1991. The role of oxygen in photooxidation of organic molecules on semiconductor particles. *J. Phys. Chem.* 95: 5261–5267.

Goswami, D.Y. 1997. A review of engineering developments of aqueous phase solar photocatalytic detoxification and disinfection processes. *J. Sol. Energy Eng. Trans. ASME* 119(2): 101–107.

Hamilton, J.W.J., J.A. Byrne, C. McCullagh, and P.S.M. Dunlop. 2008. Electrochemical Investigation of Doped Titanium Dioxide. *Int. J. Photoenergy*, Vol. 2008, Article no. 631597: 1–8.

Herrmann, J.M. 2010. Photocatalysis fundamentals revisited to avoid several misconceptions. *Appl. Catal. B Environ.* 99: 461–468.

Herrmann, J.M., C. Guillard, and C. Pichat. 1993. Heterogeneous photocatalysis: An emerging technology for water treatment. *Catal. Today* 17(1–2): 7–20.

Hoffmann, M.R., S.T. Martin, W.Y. Choi, and D.W. Bahnemann. 1995. Environmental Applications of Semiconductor Photocatalysis. *Chem. Rev.* 95(1): 69–96.

Malato, S., P. Fernandez-Ibanez, M.I. Maldonado, J. Blanco, and W. Gernjak. 2009. Decontamination and disinfection of water by solar photocatalysis: Recent overview and trends. *Catal. Today* 147: 1–59.

Matthews, R.W. 1986. Photo-oxidation of organic material in aqueous suspensions of titanium dioxide. *Water Res.* 20(5): 569–578.

McMurray, T.A., P.S.M. Dunlop, and J.A. Byrne. 2006. The photocatalytic degradation of atrazine on nanoparticulate TiO_2 films. *J. Photochem. Photobiol. A Chem.* 182: 43–51.

Mehos, M.S. and C.S. Turchi. 1993. Field testing solar photocatalytic detoxification on TCE-contaminated groundwater. *Environ. Prog.* 12(3): 194–199.

Mills, A., R.H. Davies, and D. Worsley. 1993. Water purification by semiconductor photocatalysis. *Chem. Soc. Rev.* 22(6): 417–425.

Mills, A. and S. Le Hunte. 1997. An overview of of semiconductor photocatalysis. *J. Photochem. Photobiol. A Chem.* 108: 1–35.

Miranda-García, N., M.I. Maldonado, J.M. Coronado, and S. Malato. 2010. Degradation study of 15 emerging contaminants at low concentration by immobilized TiO_2 in a pilot plant. *Catal. Today* 151: 107–113.

Okamoto, K., Y. Yamamoto, H. Tanaka, and A. Itaya. 1985. Heterogeneous photocatalytic decomposition of phenol over TiO_2 powder. *Bull. Chem. Soc. Jpn.* 58: 2023–2028.

Oregan, B. and M. Gratzel. 1991. A low-cost, high-efficiency solar cell based on dye-sensitized colloidal TiO_2 films. *Nature* 353: 737–740.

Pelizetti, E., M. Borgorello, C. Minero, E. Pramauro, E. Borgarello, and N. Serpone. 1998. Photocatalytic degradation of polychlorinated dioxins and polychlorinated biphenyls in aqueous suspensions of semiconductors irradiated with simulated solar light. *Chemosphere* 17: 499–510.

Pelizzetti, E. and C. Minero, 1993. Mechanism of the photo-oxidative degradation of organic pollutants over TiO_2 particles. *Electrochim. Acta* 38 (1): 47–55.

Prairie, M.R., B.M. Stange, and L.R. Evans. 1993a. TiO_2 Photocatalysis for the destruction of organics and the reduction of heavy metals. In *Photocatalytic Purification and Treatment of Water and Air*, eds. Ollis, D.F. and H. Al-Ekabi. Amsterdam: Elsevier Science, pp. 353–363.

Prairie, M.R., L.R. Evans, B.M. Stange, and S.L. Martinez. 1993b. An investigation of TiO_2 photocatalysis for the treatment of water contaminated with metals and organic chemicals. *Envrion. Sci. Technol.* 27: 1776–1782.

Ryu, J. and W. Choi. 2008. Substrate-Specific Photocatalytic Activities of TiO_2 and Multiactivity Test for Water Treatment Application. *Environ. Sci. Technol.* 42: 294–300.

Sekler, M.S., Y. Levi, B. Polyak, A. Novoa, P.S.M. Dunlop, J.A. Byrne, and R.S. Marks. 2004. Monitoring genotoxicity during the photocatalytic degradation of p-nitrophenol. *J. Appl. Toxicol.* 24: 395–400.

Serpone, N. 1994. A decade of heterogeneous photocatalysis in our laboratory: Pure and applied studies in energy production and environmental detoxification. *Res. Chem. Inter.* 20(9): 953–992.

Serpone, N., P. Maruthamuthu, P. Pichat, E. Pelizzetti, and H. Hidaka. 1995. Exploiting the interparticle electron transfer process in the photocatalysed oxidation of phenol, 2-chlorophenol and pentachlorophenol: Chemical evidence for electron and hole transfer between coupled semiconductors. *J. Photochem. Photobiol. A Chem.* 85: 247–255.

Tryk, D.A., A. Fujishima, and K. Honda. 2000. Recent topics in photoelectrochemistry: Achievements and future prospects. Electrochim. *Electrochim. Acta* 45(15–16): 2363–2376.

Vautier, M., C. Guillard, and J.M. Herrmann. 2001. Photocatalytic degradation of dyes in water: Case study of Indigo and Indigo Carmine. *J. Catal.* 201: 46–59.

Yan, X., T. Ohno, K. Nishijima, R. Abe, and B. Ohtani. 2006. Is methylene blue an appropriate substrate for a photocatalytic activity test? A study with visible-light responsive titania. *Chem. Phys. Lett.* 429: 606–610.

Zhang, Y., J.C. Crittenden, D.W. Hand, and D.L. Perram. 1994. Fixed-bed photocatalysts for solar decontamination of water. *Environ. Sci. Technol.* 28: 435–442.

4

Advanced Oxidation Processes: Basics and Applications

Rakshit Ameta, Anil Kumar, P. B. Punjabi, and Suresh C. Ameta

CONTENTS

4.1 Introduction

Water is one of the essential things of life, and any undesired addition of chemical substances leads to its contamination and makes it unfit for utilization. These days, the world is in a cancerous grip of environmental pollution in one or the other form, out of which, water pollution is of prime concern. Water pollution has become a major problem at the global level.

Mostly, anthropogenic activities are responsible for this pollution. In particular, effluents are discharged directly or indirectly by the industries into the nearby water resources without proper treatment.

Anthropogenic activities include rapidly growing industrialization, a series of new constructions, manyfold increases in transportation, aerospace movements, developmental and enhancement in technologies, that is, nuclear power, pharmaceutical, pesticides, herbicides, agriculture, and so on. These are all the most desirable activities for human development and welfare, but they also lead to the generation and release of objectionable materials into the environment. Thus, they pollute the whole environment, making our life on this beautiful earth quite miserable. The situation, if not controlled in a timely manner, would become a malignant problem for the survival of mankind on the earth. Many rivers are being polluted by effluent water from industries and domestic sectors. This creates a problem for the aquatic life by turning water into a resource of no use. So, it is of utmost necessity to solve this problem of water pollution.

The most important challenge in the twenty-first century is to combat against the ever-increasing environmental pollution. To have a neat, clean, healthy, and green environment, there is an urgent need to search for such an approach, which may be applicable at room temperature, safe to handle, economic, and eco-friendly. And above all, the main requirement of the treatment is that it should not be harmful to the environment in any manner.

Although conventional oxidation technologies are available for the oxidation of pollutants or disinfection of pathogenic contaminants using a variety of oxidants such as chlorine, peracetic acid, permanganate, hydrogen peroxide (H_2O_2), and ozone, there is another group of chemical oxidative processes called advanced oxidation processes (AOPs) or advanced oxidation technologies (AOTs). The concept of AOPs was originally established by Glaze et al. (1987). It is defined as "oxidation processes, which generate highly reactive radicals (especially hydroxyl radicals) in sufficient quantity to affect the water treatment." These processes are capable of degrading almost all organic contaminants. It is clear from standard redox potential data that hydroxyl radical is the strongest known oxidant (2.80 V), second to fluorine (3.03 V).

Therefore, the complete mineralization of most of the organic matters is possible, when the hydroxyl radicals are the main oxidizing species in the solution. This is one of the major advantages of AOPs, since other chemical oxidation processes mostly lead to partial oxidation of the target compounds, and thus, the generation of new hazardous compounds is possible. The other advantage of AOPs is the generation of negligible amounts of residues and their applicability, in case of very low concentrations of pollutants.

4.1.1 Various AOPs

Although a number of techniques are available under AOPs (more than 10), the main groups of AOPs are four. These are (i) Fenton and photo-Fenton,

(ii) ozonolysis, (iii) photocatalysis, and (iv) sonolysis-based processes. These oxidation processes can produce in situ reactive free radicals, mainly hydroxyl radicals. A hydroxyl radical is a nonselective oxidant, which can oxidize a wide range of organic molecules.

A hydroxyl radical has some interesting characteristics, which make it quite important in AOPs. These are:

1. It is short-lived
2. It can be easily produced
3. It is a powerful oxidant
4. It is electrophilic in behavior
5. It is ubiquitous in nature
6. It is highly reactive
7. It is nonselective

The reactivity of hydroxyl radical (2.06) is next to that of fluorine (2.23), followed by that of atomic oxygen (1.78), H_2O_2 (1.31), and then permanganate (1.24). It is the high redox potential of hydroxyl radical that makes it a powerful oxidant. Thus, hydroxyl radicals have emerged not only as an effective but also as an economic and eco-friendly species.

Hydroxyl radicals can react in water by four different routes: (i) addition, (ii) hydrogen abstraction, (iii) electron transfer, and (iv) radical interaction.

The treatment of wastewaters can be carried out using these hydroxyl radicals. The contaminants are degraded to smaller or less harmful fragments and, in the majority of cases, complete mineralization of the pollutants has been achieved. Even persistent organic pollutants (POPs) can be degraded to the desirable extent using AOPs involving hydroxyl radicals as an active oxidizing agent.

The complete mineralization of an organic pollutant leads to the formation of carbon dioxide (CO_2), water, and/or some inorganic ions, depending upon the molecular composition of that pollutant (Equation 4.1).

$$C_aH_bX + mOH \rightarrow aCO_2 + \frac{b}{2}H_2O + X^{n-} \tag{4.1}$$

All the carbon and hydrogen atoms are completely oxidized to CO_2 and water, while the X gives X^{n-}. X^{n-} may be some inorganic ions such as Cl^-, Br^-, NO_2^-, NO_3^-, SO_3^{2-}, and SO_4^{2-} (depending on experimental conditions). This inorganic ion may be removed by ion-exchanger.

Degradation and detoxification of formalin wastewaters by AOPs has been observed by Kajitvichyanukul et al. (2006). A comparison of different AOPs for phenol degradation was made by Esplugas et al. (2002). Priya et al. (2008) achieved complete photodegradation of phenol in a reasonable time, that is, less than 5 h, when the concentration of phenol was ≤100 ppm. A comparison of various AOPs has also been given by Saritha et al. (2007) for the degradation of 4-chloro-2-nitro-phenol. The decolorization and mineralization of acid orange-6 azo dye were observed by Hsing et al. (2007) using AOPs. Kawaguchi (1992) reported the photooxidation of phenol in aqueous solution in the presence of H_2O_2. The photodegradation of phenol resulted in the stoichiometric conversion of phenol with practically complete mineralization.

4.2 UV/H_2O_2 Processes

H_2O_2 is a strong oxidant (Legrine et al. 1993) that can be used to degrade the organic contaminants present in wastewaters; however, the use of individual H_2O_2 is not that efficient in oxidizing more complex and recalcitrant materials. The use of H_2O_2 becomes more effective when it is used in combination with some other reagents or energy sources capable of dissociating H_2O_2 to generate hydroxyl radicals. H_2O_2 is quite commonly used as an oxidant as it destroys the organic compounds in contaminated wastewaters containing chloro-organics, dyes, pesticides, phenols, and many other organic compounds at very low concentrations.

Ultraviolet (UV) light is an electromagnetic radiation with a wavelength shorter than that of visible light and longer than that of x-rays. It can cause chemical reactions, but it is classified as nonionizing radiation.

It decomposes to generate hydroxyl radicals (Equation 4.2) in the presence of UV irradiation as

$$H_2O_2 \xrightarrow{hv} 2^{\bullet}OH. \tag{4.2}$$

The elimination of organic pollutant is enhanced in the presence of an oxidant during UV irradiation. This substance must absorb UV light and react with water, forming highly reactive $^{\bullet}OH$ radicals.

Low-pressure mercury vapor UV lamps (254 nm) are normally used, but then, a high concentration of H_2O_2 is needed in the medium to generate sufficient hydroxyl radicals. In this way, the use of UV/H_2O_2 process is less effective because higher concentrations of H_2O_2 may also scavenge the hydroxyl radicals. In some AOP technologies, broadband UV lamps (medium pressure) and xenon flash lamps are used to overcome this limitation.

A hydroxyl radical can react with H_2O_2 and a series of reactions are then possible. It is known that H_2O_2 may act as a promoter as well as a scavenger of hydroxyl radicals (Glaze et al. 1987). H_2O_2 can generate $^{\bullet}OH$ radicals under UV exposure on the one hand, while it reacts with the $^{\bullet}OH$ radicals to form hydroperoxyl ($HO_2^{\bullet}$) radicals (Chrittenden et al. 1999) on the other.

The scavenging action of hydroxyl radicals of H_2O_2 in higher concentrations inhibits further generation of $^{\bullet}OH$ radicals, thus reducing the oxidative capability of H_2O_2.

Hydroxyl and hydroperoxyl radicals can attack organic compounds to degrade them. The hydroperoxyl radicals have a relatively lower reduction potential (1.7 V) as compared with that of hydroxyl radicals (2.80 V). As a result, hydroperoxyl radical generation is not a very interesting step in the process of AOPs. The low reduction potential of a hydroperoxyl radical indicates that this species is highly unstable in water and it undergoes disproportionation easily rather than degrading organic compounds. However, the stability of a hydroperoxyl radical varies slightly with the nature of the solvent used (Figure 4.1).

The participation of the hydroxyl radicals has been confirmed as an active oxidizing species by using their specific scavengers. It has been shown that an increase in the initial concentration of H_2O_2 increases the rate of degradation of the contaminants up to a particular (maximum) value, but later on, it declines, when very high H_2O_2 concentration levels are reached. This decrease in the rate is due to the preferential reaction of hydroxyl radicals with excess H_2O_2 as compared with the organic substrates, which leads to the formation of the $HO_2^{\bullet}$ radicals (Lapez et al. 2000).

The main advantages of using this process (H_2O_2/UV) are (i) the H_2O_2 (oxidant) is totally soluble in water, (ii) there is no limitation of mass transfer, (iii) it is an effective source of hydroxyl radicals, and (iv) there is no need for a separation process after the treatment is over (Litter 2005; Gogate and Pandit 2004).

The degradation rate also depends on the pH. As the pH increases, there will be an increase in the rate of degradation, but after attaining an optimum value of pH, the rate of degradation starts decreasing. An increase in the pH will improve the formation of hydroperoxide anion (HO_2^-) and, in turn, it will generate more hydroxyl radicals. As HO_2^- has a molar extinction coefficient higher than that of H_2O_2 (Beltran et al. 1996), the rate of reaction of

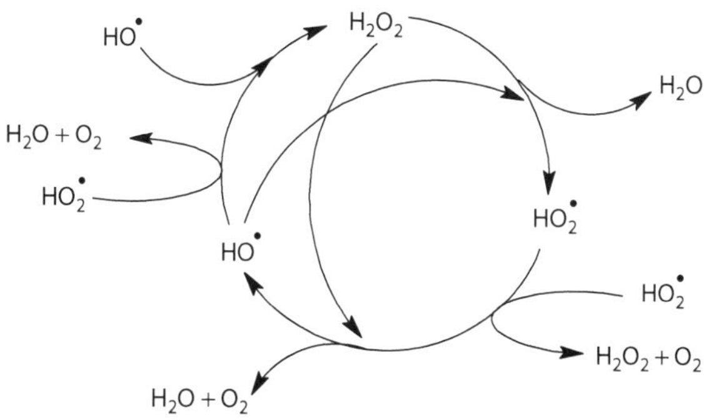

FIGURE 4.1
Different processes in H_2O_2 degradation.

HO_2^- ions with $^{\bullet}OH$ radicals is faster when compared with H_2O_2 (Buxton and Elliot 1986). In the higher pH alkaline range, the CO_2 (degradation product) is easily converted to CO_3^{2-} and HCO_3^- ions in water. These ions—HO_2^-, CO_3^{2-}, and HCO_3^-—are also scavengers of hydroxyl radicals, and as such, the rate of $^{\bullet}OH$ radical scavenging also increases rapidly with the increasing pH of the solution (Equations 4.3 through 4.6).

$$H_2O_2 \rightarrow HO_2^- + H^+ \quad (4.3)$$

$$HO_2^- + {}^{\bullet}OH \rightarrow HO_2^{\bullet} + OH^- \quad (4.4)$$

$$CO_3^{2-} + {}^{\bullet}OH \rightarrow CO_3^{\bullet -} + HO^- \quad (4.5)$$

$$HCO_3^- + {}^{\bullet}OH \rightarrow CO_3^{\bullet -} + H_2O. \quad (4.6)$$

To overcome the problems of scavenging of active oxidizing species, the pH of the solution was decreased to force the carbonate and bicarbonate ions to remain in the form of carbonic acid. Under these circumstances, the formation of HO_2^- is also monitored.

Leitner and Dore (1997) observed the degradation of glycolic acid in the presence of UV/H_2O_2, where oxalic and formic acids were obtained as the final products. It was proposed that the hydroxyl radicals abstract carbon-bound hydrogen atoms from the glycolic acid, forming a radical. It rapidly adds oxygen molecules, forming peroxyl radicals. Here, the $HO_2^{\bullet}$ radical is eliminated, initiating the degradation chain reaction. The glyoxylic acid formed from the oxidation of glycolic acid is later on oxidized to oxalic acid in a similar manner. It has been reported that oxalic acid is finally converted to formic acid.

Dimethyl sulfoxide (DMSO) is an organic pollutant, which can be mineralized by the $^{\bullet}OH$ radicals generated by the UV/H_2O_2 process (Lee et al. 2004). $^{\bullet}OH$ radicals (as the main oxidizing species) are responsible for the degradation of not only DMSO but also its intermediates (methansulfinate, methansulfonate, formaldehyde, formate ion, etc.), and the final products are CO_2 and sulfate, since $^{\bullet}OH$ is a much more powerful oxidant than the other oxidants of the UV/H_2O_2 system, that is, $HO_2^{\bullet}$ and H_2O_2.

Stepnowski et al. (2002) used H_2O_2 in the presence and absence of UV radiations to degrade oil refinery wastewater that was pretreated with flotation and coagulation. This process was also used by Philippopoulos and Poulopoulos (2003) to purify oily wastewater from a lubricant-producing unit. Hu et al. (2008) carried out the degradation of methyl *tert*-butyl ether (MTBE) (a gasoline additive), which is one of the most common contaminants in underground waters due to leakages in gasoline station storage tanks. It was observed that the H_2O_2/UV process can remove 98% of the MTBE, and the removal percentage was found to increase with an increase in the concentration of H_2O_2. But after attaining a maximum value, this removal percentage started decreasing, which may be due to the competitive reactions of H_2O_2 and the hydroxyl radicals, thus reducing the hydroxyl radicals.

Shu et al. (1994) evaluated the effect of the pH on the UV/H_2O_2 process for the treatment of synthetic wastewater containing azo dyes using acid black-1 as the model compound. It was observed that the degradation was maximum in the pH range 3.0–5.2, while Unkroth et al. (1997) used an excimer laser in place of mercury lamps for the treatment of commercial coloring agents (reactive orange-16, reactive black-5, vat red-10, and vat blue-6). When this laser irradiation (193 nm) was used in the presence of H_2O_2, almost complete oxidation of the dyes was achieved with much less energy (almost 2–7 times).

Cater et al. (2000) also utilized this system in the treatment of MTBE in contaminated water. Beltran et al. (1993) also reported oxidation of atrazine in water by H_2O_2/UV. Sorensen and Frimmel (1997) used this process in the degradation of EDTA and some aromatic sulfonates. Andreozzi et al. (2000) investigated the oxidation of *N*-methyl-*p*-aminophenol (metol) in an aqueous solution using H_2O_2 in the presence of UV radiations. Bose et al. (1998a,b) also observed the degradation of an explosive (RDX) using such an AOP.

Some researchers have also used the combination of H_2O_2 with ozone for the oxidation of some pollutants such as tri- and perchloroethene (Sunder and Hempel 1997), 1, 4-dioxane (Adams et al. 1994), and MTBE (Leitnar et al. 1994).

4.3 Fenton and Photo-Fenton Processes

The benefits of this reagent are the following:

1. It can degrade a wide variety of pollutants to biodegradable or almost harmless products.
2. It is eco-friendly in nature, as the remaining (residual) reagent is not hazardous.
3. It is also cost-effective.

Fenton (1876, 1894) reported that a solution of H_2O_2 and ferrous ions in acidic medium has a reasonably high oxidizing power, and the application of Fenton's reagent in destroying toxic organic compounds by oxidation processes was done in the early 1960s. After a controversial history about the reaction mechanism of Fenton's reaction, the classic Fenton's reaction was interpreted by Haber and Weiss (1934). This reaction consists of a combination of H_2O_2 and ferrous ions (Fe^{2+}), in aqueous acidic medium, which leads to the decomposition of H_2O_2 into a hydroxyl ion and a hydroxyl radical and the oxidation of Fe^{2+} to Fe^{3+}. The Fenton's reaction is as follows (Equation 4.7):

$$Fe^{2+} + H_2O_2 \rightarrow Fe^{3+} + {}^{-}OH + {}^{\bullet}OH \quad (4.7)$$

Bossmann et al. (1998) reported that these Fe^{2+} and Fe^{3+} species correspond to the $[Fe(OH)(H_2O)_5]^{n+}$ aqueous complexes, which form the $[Fe(OH)(H_2O_2)(H_2O)_4]^+$ and $[Fe(OH)(H_2O)_5]^{2+}$ complexes in the presence of H_2O_2. Fe^{3+} ions that are formed can be reduced back to Fe^{2+} ions again, forming hydroperoxyl radicals ($HO_2^{\bullet}$) by reacting with H_2O_2 (Walling and Weil 1994). The Fe^{3+} ions also react with the $HO_2^{\bullet}$ to reduce to Fe^{2+} ions (Equations 4.8 through 4.10).

$$Fe^{2+} + H_2O_2 \rightarrow Fe^{3+} + {}^{-}OH + {}^{\bullet}OH \quad (4.8)$$

$$Fe^{3+} + H_2O_2 \leftrightarrow [Fe \cdots OOH] \rightarrow Fe^{2+} + HO_2^{\bullet} + H^+ \quad (4.9)$$

$$Fe^{3+} + HO_2^{\bullet} \rightarrow Fe^{2+} + H^+ + O_2. \quad (4.10)$$

Safarzadeh-Amiri et al. (1997) and Wang (2008) have observed that the initial degradation rate is much lower using Fe^{3+} than Fe^{2+}. In the absence of light and any complexing ligands other than water, the mechanism of H_2O_2 decomposition in acidic aqueous solution involves the formation of hydroxyperoxyl and hydroxyl radicals.

$[Fe(H_2O)_6]^{2+}$-like complexes are formed at very low pH (<2.5), but these react quite slowly with H_2O_2 as compared with $[Fe(OH)(H_2O)_5]^+$, thus generating fewer numbers of hydroxyl radicals and thereby resulting in a decrease in the efficiency of the system (Laat and Gallard 1999). On the other hand, hydroxides of iron-like ferric and ferrous hydroxides (>2.5) are precipitated at higher pH. These hydroxides do not react with H_2O_2 and, as such, there is no Fenton's reaction. If the concentration of H_2O_2 is lower than that of ferrous ions, then the hydroxyl radicals may react with the excess ferrous ions and, as a consequence, the attack of hydroxyl radicals on organic substrates is also decreased (Neyens and Baeyens 2003) (Equation 4.11).

$$Fe^{2+} + {}^{\bullet}OH \rightarrow Fe^{3+} + {}^{-}OH. \quad (4.11)$$

The Fenton's reaction alone is not capable of degrading most of the organic compounds and mineralizing them efficiently. The Fenton reaction stops as soon as all the Fe^{2+} ions present in the solution are oxidized to Fe^{3+} ions. Thus, the generation of hydroxyl radicals is also stopped, resulting in no further degradation of the organic pollutants.

It has been reported that the Fenton reaction rates can be enhanced manyfold if the solution is irradiated with UV/visible light. This type of homogeneous photo-assisted reaction is termed as photo-Fenton reaction. The major advantage of photo-Fenton reaction is that it will occur in the presence of radiation of wavelengths up to 600 nm. Therefore, it may utilize a larger portion of the solar insolation. During this reaction, Fe^{3+} ions are not accumulated in the system and Fe^{2+} ions are regenerated so that the reaction does not stop as in the case of a Fenton's reaction (Equation 4.12).

$$Fe^{3+} + H_2O \xrightarrow{hv} Fe^{2+} + {}^{\bullet}OH + H^+. \tag{4.12}$$

So, more hydroxyl radicals are generated in this case. It is known that the addition of H_2O_2 enhances the photocatalytic reaction due to the inhibition of electron–hole recombination and production of additional $^{\bullet}OH$ radicals through a reaction with the conduction band electrons or with the superoxide radicals, especially at high H_2O_2 concentrations (Fujishima et al. 2000; Augugliaro et al. 1990; Litter 1999) (Equations 4.13 and 4.14).

$$H_2O_2 + e^- \rightarrow {}^{\bullet}OH + {}^{-}OH \tag{4.13}$$

$$O_2^{\bullet-} + H_2O_2 \rightarrow O_2 + {}^{\bullet}OH + {}^{-}OH \tag{4.14}$$

Faust and Hoigne (1990) suggested that the dominant species in the photo-Fenton process in an acidic medium (between pH 2.5 and 5) is the ferric complex $[Fe(OH)]^{2+}$. It is a simplified representation of the aqueous complex $[Fe(OH)(H_2O)_5]^{2+}$. The photolysis of this complex keeps generating more hydroxyl radicals. The newly generated ferrous ions react again with H_2O_2, generating a second $^{\bullet}OH$ radical and a ferric ion, and, in this manner, this process becomes a cycle in nature. In the reaction medium, other photoreactive species such as $[Fe_2(OH)_2]^{4+}$ and $[Fe(OH)]^{2+}$ are also present, which form hydroxyl radicals (Figure 4.2).

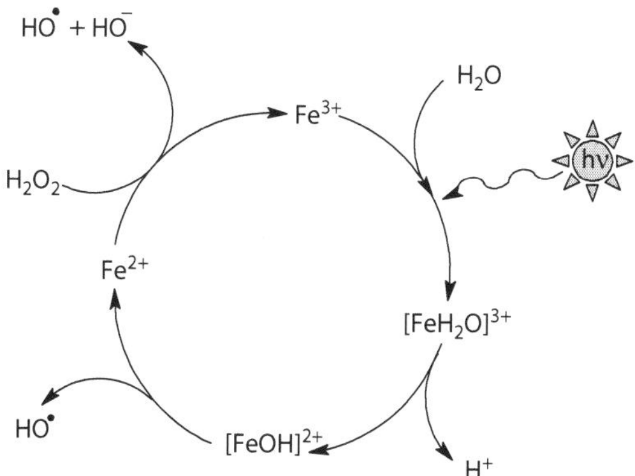

FIGURE 4.2
Photo-Fenton process.

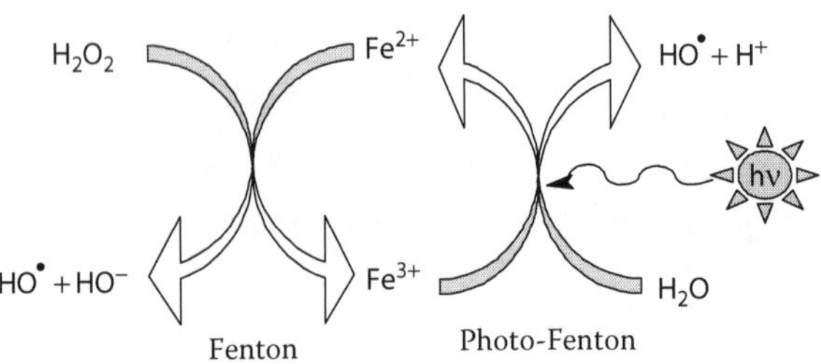

FIGURE 4.3
Difference in Fenton and photo-Fenton reactions.

Thus, more hydroxyl radicals are generated by the reaction of the regenerated ferrous ions with H_2O_2 in solution. The use of light radiation reduces the required concentration of ferrous ions significantly as compared with the dark Fenton reaction. The photo-Fenton process leads to the reduction in the pH of the medium as protons are generated (Figure 4.3).

Fenton and photo-Fenton reactions not only depend on the concentration of H_2O_2 and Fe^{2+} ions, but also on the pH of the medium. Chamarro et al. (2001) utilized the Fenton process for the degradation of some phenols (phenol, 4-chlorophenol, and 2,4-dichlorophenol) as well as nitrobenzene. This process was found to eliminate the toxic substances and also increased the biodegradability of the treated water. It was observed that the biodegradability of dichlorodiethyl ether can be enhanced by modifying the Fenton's reagent. The treatment of textile water by means of Fenton and photo-Fenton processes has also been reported with their effectiveness for color removal and chemical oxygen demand (COD) reduction (Balanosky et al. 1999; Kang et al. 2000; Perez et al. 2002).

Nadtochenko and Kiwi (1998) and De Laat et al. (2004) have reported that knowledge of the physicochemical characteristics of the wastewater is also important (before treatment) as some inorganic ions (such as Cl^-, SO_4^{2-}, $H_2PO_4^-/HPO_4^{2-}$) already present in the wastewater or added as reagents ($FeSO_4$, $FeCl_3$, HCl, H_2SO_4) may also interfere in the reaction and inhibit the whole degradation process. It has been suggested that the complexation reactions of the inorganic ions with Fe^{2+} or Fe^{3+} ions or the hydroxyl radicals leading to the formation of less reactive inorganic radicals or radicals anions ($Cl^{\bullet-}$, $Cl_2^{\bullet-}$ and $SO_4^{\bullet-}$) may be the major cause for the inhibition.

Iwahashi et al. (1999) reported that quinolic, α-piconilic, fusaric, and 2,6-pyridine dicarboxylic acids enhance the Fenton's reaction in phosphate buffer. Beltran-Heredia et al. (2001) investigated the degradation of olive mill wastewater by combining Fenton's reagent and ozonation with an aerobic biological treatment. Fenton's oxidation (with ethanol) was also utilized to convert anthracene (PAH) into its corresponding quinone, 9,10-anthraquinone, which is biodegradable (Lee et al. 1998).

Recently, Kumar et al. (2008a) reported the degradation of naphthol green B using photo-Fenton's reagent. Carbon tetrachloride has also been degraded by using modified Fenton's reagent by Teel and Watts (2002).

Sun and Pignatello (1993) used Fe(III)/H_2O_2/UV system for the total mineralization of 2,4-dichlorophenoxyacetic acid. The degradation of polychlorinated dibenzo-*p*-dioxin and dibenzofuran contaminants in 2,4,5-trichlorophenoxyacetic acid was also reported by Pignatello and Huang (1993). They also reported the complete oxidation of metolachlor and methyl parathion in an aqueous medium by such a system (Buxton et al. 1988).

Fenton's and photo-Fenton's reagents have also been used for the degradation of some nitro compounds such as nitrobenzene and nitrophenol (Lipczynska-Kochany 1992). Some interesting observations were reported on the immobilization of such a reagent and its use for the degradation of organic contaminates (Maletzky and Bauer 1999; Dhananjeyan et al. 2001). Efforts were also made to convert wastewaters from textile industries into some biodegradable components (Rodriguez et al. 2002).

Lindsey and Tarr (2000a,b) reported the inhibition of hydroxyl radical degradation of some aromatic hydrocarbons by fulvic acid as well as some dissolved natural organic materials. Ameta et al. (2006) reported a comparative study of the degradation of resorcinol with Fenton and Fenton-like reagents using some other transition metal ions such as Cu(II) and Co(II), while Mogra et al. (2003) observed the photochemical degradation of *o*-chlorobenzoic acid by photo-Fenton's reagent.

With H_2O_2, the presence of oxygen accelerated the phenol photooxidation (Auguliaro et al. 1988, 1990). This acceleration was explained by the formation of $HO_2^\bullet$ radicals (Equations 4.15 through 4.18):

$$H_2O_2 + h\nu \rightarrow H_2O_2^\bullet \tag{4.15}$$

$$H_2O_2^\bullet + O_2 \rightarrow 2HO_2^\bullet \tag{4.16}$$

$$HO_2^\bullet + C_6H_5OH \rightarrow H_2O_2 + {}^\bullet C_6H_4OH \tag{4.17}$$

$${}^\bullet C_6H_4OH + O_2 \rightarrow \text{Organic peroxide.} \tag{4.18}$$

Badawy et al. (2006) successfully achieved the complete removal of organophosphorus pesticides from wastewater using AOPs. Andreozzi et al. (2000) used AOPs for the treatment of mineral oil–contaminated wastewater.

Araujo et al. (2002) used Fenton's reagent for the degradation of residual Kraft black lignin, while Perez et al. (1983) reported this system to be comparatively economical in the treatment of effluents from paper and pulp industries. Kumar et al. (2008b) used photo-Fenton's reagent for the degradation of rhodamine B. Hsueh et al. (2005) also used Fenton and Fenton-like systems for the degradation of orange G, reactive black 5, and red MX-5B, while Scheeren et al. (2002) made a comparative study of the degradation of acid orange-7 dye by different AOPs. The oxidative degradation of lignin has also been investigated by Gold et al. (1983).

Some successful attempts have been made in immobilizing iron species on a Nafion membrane and using it for the degradation of *p*-chlorophenol (Maletzky and Bauer 1999), while some iron mineral–like lepidocrocite (Chou and Huang 1999), hematite (Natts et al. 1997), and goethite (Lin and Gurol 1998) have also been used in this process in one or the other manner.

Fenton and photo-Fenton processes are promising advanced oxidation methods, which help in the photooxidation of various organic compounds such as dyes (Chen et al. 2003; Kumar et al. 2008c; Modrishahla et al. 2007), pesticides (Fallmann et al. 1999; Lapertot et al. 2007), herbicides (Farre et al. 2007; Flox et al. 2007), fungicides (Conceicao et al. 1994), and explosives (TNT, RDX, and HMX) (Lee et al. 2002). Degradations of 2,4-dichlorophenoxy acetic acid (Kong and Lemley 2006), carbonate insecticide (Wang and Lemley 2003), herbicide chlorotoluron (Abdessalem et al. 2008), p,p′-DDT, and p,p′-DDE (Villa and Nogueira 2006) were reported using heterogeneous Fenton's reagents.

4.3.1 Electro-Fenton and Photoelectro-Fenton Processes

Different indirect electrooxidation methods have also been explored for the degradation of organic pollutants (Do and Chen 1994; Alberto and Pletcher 1999; Oturan et al. 1999; Oturan 2000; Aaron and Oturan 2001). In these environmentally benign and clean electrochemical techniques, H_2O_2 is continuously generated in an acidic, contaminated solution from the two-electron reduction of O_2 at different cathodes (Equation 4.19).

$$O_2 + 2H^+ + 2e^- \rightarrow H_2O_2. \tag{4.19}$$

In the electro-Fenton method, a Pt anode and Fe^{2+} are used in the acidic medium, so that organic contaminants are efficiently degraded by the $^{\bullet}OH$ radicals generated from the decomposition of H_2O_2 by Fe^{2+} as well as the hydroxyl radicals formed and adsorbed at the Pt anode surface from the oxidation of water (Comninellis and Pulgarin 1993; Comninellis and Battisti 1996) (Equation 4.20).

$$H_2O \rightarrow {}^{\bullet}OH_{abs} + H^+ + e^-. \tag{4.20}$$

However, there are still some problems associated with the use of the electro-Fenton reaction. These are as follows: (i) the production of H_2O_2 is quite slow because of the low solubility of oxygen in water (Savall 1995) and (ii) the current efficiency is also low (at $pH < 2.9$) (Pignatello et al. 2006).

This phenomenon is due to the formation of $Fe(OH)_3$ sludge, which is still a problem.

The efficiency of the electro-Fenton methods can be enhanced with irradiation, and these methods can be used for wastewater remediation. Such systems are called photoelectro-Fenton processes (Pratap and Lemley 1998; Sir'es et al. 2007; Boye et al. 2002; Flox et al. 2007). In the electro-Fenton process, Fe^{2+} is added to the solution, and the pollutants can be degraded by the $^{\bullet}OH$ radicals, but in the photoelectro-Fenton method, the simultaneous irradiation of the solution with light accelerates the rate of mineralization of a pollutant as it favors the regeneration rate of Fe^{2+} ions. Pratap and Lemley (1998) reported the electro-Fenton degradation of pesticides such as atrazine and metolachlor, while Flox et al. (2007) used the photoelectro-Fenton process for the destruction of a herbicide named mecoprop. Sir'es et al. (2007) investigated the degradation of clofibric acid by electro-Fenton and photoelectro-Fenton processes. The herbicides 4-chlorophenoxyacetic acid and 2,4-dichlorophenoxyacetic acids (2,4-D) were mineralized by such an advanced electrochemical oxidation process. Sir'es et al. (2006) also reported the degradation of paracetamol (an antipyretic) electrochemically in the presence of Fe^{2+} as a catalyst.

Electro-Fenton and photoelectro-Fenton processes have also been used for the detoxification of acidic wastewaters, where H_2O_2 is electrogenerated. These procedures have been used efficiently for the mineralization of aniline (Brillas et al. 1998), 4-chlorophenol (Brillas et al. 1998), and 2,4-dichlorophenoxyacetic acid (Brillas et al. 2000).

4.4 UV-Vis/Ferrioxalate/H_2O_2 Processes

There is an urgent need for developing AOPs that can generate $^{\bullet}OH$ radicals with near-UV (300–400 nm) or even visible light (>400 nm) so that a system can also utilize the solar radiation as the light source to degrade organic effluents (Duran et al. 2008; Malato et al. 2002). The homogeneous solar photo-Fenton reaction is one of the most eco-friendly as well as economically viable systems, but in this case, only photons below 350 nm are used by H_2O_2. This amounts to a very small portion (hardly 3% of solar irradiation), while ferrioxalate is a photosensitive complex that can be used to utilize a larger fraction of the solar spectrum ranging up to 450 nm, which is almost six times more use of solar irradiation (≈18%). Thus, the oxidation capacity of the solar photo-Fenton process is also enhanced (Malato et al. 2004; Bauer et al. 1999; Nogueira et al. 2002; Lucas and Peres 2007). The photolysis of ferrioxalate generates more H_2O_2, which in turn yields more hydroxyl radicals for the degradation of organic contaminants.

The treatment of water with UV/H_2O_2 is less effective due to poor absorption of the UV photons by H_2O_2, but here the UV-vis/ferrioxalate/H_2O_2

process is useful. Ferrioxalate absorbs light strongly at longer wavelength and that too with a high molar absorption coefficient. It generates hydroxyl radicals with better quantum yield; however, the quantum yield varies with different ligands.

Ferrioxalate has been used in the photo-Fenton reactions, but very little work has been done on the ferrioxalate-assisted photo-Fenton system using the ferrous-initiated process. The use of ferrous sulfate is advantageous, as it is less corrosive than the corresponding ferric salts, inexpensive, and relatively more soluble than ferric compounds. Ferrous ions are rapidly converted into ferric ions in the presence of H_2O_2, and the degradation of organic continuants is catalyzed by ferrioxalate.

At a pH lower than 3–4, the main species are $[Fe(C_2O_4)_2]^-$ and $[Fe(C_2O_4)_3]^{3-}$, which are highly photoactive. In the presence of UV light, these species will convert ferric ions into ferrous ions, producing more $^{\bullet}OH$ radicals. At pH values higher than 4–5, the main species are $[Fe(C_2O_4)]^+$ and $[Fe(C_2O_4)]^+$, which are relatively less photoactive. Above this pH range, the Fe^{3+} complex starts coagulating, which will reduce the efficiency of the catalyst. Balmer and Sulzberger (1999) observed the degradation of atrazine, while Li et al. (2007) reported the photodegradation of bisphenol A.

In this process, the initial total organic carbon (TOC) can be higher as the organic carbon is added in the form of oxalate; however, the oxalate is mineralized rapidly through photodecarboxylation and, simultaneously, the organic pollutant is oxidized. The role of such a system in water has been discussed by different workers (Zhou and Hoigne 1992; Faust and Zepp 1993; Hislop and Bolton 1999). Zepp et al. (1992) and Safarzadeh-Amiri et al. (1996) also reported an improvement in photo-assisted Fenton processes.

4.5 Ozone-Based Processes

Ozone (O_3) is a strong oxidant with an oxidation potential more than that of H_2O_2 (about 1.2 times), and it is almost 10 times more soluble in water than pure oxygen. Thus, it can also be used to degrade organic contaminants. It reacts with organic pollutants and ultimately decomposes them to oxygen. Ozone can oxidize organic contaminants directly or indirectly through the formation of hydroxyl radicals.

4.5.1 Ozone/UV Processes

The photolysis of an aqueous ozone produces hydroxyl radicals directly as in the case of the gas-phase reaction, but the photolysis of ozone at 254 nm in an aqueous acetic acid solution led to the production of H_2O_2. As the extinction coefficient of O_3 at 254 nm is much higher than that of H_2O_2 (more than

150 times), it can be used effectively for the oxidative degradation of organic pollutants in water. H_2O_2 generates 0.09 hydroxyl radicals per incident photon as compared with the two hydroxyl radicals by ozone molecule (Munter et al. 1995).

The photolysis of aqueous ozone produces H_2O_2 also, which in turn produces hydroxyl radicals. These are the active species responsible for the destruction of organic compounds. After the oxidation is complete, some of the intermediate products in the solution may be toxic or sometimes even more toxic than the initial compounds, and therefore, complete oxidation should be carried out by supplementing O_3 with UV radiation. Therefore, the O_3/UV process becomes more effective as the organic pollutants can be degraded through the absorption of the UV irradiation as well as through the reaction with hydroxyl radicals (Trapido and Kallas 2000; Gurol and Vatistas 1987).

Phenolic compounds (phenol, *p*-cresol, 2,3-xylenol, 3,4-xylenol, and *p*-nitrophenol) are easily oxidized by ozone and completely mineralized to CO_2 and H_2O (Rein 2001; Metcalf & Eddy Inc. 2003). Complete mineralization of some other low-molecular-weight organic compounds such as glyoxal, glyoxylic acid, oxalic acid, and formic acid was carried out by Takahashi (1990) using the O_3/UV system. Peyton et al. (1982) observed the elimination of C_2Cl_4 from water using O_3/UV and compared it with only ozonation and photolysis. Hung-Yee and Ching-Rong (1995) have reported that this process is the most effective method for the decolorization of azo dyes as compared with the process of oxidation by UV or ozonation alone.

Guittonneau et al. (1990) compared and opined that the O_3/UV process is more efficient than the UV/H_2O_2 system for the degradation of *p*-chloronitrobenzene. Beltran et al. (1998) have studied the effect of ozone feed rate, pH, and hydroxyl radical scavengers in the removal of nitrobenzene by this process. The results were compared with those of UV radiation and single ozonation processes.

4.5.2 O_3/H_2O_2 Processes (Peroxone)

Peroxone is a combination of ozone and H_2O_2, and it is a newer process as well as an AOP that can be used for the treatment of polluted soils, groundwater, and wastewaters. In this process, highly reactive species (hydroxyl radicals) that are responsible for the oxidation of most of the organic pollutants in a solution are formed.

The formation of a complex by mixing O_3 and H_2O_2 was shown by Engdahl and Nelander (2002), and such complex intermediates might be involved in this peroxone process. A significant concentration of H_2O_3 is produced. Wentworth et al. (2001) also reported that this H_2O_3 species reacts with another H_2O_3 to form [(HOO)(HOOO)-7r] (a head-to-tail seven-member ring complex), which finally leads to the formation of H_2O_2.

This seven-membered ring then attacks aromatic systems such as benzene, leading to the formation of phenol, oxygen, and H_2O_2. After the attack of

$^{\bullet}OH$ radical, the H can be extracted from the ring by the $HO_2^{\bullet}$ radical forming H_2O_2 and phenol.

Sufficient oxidation and exposure to UV energy give the final reaction by-products such as CO_2, water, and the corresponding inorganic salts.

4.5.2.1 *UV/O$_3$/H$_2$O$_2$ Processes*

The efficiency of the peroxone process increases upon UV irradiation, and thus the UV/peroxone process can be used to treat wastewater contaminated with a variety of pollutants such as polycyclic aromatic hydrocarbons (PAHs), chlorinated solvents, MTBE, pesticides, trinitrotoluene (TNT), and other organic contaminants.

Mokrini et al. (1997) reported the oxidation of phenol and benzoic acid by means of this process at different pH values and indicated that both are degraded more rapidly by ozone only at higher pH range (9–12), but ozonation combined with H_2O_2 and/or UV can be carried out rapidly even in lower pH range (3–7). Trapido et al. (2001) also reported that the combination of ozone with UV radiation and H_2O_2 is more effective for the degradation of nitrophenols than single ozonation or the binary combinations (O_3/H_2O_2), while Contreras et al. (2001) observed that the addition of H_2O_2 to the UV/O_3 system also improves the rate of TOC removal (slightly) in solutions of nitrobenzene.

4.6 Photocatalysis

A photocatalyst is defined as a species that induces photochemical reactions, and the term photocatalysis is used for those chemical reactions that occur in the presence of light and the photocatalyst. In this type of chemical reactions, an electron–hole pair is generated on exposing a photocatalyst to light of suitable wavelength. This electron can be used for reducing a substrate, whereas the hole may be utilized for oxidation. A photocatalyst is basically a semiconductor.

Photocatalysis is based on the activation of a semiconductor material (CdS, TiO_2, ZnO, WO_3, etc.) by the action of radiation with an appropriate wavelength. The activation is achieved with the absorption of a photon by the semiconductor particle possessing enough energy to promote an electron (e^-) from its valence band (VB) to the conduction band (CB), thus creating holes (h^+) in the VB. These holes will act as oxidizing species themselves or will generate $^{\bullet}OH$ radicals (Figure 4.4) (Equations 4.21 through 4.23).

$$TiO_2 \xrightarrow{hv} TiO_2\left(e^-_{CB} + h^+_{VB}\right) \tag{4.21}$$

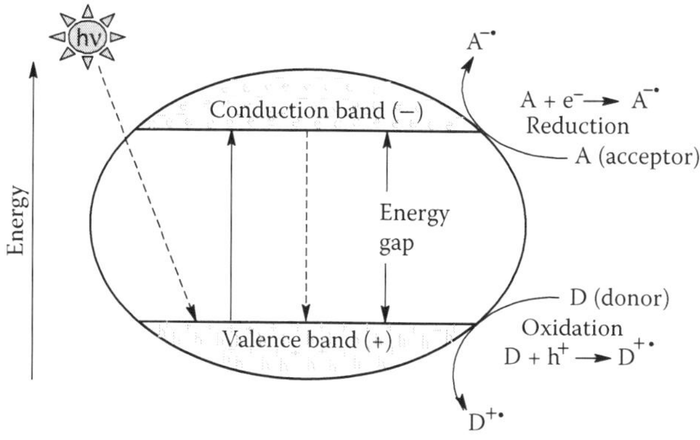

FIGURE 4.4
Principle of photocatalysis.

$$h^+ + H_2O \rightarrow {}^\bullet OH + H^+ \tag{4.22}$$

$$h^+ + {}^-OH \rightarrow {}^\bullet OH. \tag{4.23}$$

Matthews (1998) used ZnO for the photocatalytic oxidation of aromatic hydrocarbons containing phosphorus, sulfur, nitrogen, and halogen. The final products were identified as CO_2, H_2O, PO_4^{3-}, SO_4^{2-}, NO_3^-, and X^-. Cun et al. (1998) also used ZnO for the photocatalytic oxidation of aniline.

ZnO was used for the decontamination of effluents from pharmaceutical industries (discharges from cephalosporin- and penicillin-producing industries) (Shanhu et al. 2002). It was also found that Gram-negative *Escherichia coli* and Gram-positive *Lactobacillus helveticus* (Lui and Thomas 2003) can also be inactivated by ZnO and TiO_2 exposed to UV light. Gouvea et al. (2000) were able to remove kraft effluents completely using ZnO photocatalyst doped with silver ions.

Photocatalysts are not only useful in removing or decomposing organic materials (pollutants), but can also be used for the recovery of high-purity heavy/noble metals such as gold and silver from wastewater. The silver ions present in the used photo film–developing solution were completely removed within just 15 min using ZnO nanopowder (Equation 4.24).

$$4Ag^+ + 2H_2O \rightarrow O_2 + 4H^+ + 4Ag \tag{4.24}$$

The photocatalytic oxidation of organic substances such as isopropanol (Kuriacose and Markham 1962), phenol (Markham et al. 1954), and acetamides (Markham et al. 1962) over ZnO has also been investigated. A cellulose-bleaching effluent can also be degraded by photocatalysis using

TiO_2 and ZnO supported on glass Raschig rings (Yeber et al. 2000). The effluent was completely decolorized, and the total phenol contents were reduced by 85% within 120 min in the presence of both these photocatalysts.

Borgarillo et al. (1982) reported the oxidation of H_2S (a by-product of the coal-processing and petroleum industry) to sulfur at the CdS semiconductor surface. A semiconductor carried by carbon nanotubes (SC/CNTs) is also a promising candidate for the photocatalytic degradation of organic pollutants. The capability of SC/CNTs in degrading organic molecules is higher than that of semiconductors alone. CdS mixed with TiO_2 suspensions increases the photodegradation of 2-chlorophenol drastically (Doong et al. 2001).

Heat treatment of CdS with KCl (Kobayakawa et al. 1993) also enhances its photocatalytic activity. Wang et al. (1992) carried out the photocatalytic reduction of environmental pollutant Cr(VI) over CdS powder under visible light illumination.

Photocatalysts have been found useful not only in the oxidation and in the degradation of some organic molecules, but also in the synthesis of organic compounds, isomerization/transformation, wastewater treatment, self-cleaning, sterilization, and so on.

More than 10,000 different synthetic dyes and pigments (about 10^9 kg) are produced every year and used in dyeing and printing industries. It is estimated that about one-tenth of these dyes are lost in industrial wastewaters and pollute the water resources. Mu et al. (2004) investigated the photocatalytic degradation of orange-II in the presence of Mn^{2+} and TiO_2 in solution, while the photocatalytic degradation of ethyl violet in aqueous TiO_2 suspension was reported by Chen et al. (2006). Photocatalytic degradation of water contaminants can be enhanced by decreasing the particle size as it increases its surface area, thereby increasing the number of active sites available on the surface of the photocatalyst. The photocatalytic degradation of Congo red catalyzed by nano-sized TiO_2 was observed by Wahi et al. (2005). Photodegradation, decolorization, and demineralization of malachite green in an aqueous suspension of TiO_2 nanoparticles under aerated conditions have been carried out by Kominami et al. (2003).

Mrowtez and Selli (2004) observed the effect of iron species in the photocatalytic degradation of an azo dye in TiO_2 suspension. Similarly, Rao et al. (2003) reported the influence of metallic species on TiO_2 for the photodegradation of dyes and dye intermediates. The photodegradation of eosin Y, an anionic xanthene fluorescent dye, has also been investigated in an aqueous heterogeneous solution containing TiO_2 (Poulios et al. 2003). Zhiyong et al. (2003) synthesized and observed the stabilization of TiO_2 on thin polyethylene films (LDPE). It was reported that the LDPE-TiO_2 films were able to mediate the complete decolorization of orange-II, and it was seven times more active (catalytically) than TiO_2 suspension only.

Kraft black liquor is an important effluent from pulp and paper industries. It is easily bleached photocatalytically using ZnO as a photocatalyst (Mansilla et al. 1994). The photocatalytic degradations of brilliant green and bromopyrogallol over ZnO were investigated by Ameta et al. (1997) and Sharma et al. (1998), respectively. The presence of surface charge on the ZnO photocatalyst affects favorably the rate of photocatalytic degradation of methyl orange and crystal violet (Bhandari et al. 2006). Ye et al. (2006) reported the synthesis and use of ZnO nanoplatelets (as thin as 10 nm) in the photodegradation of methylene blue and eosin Y with high efficiency. It was observed that COD is also reduced along with the removal of color.

Ameta et al. (1998) reported an enhancement in ZnO-mediated photodegradation of basic blue 24 dye in the presence of a surfactant. Patel et al. (2006) used photocatalysis for desalination. Hasnat et al. (2005) made a comparative study of photocatalytic degradation of a cationic dye and an anionic dye. Photoreduction of methyl orange dye in the presence of CdS was investigated by Mills and Williams (1987) using EDTA as an electron donor. Takizawa et al. (1978) compared the photocatalytic degradation of CdS-suspended aerated aqueous solution of methylene blue and rhodamine B, while Zhang and Shen (1996) reported on the photocatalytic reduction of methyl yellow on CdS nanoparticles mediated in reverse micelle.

Although chlorine (in some form or the other) is used for the treatment of water, it is equally true that an attachment of chlorine atom in a moiety of an organic molecule creates many harmful effects on the human body as well as on aquatic life. Photocatalysis is the best method to decompose these chloro compounds.

The degradation of different chloro-organics has been achieved, that is, chlorobenzenes (Ollis et al. 1984) and monochloroacetic acid (Kopf et al. 2000). Polymers such as polyvinyl chloride (PVC) and polyurethane can also be photodegraded (Osawa et al. 1994). Chlorophenols are toxic in nature and are resistant to biodegradation; therefore, alternative treatment methods for the destruction of chlorophenol wastes are necessary. Serpone et al. (1995) reported the photocatalyzed oxidation of phenol, 2-chlorophenol, and pentachlorophenol (PCP) on coupled semiconductors such as CdS/TiO_2. It was observed that the use of coupled semiconductors leads to an enhancement in the rate of photocatalytic reactions as compared with TiO_2 alone.

The photodegradation of dichloromethane, tetrachloroethylene, and 1,2-dibromo-3-chloropropane in an aqueous suspension of TiO_2 under sunlight was observed by Halmann et al. (1992), while the photocatalytic degradation of $CHCl_3$, $CHBr_3$, CCl_4, and CCl_3COOH was investigated by Choi and Hoffmann (1997) in an aqueous TiO_2 suspension. In the absence of dissolved oxygen, $CHCl_3$ and $CHBr_3$ degraded into carbon monoxide and halide ions, while in the presence of oxygen, $CHCl_3$ degraded to give CO_2 and halide ions. Trichlorobenzene can be photocatalytically decomposed using TiO_2 supported on a nickel poly(tetrafluoroethylene) composite plate (Uchida et al. 1995).

An exponential increase in population has resulted in the requirement of not only a larger quantity of food but also better quality. In an effort to increase the quantity of food and also to reduce the loss of crop by the attack of termites, pests, insects, herbs, and so on, farmers started using various chemicals. They also prefer such chemical treatments for various purposes such as early ripening of fruits and flowers, prolonging the freshness of vegetables, and checking the dropping of fruits, flowers, and vegetables. Use of these chemicals (pesticides, insecticides, herbicides, and acetylene) is harmful to human beings, animals, and the environment also. Most of the pesticides and herbicides are nonbiodegradable.

The insecticide chlordane is used for the control of termites, ants, wasps, and cockroaches. Its effective decomposition was achieved by photocatalytic degradation in aqueous dispersions of TiO_2 (Hustert et al. 1983). The photodegradation of trimethyl phosphate was carried out and different products were obtained in stepwise degradation, that is, dimethyl phosphate, methyl phosphate, and orthophosphate, as well as formaldehyde, formic acid, CO, and CO_2.

The degradation of metasystox can be achieved in only 3 h of solar irradiation, which involves almost complete elimination of the active compound, methyloxy-demeton (Arques et al. 2007). The degradation of carbendazim pesticide was carried out by the combination of photocatalysis and ozonation (Rajeshwari and Kanmani 2009). It was also demonstrated that this combination of TiO_2 and ozonation will overcome the disadvantage of the selectivity of ozonation as well as the lower efficiency of TiO_2 photocatalysis. Recently, the homogeneous and heterogeneous photocatalytic degradations of imidacloprid (a systematic chloronicotinoid insecticide) has been investigated in an aqueous solution using artificial UV-A or visible illumination (Kitsiou et al. 2009). Ammonium, nitrate, and chloride ions were detected as final mineralization products.

Muszkat et al. (1995) observed TiO_2-photocatalyzed degradation of pesticides in the rinse water of agricultural sprayers and also in heavily polluted well water under sunlight. The nitrogen-doped TiO_2 crystalline nanopowder was used for the photocatalytic degradation of the herbicides RS-2-(4-chloro-*o*-tolyloxy) propionic acid (mecoprop) and (4-chloro-2-methylphenoxy) acetic acid (MCPA; Abramovic et al. 2009) using various light sources. Commercial mixtures of pesticides (Folimat and Ronstar) and two fungicides (pyrimethanil and triadimenol) also undergo photocatalytic and biological degradations (Arana et al. 2008). It was observed that the photocatalytic treatment was able to degrade dicofol, tetradifon, pyrimethanil, triadimenol, and some components of Ronstar.

Indole is one of the most important components of industrial and agricultural activities. It is quite toxic to aquatic organisms and may act as a carcinogen on chronic exposure. Merabet et al. (2009) developed a response surface methodology (RSM) for the optimization and modeling of photocatalytic degradation of indole in UV/TiO_2.

Surfactants are important for cleaning in daily life in households as well as in industries. These are used in industries such as agrochemical, mining, and oil fields. The hydrophobic part of these molecules is free to attach to grease, fat, or oil on the surface, and hence it is used in cleansing processes.

The soap is biodegradable, but nowadays, detergents have almost replaced soaps, leaving aside bathing soaps. The detergents are not biodegradable, and thus a new kind of pollution is appearing on the forefront, that is, detergent pollution.

Surfactant hexadecyltrimethylammonium chloride is photodegradable with the help of TiO_2, H_2O_2, and UV radiations (Mukhtar-Al-Hassan et al. 2008). Two industrial-grade surfactants, sodium lauryl sulfate (SLS) and sodium dodecylbenzene sulfonate (SDDBS), were degraded photocatalytically using a black light fluorescent tube as an irradiation source in the photoreactor. It was observed that about 80% removal of these surfactants can be achieved within 60 min of irradiation (Lizama et al. 2005) only. The degradation of nonylphenol ethoxylated surfactant was also carried out (Pelizzetti et al. 1989) in a slurry reactor as well as in a vessel where the photocatalyst was anchored onto an aluminum surface to avoid the final filtration of the powder at the end of the reaction. TiO_2-based photocatalytic degradation of an anionic detergent (lauryl benzene sulfonate) had been quite efficient (Fabbri et al. 2009).

Singhal et al. (1997) were also successful in degrading cetylpyridinium chloride photocatalytically. Photocatalytic treatment also proved effective in the treatment of certain other surfactants (Horikoshi et al. 2000; Mehrab and Haffield 2005; Sherrard et al. 1994; Hidaka et al. 2003; Parra et al. 2001; Taicheng et al. 2008; Mortazavi et al. 2008) such as alkylbenzene sulfonate (LAS), polyethoxylate surfactants, nonbiodegradable *p*-nitrotoluene-*o*-sulfonic acid (cp-NTS), cationic surfactants cetyltrimethylammonium bromide (CTMAB), and decabromodiphenyl ether (BDE 209), dodecyl pyridinium chloride (DPC), and anionic surfactant sodium dodecyl sulfate from wastewater.

Almost daily, a new disease seems to appear in the headlines. Advances in transportation technology have caused an increase in the movement of people across countries, oceans, and continents. It helps infectious diseases to spread rapidly and to long ranges, thus becoming serious threats for the world. Photocatalysis has also emerged as a revolutionary invention in sterilizing action. It kills the bacterial cells. Disinfection by TiO_2 is 3 times stronger than chlorine and 1.5 times stronger than O_3. Photocatalyst nano-TiO_2 can kill *Pseudomonas aeruginosa*, *E. coli* influenza virus, MRSA (methicillin-resistant *Staphylococcus aureus*), *Tuberculi bacillus*, and so on and some fungi. Thus, disinfection by nano-TiO_2 photocatalyst may be a promising technology with further researches into its practical use for the treatment of water.

Subramani et al. (2007) used TiO_2-impregnated activated carbon for photocatalytic degradation of indigo carmine, while Pouretedal et al. (2009) observed the photodegradation of methylene blue and safranin using nanoparticles of cadmium sulfide (as nanophotocatalyst) doped with manganese, nickel, and copper. Zhou et al. (2010) carried out photodegradation of rhodamine B and nitrobenzene in wastewater by bonding TiO2/single-walled carbon nanotube composites. The enhanced photocatalytic activity was attributed to smaller crystalline size, larger surface area, and especially the ester bonds.

Photodegradation of parathion by TiO_2 and zero-valent iron in an aqueous solution was observed by Doong and Chang (1998) in the presence of H_2O_2, while Liang et al. (2010) made a comparison in the degradations of diphenamid by homogeneous photolysis and heterogeneous photocatalysis. Chen and Chang (1999) reported the photocatalytic degradation of oil films floating on water by using TiO_2 supported on hollow glass microbeads.

Gikika et al. (2005) observed the photocatalytic reduction and the recovery of mercury using polyoxometallates such as $PW_{12}O_{40}^{3-}$ and $SiW_12O_{40}^{4-}$. It was observed that mercury(II) was reduced to Hg(0) via Hg(I), thereby giving a dark gray deposit. Turchi and Ollis (1990) proposed the mechanism of photocatalytic degradation of organic water contaminants involving hydroxyl radicals, while the role of electron transfer and superoxide has been observed for TiO_2-assisted photocatalytic degradation of haloquinolines in water by Cermenati et al. (2000). TiO_2 has also been used for the photocatalytic decomposition of DMMP by Obee and Satyapal (1998). Dionysiou et al. (2000) observed the effect of ionic strength and H_2O_2 on the photocatalytic degradation of 4-chlorobenzoic acid in water.

Maillard-Dupuy et al. (1994a) reported on the kinetics and products of photocatalytic degradation of pyridine in water, while Doherty et al. (1995) reported such observation for the degradation of morpholine. Phenyltrifluoromethyl ketone forms trifluoroacetic acid (a pollutant). Its degradation by photocatalysis, sonolysis, and their combination was investigated by Theron et al. (2001). Uchida et al. (1993) used activated carbon as a support for TiO_2 in the photocatalytic decomposition of propyzamide. Tanaka et al. (1986) reported photocatalytic deposition of metal ions on TiO_2 powder. The oxidation of naphthalene and its derivatives was observed by Soana et al. (2000) in the presence of photocatalyst titania. TiO_2 was also used as a photocatalyst by Maillard-Dupuy et al. (1994b) for the degradation of nitrobenzene in water.

4.7 Sonolysis

Sonochemistry is concerned mainly with the understanding of the effect of ultrasound on chemical systems and its applications to chemical reactions and processes. Ultrasonic waves have unique physical properties and a frequency higher than the human hearing range, that is, above 20 kHz.

Wood and Loomis (1927) were the first to report the physical and biological influences of sonic waves. The use of ultrasound waves to induce cavitation for the destruction of some contaminants has resulted in the development of environmental sonochemistry. This application of sonochemistry in environmental remediation to reduce the amounts of chemicals required in other conventional treatments is welcome as a clean energy source.

Sonolysis of water gives $H^\bullet$ and $^\bullet OH$ by the thermal dissociation of water vapor present in the cavities during the compression phase. H_2O_2 and $H_2(g)$ are produced via hydroxyl radicals and hydrogen atoms during sonolysis of water (Equations 4.25 through 4.30).

$$H_2O \xrightarrow[\bullet)))]{} H^\bullet + {}^\bullet OH \tag{4.25}$$

$$H^\bullet + H^\bullet \rightarrow H_2 \tag{4.26}$$

$${}^\bullet OH_{(aq)} + {}^\bullet OH_{(aq)} \rightarrow H_2O_{2(aq)} \tag{4.27}$$

$$H^\bullet + O_2 \rightarrow HO_2^\bullet \tag{4.28}$$

$$HO_2^\bullet + H^\bullet \rightarrow H_2O_2 \tag{4.29}$$

$$HO_2^\bullet + HO_2^\bullet \rightarrow H_2O_2 + O_2 \tag{4.30}$$

$^\bullet OH$ and $^\bullet H$ radicals have a very short life span and hence are detected by spin trapping with nitrone compounds (the spin traps) that convert the short-lived radicals into relatively stable nitroxide radicals.

A bubble is produced on irradiation of a liquid with ultrasound. These bubbles oscillate and grow a little more during the expansion phase of the sound wave. On the other hand, they shrink during the compression phase of the sound wave. Under certain conditions, these bubbles can undergo a violent collapse, resulting in the generation of very high pressure and temperature, forming transient supercritical water. This process is called cavitation. In other words, the cavitation is the formation, growth, and collapse of bubbles in a liquid.

Cavitation serves as a means of concentrating the diffuse energy of sound and, as a consequence of the extreme conditions created, the cleavage of the dissolved oxygen molecules and water molecules (into $^\bullet H$ atoms and $^\bullet OH$ radicals) takes place. These hydroxyl radicals can be utilized for the degradation of various organic contaminants.

Lesko et al. (2006) reported the sonochemical decomposition of phenols. Here, $^\bullet OH$ radicals attack the phenols and lead to their oxidation. The reaction of the hydroxyl radicals breaks down the benzene rings of the phenols.

Diclofenac is a widely used anti-inflammatory nonsteroidal drug that escapes conventional urban wastewater treatment because it is resistant to such treatments. Hence, it is accumulated in treated effluents as well as water resources such as lakes and rivers, which may exhibit some adverse effects on aquatic organisms. The degradation of diclofenac during sonolysis, ozonation, and their simultaneous application has been reported by Naddeo et al. (2009). Hapeshi et al. (2010) reported the use of ultrasound for the degradation of the antibiotic ofloxacin in water.

Panwar et al. (2007, 2008) proposed the sonochemical degradation of some dyes (malachite green, brilliant cresyl blue, and azure-B). Kim and Huang (2005) observed the formation of a hydroxylated product such as 3-hydroxybenzothiophene, which is finally mineralized to the end product such as CO_2 and some inorganic sulfur species.

4.7.1 Sonocatalysis

The use of sound waves to impart or enhance the catalytic activity of a chemical compound is termed as sonocatalysis. It is a potential way to speed up the chemical catalysis. The sonication effects are mostly generated by the ultrasonic cavitation in liquids and, therefore, at least one reagent must be in liquid phase. It generates a highly reactive species in a short period with high temperature and pressure, which contributes to molecular decomposition, increasing the reactivity of many chemical substances. Ultrasonic irradiation can also be used to prepare the catalyst, for example, to produce the aggregates of fine-sized particles.

In heterogeneous catalysis, the catalyst and the reagent are in different phases. The heterogeneous catalysis is limited to the phase boundary, because only here the reagent and the catalyst are present. Ultrasound is used in many processes for the effective dispersion of solids and for the fine emulsification of liquids. In sonocatalysis, various ultrasound frequencies and catalysts, such as TiO_2, ZnS, and ZnO, can be used for the oxidation of iodide ions to iodine.

Ultrasonic irradiation can affect the reactivity of a catalyst in different reactions. Masuyama et al. (1992) observed that the allylation of ketones and aldehydes by allylic alcohols is improved using ultrasonic irradiation of a palladium/tin dichloride catalyst in less polar solvents. An inverted regioselectivity was also observed as compared with the homogeneous carbonyl allylation in polar solvents. Pratt and McGovern (1964) observed that aniline and *p*-methoxyaniline are converted to azobenzene and p,p′-dimethoxyazobenzene, respectively, over activated MnO_2 using the sonocatalysis process.

4.7.2 Sonophotocatalysis

Sonoluminescence and "hot spot" are normally considered as the inherent qualities of ultrasound and are used to explain the phenomenon of sonocatalytic degradation of organic pollutants in an aqueous solution in the presence of a catalyst. The ultrasonic irradiation can result in the generation of light with a comparatively wide wavelength range. Light with wavelengths < 375 nm may excite the catalyst acting as a photocatalyst and, as a result, hole–electron pairs are generated. In general, the photogenerated electrons react with water molecules to produce $^{\bullet}OH$ radicals with high oxidative activity. However, the holes can also play an important role depending upon the different crystal forms of the catalyst. The holes can directly degrade the organic pollutants adsorbed on the surface of the catalyst and also react with water molecules, producing $^{\bullet}OH$ radicals. Thus, holes degrade the organic pollutants in an aqueous solution directly or indirectly.

Sonophotocatalysis is the combination of two AOPs, that is, sonication and photocatalysis, wherein the free radicals are generated in both these processes. The rate of reaction is increased by the combination of these two modes of irradiations (UV/solar and ultrasound).

H_2O_2 is formed by coupling photocatalysis with sonolysis. H_2O_2 can also be added externally, which increases the number of hydroxyl radicals in the reaction system through the following steps (Equations 4.31 through 4.34):

$$H_2O \xrightarrow[\bullet)))]{} H^{\bullet} + {}^{\bullet}OH\,(\text{Sonolysis}) \quad (4.31)$$

$$H_2O_2 + H^{\bullet} \rightarrow H_2O + {}^{\bullet}OH\,(\text{Dissociation of } H_2O_2) \quad (4.32)$$

$$H_2O_2 + h\nu \rightarrow 2{}^{\bullet}OH\,(\text{Photolytic dissociation}) \quad (4.33)$$

$$H_2O_2 + e_{cb}{}^{-\bullet}OH + {}^{-}OH\,(\text{Reduction of } H_2O_2) \quad (4.34)$$

Silva et al. (2007) used this combination for the destruction of the phenolic compounds in the effluents from agroindustries.

An extensive overview of the treatment of pollutants in wastewaters by the sonophotocatalysis oxidation process has been given by Adewuyi (2005). If the adsorption of pollutants at the specific sites is the rate-controlling step in a particular reaction, ultrasound will play an important role by increasing the number of active sites substantially. Simultaneously, the surface area available due to the defragmentation of the catalyst also agglomerates under these conditions of turbulence generated and corresponding increases in the diffusion rates of the contaminants have been observed.

Selli et al. (2008) have reported on the degradation and mineralization of 1,4-dichlorobenzene (1,4-DCB) in an aqueous solution under different conditions: (i) photolysis, (ii) photocatalysis in the presence of TiO_2, and (iii) sonolysis. It was observed that photocatalysis results in faster removal of 1,4-DCB as compared with sonolysis and photolysis. However, the highest degradation and mineralization rates were attained under sonophotocatalytic conditions. They also reported the combined use of sonolysis and photocatalysis for the degradation of MTBE in water (Selli et al. 2005).

Reddy et al. (2003) reported that when high-frequency ultrasonication was combined with photocatalysis, the total degradation rate was found to be a sum of the individual rates of photocatalysis and sonication, whereas the combined effect of low-frequency ultrasound with photocatalysis is, however, much more complex. Ultrasonic degradation involves only the $^{\bullet}OH$ radicals (Peller et al. 2001), whereas photocatalytic oxidation involves both the hydroxyl radicals and the photogenerated holes (Nosaka et al. 1998; Yang and Tamai 2001); thus, some mechanistic differences exist between photocatalytic and sonophotocatalytic degradations.

The degradation of malachite green in water by means of a combination of ultrasound irradiation and heterogeneous photocatalysis was investigated by Berberidou et al. (2007). It was observed that the extent of sonolytic degradation increased with an increase in the power of the ultrasound (75–135 W), while the presence of TiO_2 in the dark generally had little or no effect on the degradation of the dye. The process of sonolysis under argon was relatively faster than in the presence of air. TiO_2 photocatalysis led to complete degradation of malachite green in 15–60 min. The sonophotocatalytic process was faster than the respective sonolytic or photocatalytic process due to the enhanced formation of reactive radicals as well as the increase in the active surface area of the catalyst in the presence of ultrasound. Malachite green was completely degraded in less than 20 min under sonophotocatalytic treatment.

Maezawa et al. (2007) reported the treatment of wastewater containing dye by photocatalytic oxidation with sonication. Such a type of combinative sonolysis and photocatalysis for textile dye degradation was also reported by Stock et al. (2000). Mrowetz et al. (2003) investigated the degradation of organic water pollutants through sonophotocatalysis in the presence of TiO_2. Bejareno-Peraz and Suarez-Herrera (2007) used TiO_2 as a photocatalyst for sonophotocatalytic degradation of Congo red and methyl orange. Visible light–induced sonophotocatalytic degradation of reactive red dye-198 was carried out by Kaur and Singh (2007) using dye-sensitized TiO_2, while Gonzalez and Martinez (2008) observed the sonophotocatalytic degradation of basic blue 9 industrial textile dye over TiO_2 and the factors influencing the degradation. Ameta et al. (2009) carried out a comparative study of sonolytic, photocatalytic, and sonophotocatalytic degradations of toluidine blue.

4.7.3 Sono-Fenton and Sonophoto-Fenton Processes

The dark reaction of H_2O_2 with ferrous salts (known as Fenton's reagent) and the photo-assisted decomposition of H_2O_2 (photo-Fenton) are known sources of hydroxyl radicals for the destruction of organic pollutants (Eisenhauer 1964;

Murphy et al. 1989; Safarzadeh-Amiri et al. 1996; Mazellier et al. 1997; Sehgal and Wang 1981; Mason and Lorimer 1988).

Sono-Fenton and photo-Fenton processes are applied separately as well as in combined systems, in order to evaluate the possible synergic effects produced by using such coupled systems. The cavitation effect of ultrasound reduces the particle size, resulting in a higher number of available active sites due to an increased surface area, which assists the subsequent photo-Fenton system. Such encouraging results open up new avenues in the field of AOTs for wastewater treatment.

Sound waves travelling through water with frequencies >15 kHz force the growth and subsequent collapse of small bubbles of gas in response to the passage of expansion and compression waves. The greatest coupling occurs when the natural resonance frequency of a bubble is equal to the ultrasonic frequency (e.g., 20 kHz equals a bubble diameter of 130 pm). The chemical effects of sound waves are realized during and immediately after the collapse of a vapor-filled cavitation bubble (Eisenhauer 1964; Murphy et al. 1989; Suslick 1988). Transient temperatures approaching 5000°C (Suslick et al. 1986, 1991) in the interior of the collapsing cavitation bubbles and pressures of several hundred atmospheres have been measured. As a result, temperatures around 2000°C have been observed for the interfacial region surrounding a collapsing bubble (Suslick et al. 1986, 1991). Collapse of bubbles occurs within 100 ns in the sonophoto-Fenton reaction. $^{\bullet}OH$ radicals are produced in two reactions. Ultrasonic radiations can induce dissociation of H_2O into $^{\bullet}OH$ and $H^{\bullet}$ radicals, and this $H^{\bullet}$ radical can induce dissociation of H_2O_2 into $^{\bullet}OH$ and water. H_2O_2 also generates two $^{\bullet}OH$ radicals in the presence of ultrasound. This extra generation of $^{\bullet}OH$ radicals increases the rate of sonophoto-Fenton reaction manyfold.

Segura et al. (2009) investigated the heterogeneous sonophoto-Fenton process for the degradation of phenolic aqueous solutions, while the sono-Fenton process is also efficient for the degradation of 2,4-dichlorophenol as compared with the degradation using Fenton process. Vaishnav et al. (2010) made a comparative study of photo-Fenton and sonophoto-Fenton degradations of methylene blue.

Ultrasound is a unique method for the production of radical species. When H_2O_2 is irradiated with ultrasound, it dissociates into $H^{\bullet}$ and $^{\bullet}OH$ radical species, which can be further utilized for the degradation of acid blue-25 in aqueous media (Ghodbane and Hamdaoui, 2009).

4.7.4 Sonolytic Ozonation (Ultrasound/O_3) Processes

Sonolytic ozonation normally means the combined effect of ultrasound and ozone, and this combination is also a part of AOPs. In this process, the O_3/O_2 mixture is bubbled in sonicated liquid, where the thermal decomposition of O_3 (gaseous form) in the cavitation bubbles leads to enhanced formation of $^{\bullet}OH$ radicals and H_2O_2. In the presence of ultrasound, ozone is decomposed thermolytically in the vapor phase of a cavitation bubble (Equation 4.35) as

$$O_{3(g)} \rightarrow O_{2(g)} + O\left(^3P\right)_{(g)} \tag{4.35}$$

The cavitation bubbles, capable of decomposing the ozone under mild conditions in a given time, are more in number when compared with those reaching a temperature high enough to dissociate the water molecule. The ground state $^{\bullet}O$ atoms produced here are not reactive and generally recombine to yield oxygen molecules, but they also increase the number of $^{\bullet}OH$ radicals forming H_2O_2 by recombination (Equation 4.36).

$$O\left(^3P\right)_{(g)} + H_2O_{(g)} \rightarrow 2^{\bullet}OH_{(g)} \tag{4.36}$$

The combination of O_3 and ultrasound may be an effective oxidation system, since two $^{\bullet}OH$ radicals are formed with each O_3 molecule consumed, but, on the other hand, the $^{\bullet}OH$ production is also checked as ozone may also react with atomic oxygen (Equation 4.37).

$$O_3 + O\left(^3P\right) \rightarrow 2O_2 \tag{4.37}$$

An increased H_2O_2 production by sonolytic ozonation over sonolysis of oxygen has been reported (Hart and Henglein 1986). The sonolytic ozonation is capable of mineralizing 4-nitrophenol completely in the bulk aqueous phase during acoustic cavitation (Hua and Hoffmann 1997).

As H_2O_2 is formed during the sonolysis of water, the coupled reactions of ozone and H_2O_2 will also contribute to the degradation of the organic pollutant (Equations 4.38 and 4.39).

$$O_3 + H_2O_2 \rightarrow {}^{\bullet}OH + HO_2^{\bullet} + O_2 \tag{4.38}$$

$$O_3 + HO_2^{\bullet} \rightarrow {}^{\bullet}OH + 2O_2 \tag{4.39}$$

Terzian et al. (1991) reported that *o*-chloranil and tetrachlorocatechol are formed as by-products when PCP was degraded by the attack of $^{\bullet}OH$ radicals on the aromatic ring. The $^{\bullet}OH$ radical reacts with the aromatic ring either by e^- abstraction or by addition.

At temperatures greater than 400°C, the $^{\bullet}OH$ radicals reacts with phenols and other aromatics mainly by H-atom abstraction (He et al. 1988). This reaction seems to proceed via $^{\bullet}OH$ addition at the ortho and para positions, because resonance stabilization is provided by the phenoxyl substituent, with the ortho position favored (based on statistical considerations). *o*-Chloranil and the semiquinone radical are then formed by oxidation.

o-Chloranil and tetrachlorocatechol are formed by the disproportionation of the semiquinone radical. Aromatic ring opening then takes place and *o*-chloranil yields simple organic acids such as oxalate and chlorinated organic acids.

4.8 Other Important AOPs

Apart from these AOPs, there are some more AOPs that are used for the remediation of water. These are described briefly.

4.8.1 Microwave/H_2O_2 Processes

AOPs have been successful in treating most of the organic compounds present in polluted water, and the major reason for the development of AOPs is the inability of the conventional processes to treat highly contaminated toxic water. The applications of microwave energy in the presence of H_2O_2 to enhance chemical reactions are also an alternative method for wastewater treatment (Caddick 1995; Airton et al. 2000). Microwave irradiation has widely been used as a more effective, easier, and cost-effective technique as compared with the conventional methods of water treatment. Zhihui et al. (2005) reported that microwave irradiation can greatly enhance the efficiencies of AOPs in the degradation of *p*-chlorophenol; however, it was observed that using UV power (55 W), the percent degradation of *p*-chlorophenol was 99.65% within 45 min, while using MW power (600 W), the percent degradation of *p*-chlorophenol was 93% within 180 min (Movahedyan et al. 2009).

4.8.2 Irradiation by γ-Ray, X-Ray, and Electron Beam

These are some of the important AOPs, where the pollutants are degraded by irradiating with γ-ray, x-ray, or electron beam.

Degradation of phenol in an aqueous solution was carried out by irradiating it with γ-ray and electron beam. It was observed that addition of TiO_2 has no significant effect on phenol decomposition; however, TOC removal was drastically increased (Chitose et al. 2003).

Halobenzenes are quite commonly available contaminants. Recently, Gu et al. (2009) reported the degradation of bromobenzene by irradiating it with an electron beam, while the radiolytic degradation of dicamba (3,6-dichloro-2-methoxybenzoic acid; a herbicide) has been observed by Drzewicz et al. (2004). The application of this method was also examined in some commercial agrochemical preparations. The radiolytic degradation of a polymer poly(vinyl alcohol) (PVA) was also investigated in an aqueous solution (Zhang et al. 2005). The effects of different parameters such as the pH and the concentration of

PVA were observed. The degradation of PVA was initiated by •OH and H• radicals, which were followed by random chain scission, resulting in the formation of aldehydes, ketones, enols, and acids.

Ionizing radiation technology has been utilized for the treatment of municipal wastewater by Al-Ani and Al-Khalidy (2006). It can be used as an alternate technology to reduce treatment cost. They observed that organic contaminants were degraded and reduced to about 12% by using γ-ray with a dose of 100–500 krad. Choi et al. (2010) reported an enhancement in the degradation of alachlor (a pesticide) in aqueous solutions by irradiating it with γ-radiation in the presence of H_2O_2. Zhu et al. (2010) observed that thiocyanate in an aqueous solution can be degraded by electron beam radiolysis. Sulfur and carbon were converted to sulfate and carbonate, respectively, in an alkaline medium.

4.8.3 Supercritical Water Oxidation

This method takes advantage of the unique properties of water and is applicable above its critical point, that is, $T_C = 374°C$ and $P_C = 22.1$ MPa. It has other added advantages such as compact size, competitive cost, high destruction efficiencies, and low NO_x and SO_x. The degradation under these conditions leads to the degradation of organic pollutants to CO_2, water, and corresponding salts and only negligible quantities of carbon monoxide and oxides of sulfur and nitrogen are formed.

Supercritical water oxidation (SCWO) involves the oxidation and degradation of organic contaminants when oxygen is injected into the SCW containing these pollutants. The use of SCWO can be classified as a clean technology. It permits the degradation of the chlorinated hydrocarbons such as polychlorinated biphenyl (PCB). The field of SCWO has been extensively reviewed by Yeshodhran (2002). It was concluded that this process can be recommended for the disposal of hazardous wastes such as halogenated organics, chemical weapons, explosives, and propellants. Treatment of different wastes by SCWO has been reported by Gidner et al. (2001).

The destruction of toxic organic wastewaters associated with sludge volume reduction is a global problem. There is a good possibility of destroying the organic materials in the waste by this method and leave only the inorganic materials in the effluent, which could be easily removed or recovered. Pinto et al. (2006) reported on the supercritical oxidation of quinoline and observed the effect of different parameters such as process temperature, ratio of oxidant to organic material, system pressure, and residence time. It was observed that the color of wastewater was completely removed at temperature of 450°C and above. Here, TOC removal efficiency was also determined.

Li et al. (1993) reported the treatment of dinitrotoluene process wastewater by SCWO. SCWO of nitrobenzene has also been reported by Zhang and Hua (2003). They observed almost >95% mineralization at 600°C temperature and 356 atm. pressures. SCWO of 1,4-DCB has also been carried out by

Jin et al. (1992). Shimookawa et al. (2010) used a hybrid process of TiO_2 photocatalysis and SCWO for the degradation of chlorobenzene. The synergic decomposition of chlorobenzene was more than 80%.

4.9 Conclusion

Water pollution is rapidly increasing day by day, and, therefore, there is a pressing demand to search for newer methods for the treatment of wastewaters. AOPs have emerged as a promising technologies in this direction and have proved their worth in the treatment of various contaminants in the industrial effluents/wastewaters. Most of the organic pollutants can be mineralized by one or the other AOT or by their different combinations.

References

Aaron, J.J. and M.A. Oturan. 2001. New photochemical methods for the degradation of pesticides aqueous media: Environment applications. *Turk. J. Chem.* 25: 509–520.

Abdessalem, A.K., N. Qturan, N. Bellakhal, M. Dachraoui, and M.A. Qturan. 2008. Experimental design methodology applied to electro-Fenton treatment for degradation of herbicide chlorlouron. *Appl. Catal. B Environ.* 78(3–4): 334–341.

Abramovic, B.F., D.V. Sojic, V.B. Anderluh, N.D. Abazovic, and M.I. Comor. 2009. Nitrogen-doped TiO_2 suspensions in photocatalytic degradation of mecoprop and (4-chloro-2 methylphenoxy) acetic acid herbicides using various light sources. *Desalination* 244: 293–302.

Adams, C.D., P.A. Scanlan, and H.D. Secrit. 1994. Oxidation and biologradability enhancement of 1,4-dioxane using hydrogen peroxide and ozone. *Environ. Sci. Technol.* 28: 1812–1818.

Adewuyi, Y.G. 2005. Sonochemistry in environmental remediation. 2: Heterogenous sonophotocatalytic oxidation processes for the treatment of pollutants in water. *Environ. Sci. Technol.* 39: 8557–8570.

Airton, K., P.Z. Patriao, and D. Nelson. 2000. Hydrogen peroxide assisted photochemical degradation of ethylenediaminetetraacetic acid. *Adv. Environ. Res.* 7: 197–202.

Al-Ani, M.Y. and F.R. Al-Khalidy. 2006. Use of ionizing radiation technology for treating municipal wastewater. *Iran J. Environ. Res. Public Health.* 3: 360–368.

Alberto, A.G. and D. Pletcher. 1999. The removal of low level organics via hydrogen peroxide found in a reticulated vitreous carbon cathode cell. Part 2: The removal of phenols and related compounds from aqueous effluents. *Electrochim. Acta.* 44: 2483–2492.

Ameta, G., P. Vaishnav, R.K. Malkani, and S.C. Ameta. 2009. Sonolytic, photocatalytic and sonophotocatalytic dyradation of toluidine blue. *J. Ind. Council Chem.* 26(2): 100–105.

Ameta, R., C. Kumari, C.V. Bhatt, and S.C. Ameta. 1998. Use of zinc oxide particulate system as a photocatalyst: Photobleaching of basic blue 24. *Ind. Quim.* 33: 36–39.

Ameta, R., J. Vardia, C.V. Bhatt, and S.C. Ameta. 1997. Photocatalytic degradation of brilliant green over semiconducting zinc oxide particles. *Sam. J. Chem.* 1: 29–31.

Ameta, S.C., P.B. Punjabi, C. Kumari, and Yasmin. 2006. A comparative study of Fenton's, photo-Fenton's and other related reagents: Resorcinol–TiO_2 system. *J. Indian Chem. Soc.* 83: 42–48.

Andreozzi, R., V. Caprio, A. Insola, and R. Marotta. 2000. The oxidation of metal (N-methyl-p-aminophenol) in aqueous solution by UV/H_2O_2 photolysis. *Water Res.* 34: 463–472.

Andreozzi, R., V. Caprio, A. Insola, R. Marotta, and R. Sanchirico. 2000. Advanced oxidation processes for the treatment of mineral oil-contaninated wastewaters. *Water Res.* 34: 620–628.

Arana, J., C. Cabo, C.F. Rodriguez, et al. 2008. Combining TiO_2–photocatalysis and wetland reactor for the efficient treatment of pesticides. *Chemosphere* 71: 788–793.

Araujo, E, A.J. Rodriguez-Malaver, A.M. Gonzalez, et al. 2002. Fenton's reagent. Mediated degradation of residual kraft black liquor. *Appl. Biochem. Biotechnol.* 97: 91–103.

Arques, A., A. Garcia–Ripoll, R. Sanchis, et al. 2007. Detoxification of aqueous solutions containing the commercial pesticide metasystox by TiO_2-mediated solar photocatalysis. *J. Sol. Energy Eng.* 129: 74–79.

Augugliaro, V., E. Davi, L. Palmisano, M. Schiavello, and A. Sclafani. 1990. Influence of hydrogen peroxide on the kinetics of phenol photodegradation in aqueous titanium dispersion. *Appl. Catal.* 65: 101–116.

Auguliaro, V., L. Palmisano, A. Sclafani, C. Minero, and E. Pelizzetti. 1988. Photocatalytic degradation of phenol in aqueous titanium dioxide dispersion. *Toxicol. Environ. Chem.* 16: 89–109.

Badawy, M.I., M.Y. Ghaly, and T.A. Gad-Allah. 2006. Advanced oxidation processes for the removal of organphorphorous pesticides from wastewaters. *Desalination* 194: 166–175.

Balanosky, E., J. Fernandez, J. Kiwi, and A. Lopez. 1999. Degradation of membrane concentrates of the textile industry by Fenton like reactions in iron free solutions at biocompatible pH values (pH 7–8). *Water Sci. Technol.* 40: 417–424.

Balmer, M.E. and B. Sulzberger. 1999. Atrazine degradation in irradiated iron/oxalate systems: Effects of pH and oxalate. *Environ. Sci. Technol.* 33: 2418–2424.

Bauer, R., G. Waldner, H. Fallmann, et al. 1999. The photo-Fenton reaction and the TiO_2/UV process for waste water treatments – Novel developments. *Catal. Today* 53: 131–144.

Bejareno-Peraz, N.J. and M.F. Suarez-Herrera. 2007. Sonophotocatalytic degradation of congo red and methyl orange in the presence of TiO_2 as a photocatalyst. *Ultrason. Sonochem.* 14: 589–595.

Beltran, F.J., J.M. Encinar, and M.A. Alonso. 1998. Nitroaromatic hydrocarbon ozonation in water. 2. Combined ozonation with hydrogen peroxide or UV radiation. *Ind. Eng. Chem. Res.* 37: 32–40.

Beltran, F.J., G. Overjero, and B. Acedo. 1993. Oxidation of atrazine in water by hydrogen peroxide combined with UV radiation. *Water Res.* 27: 1013–1021.

Beltran, F.J., G. Ovejero, and J. Rivas. 1996. Oxidation of polynuclear aromatic hydrocarbons in water. 3: UV radiation combined with hydrogen peroxide. *Ind. Eng. Chem. Res.* 35: 883–890.

Beltran-Heredia, J., J. Torregrossa, J. Garcid, J.R. Dominguez, and J.C. Tiemo. 2001. Degradation of olive mill wastewater by the combination of Fenton's reagent and ozonation processes with an aerobic biological treatment. *Water Sci. Technol.* 44: 103–108.

Berberidou, C., I. Poulios, N.P. Xekoukoulotakis, and D. Mantzavinos. 2007. Sonolytic, photocatalytic and sonophotocatalytic degradation of malachite green in aqueous solutions. *Appl. Catal. B. Environ.* 74: 63–72.

Bhandari, S., J. Vardia, R.K. Malkani, and S.C. Ameta. 2006. Effect of transition metal ion an photocatalytic activity of ZnO in bleaching some dyes. *Toxicol. Environ. Chem.* 88: 35–44.

Borgarillo, F., K. Kalyansundaram, M. Gratzel, and E. Pelizzetti. 1982. Visible light induced generation of hydrogen from H_2S in CdS dispersions: Hole transfer catalysis by RuO_2. *Helv. Chim. Acta.* 65: 243–248.

Bose, P., W.H. Glaze, and S. Maddox. 1998a. Degradation of RDX by various advanced oxidation processes. I: Reaction rates. *Water Res.* 32: 997–1004.

Bose, P., W.H. Glaze, and S. Maddox. 1998b. Degradation of RDX by various advanced oxidation processes. II: Organic by products. *Water Res.* 32: 1005–1018.

Bossmann, S.H., E. Oliveros, S. Gob, et al. 1998. New evidence against hydroxyl radicals as reactive intermediates in the thermal and photochemically enhanced Fenton reactions. *J. Phys. Chem. A.* 102: 5542–5550.

Boye, B., M.M. Dieng, and E. Brillas. 2002. Degradation of herbicide 4-chlorophenoxyacetic acid by advanced electrochemical oxidation methods. *Environ. Sci. Technol.* 36: 3030–3035.

Brillas, E., J. Calpe, and J. Casodo. 2000. Mineralization of 2,4-D by advanced electrochemical oxidation process. *Water. Res.* 34: 2253–2262.

Brillas, E., E. Mur, R. Sauleda, et al. 1998. Aniline mineralization by AOP's: Anodic oxidation, photocatalysis, electro-Fenton and photoelectron-Fenton processes. *Appl. Catal.* 16: 31–42.

Brillas, E., R. Sauleda, and J. Casado. 1998. Degradation of 4-chlorophenol by anodic oxidation, electro-Fenton, photoelectron-Fenton and peroxi-coagulation process. *J. Electrochem. Soc.* 145: 750–765.

Buxton, G.V. and A.J. Elliot. 1986. Rate constant for reaction of hydroxyl radicals with bicarbonate ions. *Radiat. Phys. Chem.* 27: 241–243.

Buxton, G.V., C.L. Greenstock, W.P. Helman, and A.B. Ross. 1988. Critical review of rate constants for reactions of hydrated electrons, hydrogen atoms and hydroxyl radical ($^{\bullet}OH/O^{-\bullet}$) in aqueous solution. *J. Phys. Chem. Ref. Data.* 17: 513–886.

Caddick, S. 1995. Microwave assisted organic reactions. *Tetrahedran* 38: 10403–10432.

Cater, S.R., M.I. Stefan, J.R. Bolton, and A. Safarzadeh-Amiri. 2000. Treatment of methyl tert-butyl ether in contaminated waters. *Environ. Sci. Technol.* 34: 659–662.

Cermenati, L., A. Albini, P. Pichat, and C. Guillard. 2000. TiO_2 photocatalytic degradation of haloquinolines in water: Aromatic products GM-MS identification, role of electron transfer and superoxide. *Res. Chem. Intermed.* 26: 221–234.

Chamarro, E., A. Marco, and S. Esplugas. 2001. Use of Fenton reagent to improve organic chemical biodegradability. *Water Res.* 35: 1047–1051.

Chen, C.C., C.S. Lu, and Y.C. Chung. 2006. Photocatalytic degradation of ethyl violet in aqueous solution mediated by TiO_2 suspensions. *J. Photochem. Photobiol. A. Chem.* 181: 120–125.

Chen, J., M. Liu, J. Zhang, Y. Xian, and L. Jin. 2003. Electrochemical degradation of bromopyrogallol red in presence of cobalt ions. *Chemosphere* 53: 1131–1136.

Chen, S. and X.L. Cheng. 1999. Photocatalytic degradation of oil films floating on water by TiO_2 supported hollow glass microbeads. *Zhongguo Huanjing kexue* 19(1): 47–50.

Chitose, N., S. Ueta, S. Seino, and T.A. Yamamota. 2003. Radiolysis of aqueous phenol solutions with nanoparticales. 1: Phenol degradation and TOC removal in solutions containing TiO_2 induced by UV, γ-ray and electron beams. *Chemosphere* 50(8): 1007–1013.

Choi, D., O.M. Lee, S. Yu, and S.-W. Jeorg. 2010. Gamma radiolysis of alachlor aqueous solution in the presence of hydrogen peroxide. *J. Hazard. Mater.* 184: 308–312.

Choi, W. and M.R. Hoffmann. 1997. Novel photocatalytic mechanisms for $CHCl_3$, $CHBr_3$, and CCl_3COO^- degradation and fate of photogenerated trihalomethyl radical on TiO_2. *Environ. Sci. Technol.* 31: 89–95.

Chou, S. and C. Huang. 1999. Applications of supported iron oxyhydroxide catalyst in oxidation of benzoic acid by hydrogen peroxide. *Chemosphere* 38: 2719–2731.

Chrittenden, J.C., S.M. Hu, D.W. Hand, and S.A. Grean. 1999. A kinetic model for H_2O_2-UV process in a completely mixed batch reactor. *Water Res.* 33: 2315–2328.

Comninellis, C. and A. De Battisti. 1996. Electrocatalysis in anodic oxidation of organics with simultaneous oxygen evolution. *J. Chem. Phys.* 93: 673–679.

Comninellis, C. and C. Pulgarin. 1993. Electrochemical oxidation of phenol for waste water treatment using SnO_2 anodes. *J. Appl. Electrochem.* 23: 108–112.

Conceicao, M., D.A. Amateus, A.M. Silva, and H.D. Burrows. 1994. Environmental and laboratory studies of the photodegradation of the pesticide fenarimol. *J. Photochem. Photobiol. A. Chem.* 80: 409–416.

Contreras, S., M. Rodríguez, E. Chamarro, S. Esplugas, and J. Casado. 2001. Oxidation of nitrobenzene by O_3/UV: The influence of H_2O_2 and Fe(II): Experience in a pilot plant. *Water Sci. Technol.* 44: 39–46.

Cun, H., L. Xingjuan, and L. Shuang. 1998. Photocatalytic degradation of aniline with ZnO. *Acta Scientiae Circumstantiae* 18: 81–85.

De Laat, J., G.T. Le, and B. Legube. 2004. A comparative study of the effects of chloride, sulfate and nitrate ions on the rates of decomposition of H_2O_2 and organic compounds by Fe(II)/H_2O_2 and Fe(III)/H_2O_2. *Chemosphere* 55: 715–723.

Dhananjeyan, M.R., J. Kiwi, P. Albers, and O. Ened. 2001. Photo-assisted immobilized Fenton degradation up to pH 8 of azo dye orange II mediated by Fe^{3+}/nafion glass fibers. *Helv. Chim. Acta.* 84: 3433–3445.

Dionysiou, D.D., M.T. Suidan, E. Bekou, I. Baudin, and J.M. Laine. 2000. Effect of ionic strength and hydrogen peroxide on the photocatalytic degradation of 4-chlorobenzoic acid in water. *Appl. Catal B. Environ.* 25: 153–171.

Do, J.S. and C.P. Chen. 1994. *In situ* oxidative degradation of formaldehyde with hydrogen peroxide electrogenerated on the modified graphites. *J. Appl. Electrochem.* 24: 936–942.

Doherty, S., C. Guillard, and P. Pichat. 1995. Kinetics and products of the photocatalytic degradation of morpholine. *J. Chem. Soc. Faraday Trans.* 91: 1853–1859.

Doong, R.A. and W.H. Chang. 1998. Photodegradation of parathion in aqueous titanium dioxide and zero valent iron solutions in the presence of hydrogen peroxide. *J. Photochem. Photobiol. A. Chem.* 116: 221–228.

Doong, R.A., H. Chen, R.A. Maithrupala, and S.M. Chang. 2001. The influence of pH and cadmium sulphide on the photocatalytic degradation of 2-chlorophenol in titanium dioxide suspension. *Water Res.* 35: 2873–2880.

Drzewicz, P., P. Gehringer, A. Bajanowska-Czajka, et al. 2004. Radiolytic degradation of the herbicide dicamba for environmental protection. *Arch. Environ. Contam. Toxicol.* 48(3): 311–322.
Duran, A., J.M. Monteagudo, and E. Amores. 2008. Solar photo-Fenton degradation of reactive blue 4 in a CPC reactor. *Appl. Catal. B. Environ.* 80: 42–50.
Eisenhauer, H.R. 1964. Oxidation of phenolic wastes. *Water Pollut. Control Fed.* 36: 1116–1128.
Engdahl, A. and B. Nelander. 2002. The vibrational spectrum of H_2O_3. *Science* 295: 482–483.
Esplugas, S., J. Gimenez, S. Contreras, E. Pascual, and M. Rodreguez. 2002. Comparison of different advanced oxidation processes for phenol degredation *Water Res.* 36: 1034–1042.
Fabbri, D., A. Crime, M. Davezza, et al. 2009. Surfactant-assisted removal of swep residues from soil and photocatalytic treatment of the washing wastes. *Appl. Catal. B. Environ.* 92: 318–325.
Fallmann, H., T. Krutzler, R. Bauer, S. Malato, and J. Blanco. 1999. Applicability of the photo-Fenton method for treating water containing pesticides. *Catal. Today* 54: 309–319.
Farre, M.J., S. Brosillon, X. Domenech, and J. Peral. 2007. Evaluation of the intermediates generated during the degradation of diuron and linuron hericides by the photo-Fenton reactions. *J. Photochem. Photobiol. A. Chem.* 189: 364–363.
Faust, B.C. and J. Hoigne. 1990. Photolysis of Fe(III)-hydroxy complexes as sources of OH radicals in clouds, fog and rain. *Atmos. Environ.* 24A: 79–89.
Faust, B.C. and R.G. Zepp. 1993. Photochemistry of aqueous iron(III)-polycarboxylate complexes: Roles in the chemistry of atmospheric and surface waters. *Environ. Sci. Technol.* 27: 2517–2522.
Fenton, H.J.H. 1876. On a new reaction of tartaric acid. *Chem. News.* 33: 190.
Fenton, H.J.H. 1894. Oxidation of tartaric acid in presence of iron. *J. Chem. Soc.* 65: 899–910.
Flox, C., J.A. Garrido, R.M. Rodriguez, et al. 2007. Mineralization of herbicide mecoprop by photoelectron-Fenton with UVA and solar light. *Catal. Today* 129: 29–36.
Fujishima, A., T.M. Rao, and D.A. Tryk. 2000. Titanium dioxide photocatalysis. *J. Photochem. Photobiol. C.* 1: 1–21.
Ghodbane, S.H. and O. Hamdaoui. 2009. Degradation of acid blue 25 in aqueous media using 1700 kHz ultrasonic irradiation; ultrasound/Fe(II) and ultrasound H_2O_2 coordination. *Ultrason. Sonochem.* 16(5): 593–598.
Gidner, A.V., L.B. Stenmark, and K.M. Carlsson. 2001. Treatment of different wastes by supercritical water oxidation. In *Twentieth IT3 Conference*, Philadelphia, USA, May 14–18.
Gikika, E., A. Traupis, A. Hiskia, and E. Papaconstantinou. 2005. Phtocatalytic reduction and recovery of mercury by polyoxometallates. *Environ. Sci. Technol.* 39(11): 4242–4248.
Glaze, W.H., J.W. Kang, and D. Chapin. 1987. The chemistry of water treatment processes involving ozone, hydrogen peroxide and ultraviolet radiation. *Ozone Sci. Eng.* 9: 335–352.
Gogate, P.R. and A.B. Pandit. 2004. A review of imperative technologies for wastewater treatment II: Hydrid methods. *Adv. Environ. Res.* 8: 553–597.

Gold, M.H, H. Kutsuk, and M.A. Morgan. 1983. Oxidative degradation of lignin by photochemical and chemical radical generating systems. *J. Photochem. Photobiol. A. Chem.* 38: 647–651.

Gonzalez, A.S. and S.S. Martinez. 2008. Study of the sonophotocatalytic degradation of basic blue industrial textile dye over slurry titanium dioxide and influencing factors. *Ultrason. Sonochem.* 15: 1038–1042.

Gouvea, C.A.K., F. Wypych, S.G. Moraes, et al. 2000. Semiconductor assisted photocatalytic degradation of reactive dye in aqueous solution. *Chemosphere* 40: 433–440.

Gu, J., W. Wu, L. Tang, et al. 2009. *International Conference on Energy and Environmental Technology* (ICET '09), Guilin Guengxi, October 16–18, pp. 480–483.

Guittonneau, S., J. De Laat, J.P. Duguet, C. Bonnel, and M. Dore. 1990. Oxidation of para-chloronitrobenzene in dilute aqueous solutions by O_3 + UV and H_2O_2 + UV: A comparative study. *Ozone Sci. Eng.* 12: 73–94.

Gurol, M.D. and R. Vatistas. 1987. Oxidation of phenolic compounds by ozone and ozone/UV radiation: A comparative study. *Water Res.* 21: 895–903.

Haber, F. and J.J. Weiss. 1934. The catalytic decomposition of hydrogen peroxide by iron salts. *Proc. Roy. Soc. Lond. Ser. A.* 147: 332–345.

Halmann, M., A.J. Hunt, and D. Spath. 1992. Photodegradation of dichloromethane, tetrachloroethylene and 1,2-dibromo-3-chloropropane in aqueous suspensions of TiO_2 with natural, concentrated and simulated sunlight. *Solar Energy Mater. Solar Cells* 26: 1–16.

Hapeshi, E., A. Achilleos, A. Papaioammou, et al. 2010. Sonochemical degradation of oflaxacin in aqueous solutions. *Water Sci. Technol.* 61: 3141–3146.

Hart, E.J. and A. Henglein. 1986. Sonolysis of ozone in aqueous solution. *J. Phys. Chem.* 90: 3061–3062.

Hasnat, M.A., I.A. Siddiquey, and A. Nuruddin. 2005. Comparative photocatalytic studies of degradation of a cationic and an anionic dye. *Dyes Pigments* 66: 185–188.

He, Y.Z., W.G. Mallard, and W. Tsang. 1988. Kinetics of hydrogen and hydroxyl radical attack on phenol at high temperature. *J. Phys. Chem.* 92: 2196–2201.

Hidaka, H., T. Koike, and N. Serpone. 2003. Photocatalytic degradation of surfactants. XX. Photooxidation of sodium butylnaphthalenesulfonates. *J. Oleo Sci.* 52: 245–253.

Hislop, K.A. and J.R. Bolton. 1999. The photochemical generation of hydroxyl radicals in the UV–Vis/Ferrioxalate/H_2O_2 system. *Environ. Sci. Technol.* 33: 3119–3126.

Horikoshi, S., N. Watanabe, and H. Hidaka. 2000. Photocatalytic wastewater treatment for 4-nonylphenol of an endocrine disruptor and 4-nonylphenol polyethoxylate surfactant at the titania/water interface. *J. Jpn. Oil Chem. Soc.* 49: 631–640.

Hsing, H.J., P.C. Chiang, E.E. Chan, and M.-Y. Chen. 2007. The decolorization and investigation of acid orange 6 dye in aqueous solution by advanced oxidation processes: A comparative study. *J. Hazard. Mater.* 141: 8–16.

Hsueh, C.L., Y.H. Huang, C.C. Wang, and C.Y. Chen. 2005. Degradation of azo dyes using low iron concentration of Fenton and Fenton-like system. *Chemosphere* 58(10): 1409–1414.

Hu, Q., C. Zhang, Z. Wang, Y. Chen, K. Mao, X. Zhang, Y. Xiong, and M. Zhu. 2008. Photodegradation of methyl tert-butyl ether (MTBE) by UV/H_2O_2 and UV/TiO_2 *J. Hazard. Mater.* 154: 795–803.

Hua, I. and M.R. Hoffmann. 1997. Optimization of ultrasonic irradiation as an advanced technology. *Environ. Sci. Technol.* 31: 2237–2243.

Hung-Yee, S. and H. Ching-Rong. 1995. Degradation of commercial azo dyes in water using ozonation and UV enhanced ozonation process. *Chemosphere* 31: 3813–3825.

Hustert, K., D. Kotzias, and F. Korte. 1983. Photocatalytischer cebbau won organization verbindungen titandioxid. *Chemosphere* 12: 55–58.

Iwahashi, H., H. Kawamori, and K. Fukushima. 1999. Quinolinic acid, α-picolinic acid, fusaric acid and 2,6-pyridine dicarboxylic acid enhance the Fenton reaction in phosphate buffer. *Chem. Biol. Interact.* 118: 201–215.

Jin, L., Z. Ding, and M.A. Abraham. 1992. Catalytic supercritical oxidation of 1,4-dichlorobenzene. *Chem. Eng. Res.* 47: 2659–2664.

Kajitvichyanukul, P., M.C. Lu, C.-H. Liao, W. Wirojanagud, and T. Koottatep. 2006. Degradation and detoxification of formalin waste water by advanced oxidation processes. *J. Hazard. Mater.* 135: 337–343.

Kang, S., C. Liao, and S. Po. 2000. Decolorization of textile wastewater by photo-Fenton oxidation technology. *Chemosphere* 41: 1287–1294.

Kaur, S. and V. Singh. 2007. Visible light induced sonophotocatalytic degradation of reaction red dye 198 using dye sensitized TiO_2. *Ultrason. Sonochem.* 14: 531–537.

Kawaguchi, H. 1992. Photooxidation of phenol in aqueous solution in the presence of hydrogen peroxide. *Chemosphere* 24: 1707–1712.

Kim, K. and C.P. Huang. 2005. Sonochemical degradation of polycyclic aromatic sulfur hydrocarbons (PASHs) in aqueous solutions exemplified by benzothiophene. *J. Chin. Inst. Eng.* 28: 1107–1118.

Kitsiou, V., N. Filippids, D. Mantzavinos, and I. Poulis. 2009. Heterogenous- and homogeneous photocatalytic degradation of the insecticide imidectoprid in aqueous solutions. *Appl. Catal. B. Environ.* 86: 27–35.

Kobayakawa, K., T. Miura, A. Suzuki, Y. Sato, and A. Fujishima. 1993. Enhancement of photocatalytic activity of CdS powder for hydrogen evolution from aqueous sulfuric solution by treatment with KCl. *Sol Energy Mater. Solar Cells* 30: 201–209.

Kominami, H., H. Kumamoto, Y. Kera, and B. Ohtani. 2003. Photocatalytic decolorization and mineralization of malachite green in an aqueous suspension of titanium(IV) oxide nanoparticales under aerated conditions: Correlation between some physical properties and their photocatalytic activity. *J. Photochem. Photobiol. A. Chem.* 160: 99–104.

Kong, L. and A.T. Lemley. 2006. Kinetic modeling of 2,4-dichlorophenoxyacetic acid (2,4-D) degradation in soil slurry by anodic Fenton treatment. *J. Agric. Food Chem.* 54: 3941–3950.

Kopf, P., E. Gilbert, and S.H. Eberie. 2000. TiO_2 photocatalytic oxidation of monochloroacetic acid and pyridine: Influence of ozone. *J. Photochem. Photobiol. A. Chem.* 136: 163–168.

Kumar, A., M. Paliwal, R. Ameta, and S.C. Ameta. 2008a. A novel route for wastewater treatment: Photo-assisted Fenton degradation of naphthol green B. *Collect. Czech. Chem. Commun.* 73: 679–689.

Kumar, A., M. Paliwal, R. Ameta, and S.C. Ameta. 2008b. Photochemical treatment of rhodamin B wastewater by photo-Fenton reagent. *Ind. J. Chem. Tech.* 15: 7–11.

Kumar, A., M. Paliwal, R. Ameta, and S.C. Ameta. 2008c. Use of photo-Fenton's reagent for the photochemical bleaching of metanil yellow. *Proc. Natl. Acad. Sci.* 78: 123–128.

Kuriacose, J.C. and M.C. Markham. 1962. Conversion of isopropyl alcohol to acetone on irradiated iron oxide. *J. Catal.* 1: 498–507.

Laat, J. De and H. Gallard. 1999. Catalytic decomposition of hydrogen peroxide by Fe(III) in homogeneous aqueous solutions: Mechnaism and kinetic modelling *Environ. Sci. Technol.* 33: 2726–2732.

Lapertot, M., S. Ebrahimi, S. Dazio, A. Rubinelli, and C. Pulgarin. 2007. Photo-Fenton and biological integrated process for degradation of a mixture of pesticides. *J. Photochem. Photobiol. A. Chem.* 186: 34–40.

Lapez, J.L., F.S.G. Einschlog, M.C. Gon Zalez, A.L. Capparelli, E. Oliveros, T.M. Hashem, and A.M. Braun. 2000. Hydroxyl radical initiated photodegradation of 4-chloro-3,5-dinitrobenzoic acid in aqueous solution. *J. Photochem. Photobiol. A. Chem.* 137: 177–184.

Lee, B.D., M. Hosomi, and A. Murakami. 1998. Fenton oxidation with ethanol to degrade anthracene into biodegradable 9,10-anthraquinone. A. Pretreatment method for anthracene contaminated Soil. *Water Sci. Technol.* 38: 91–97.

Lee, S.J., H.S. Son, H.K. Lee, and K.D. Zoh. 2002. Photocatalytic degradation of explosives contaminated water. *Water Sci. Technol.* 46: 139–145.

Lee, Y., C. Lee, and J. Yoan. 2004. Kinetics and mechanisms of DMSO (dimethyl sulfoxide) degradation by UV/H_2O_2 process. *Water Res.* 38: 2579–2588.

Legrine, O., E. Oliveros, and A.M. Braun. 1993. Photochemical processes for water treatment. *Chem. Rev.* 93: 671–698.

Leitner, N.K.V. and M. Dore. 1997. Mechanism of reaction between hydroxyl radicals and glycolic, acetic and oxalic acids in aqueous solutions: Consequence on hydrogen peroxide consumption in H_2O_2/UV and O_3/UV systems. *Water Res.* 31: 1383–1397.

Leitnar, N.K.V., A.L. Papailhou, J.P. Croue, J. Peyrot, and M. Dore. 1994. Oxidation of methyl tert-butyl ether (MTBE) and ethyl/tert-butyl ether (ETBE) by ozone and combined ozone/hydrogen peroxide. *Ozone Sci. Eng.* 16: 41–54.

Lesko, T., A.J. Colussi, and M. Hoffmann. 2006. Sonochemical decomposition of phenol: Evidence of a synergistic of ozone and ultrasound for the elimination of total organic carbon from water. *Environ. Sci. Technol.*, 40: 6818–6823.

Li, F.B., X.Z. Li, X.M. Li, T.X. Liu, and J. Dong. 2007. Heterogeneous photodegradation of bisphenol-A with iron oxides and oxalate in aqueous solution. *J. Colloid Interface Sci.* 311: 481–490.

Li, L., E.F. Gloyna, and J.E. Swicki. 1993. Treatability of DNT process waste water by supercritical water oxidation. *Water Environ. Res.* 65(3): 250–257.

Liang, H.C., X.Z. Li, Y.H. Yang, and K.H. Sze. 2010. Comparison of the degradation of diphenamid by homogeneous photolysis and heterogeneous photocatalysis in aqueous solution. *Chemosphere* 80(4): 366–374.

Lin, S.S. and M.D. Gurol. 1998. Catalytic decomposition of hydrogen peroxide on iron oxides: Kinetics, mechanisms and implications. *Environ. Sci. Technol.* 32: 1417–1423.

Lindsey, M.E. and M.A. Tarr. 2000a. Inhibited hydroxyl radical degradation of aromatic hydrocarbons in the presence of dissolved fulvic acid. *Water Res.* 34: 2385–2389.

Lindsey, M.E. and M.A. Tarr. 2000b. Inhibition of hydroxyl radical reaction with aromatics by dissolved natural organic matter. *Environ. Sci. Technol.* 34: 444–449.

Lipczynska-Kochany, E. 1992. Degradation of nitrobenzene and nitrophenols in homogeneous aqueous solution. Direct photolysis versus photolysis in the presence of hydrogen peroxide and the Fenton reagent. *Water Pollut. Res. J. Can.* 27: 97–122.

Litter, M. 2005. Introduction to photochemical advanced oxidation processes for water treatment. *Adv. Environ. Chem.* 2: 325–366.

Litter, M.I. 1999. Heterogeneous photocatalysis: Transition metal ions in photocatalytic system. *Appl. Catal. B. Environ.* 23: 89–114.

Lizama, C., C. Bravo, C. Caneo, and M. Ollino. 2005. Photocatalytic degradation of surfactants with immobilized TiO_2: Comparing two reaction systems. *Environ. Technol.* 26: 909–914.

Lucas, M.S. and J.A. Peres. 2007. Degradation of reactive black 5 by Fenton/UV-C and ferrioxalate/H_2O_2/solar light processes. *Dyes Pigments* 74: 622–629.

Lui, H.L. and C.K. Thomas. 2003. Photocatalytic inactivation of *Eschenchia coli* and *Lactobacillus helveticus* by ZnO and TiO_2 activated with ultraviolet light. *Proc. Biochem.* 39: 475–481.

Maezawa, A., H. Nakadoi, K. Suzuki, T. Furusawa, Y. Suzuki, and S. Uchida. 2007. Treatment of dye wastewater by using photocatalytic oxidation with sonication. *Ultrason. Sonochem.* 14: 615–620.

Maillard-Dupuy, C., C. Guillard, H. Courbon, and P. Pichat. 1994a. Kinetics and products of the TiO_2 photocatalytic degradation of pyridine in water. *Environ. Sci. Technol.* 28: 2176–2183.

Maillard-Dupuy, C., C. Guillard, and P. Pichat. 1994b. The degradation of nitrobenzene in water by photocatalysis over TiO_2: Kinetics and products; simultaneous elimination of benzamide or phenol or Pb^{2+} cations. *New J. Chem.* 18: 941–948.

Malato, S., J. Blanco, J. Cáceres, A.R. Fernández-Alba, A. Aguera, and A. Rodríguez. 2002. Photocatalytic treatment of water-soluble pesticides by photo-Fenton and TiO_2 using solar energy. *Catal. Today* 76: 209–220.

Malato, S., J. Blanco, M.I. Maldonado, et al. 2004. Engineering of solar photocatalytic collectors. *Solar Energy* 77: 513–524.

Maletzky, P. and R. Bauer. 1999. Immobilisation of iron ions on nafion and its applicability to the photo-Fenton method. *Chemosphere* 38: 2315–2325.

Maletzky, P. and R. Bauer. 1999. Immobilisation of iron ions on nafion and its applicability to the photo-Fenton method. *Chemosphere* 38: 2315–2325.

Mansilla, H.D., J. Villasnov, G. Maturana, et al. 1994. ZnO catalysed photodegradation of Kraft black liquor. *J. Photochem. Photobiol. A. Chem.* 78: 267–273.

Markham, M.C., M.C. Hanna, and S.W. Evans. 1954. Factors influencing the oxidation of phenols, catalysed by zinc oxide and light. *J. Am. Chem. Soc.* 76: 820–823.

Markham, M.C., J.C. Kuriacose, J.A.D. Marco, and C. Giaquint. 1962. Effect of amides on photochemical processes of zinc oxide surfaces. *J. Phys. Chem.* 66: 932–936.

Mason, T.J. and P. Lorimer. 1988. *Sonochemistry: Theory, Applications and Uses of Ultrasounds in Chemistry*. Chichester: Ellis Horwood.

Masuyama, Y., A. Hayakawa, and Y. Kurusu. 1992. Ultrasound-promoted carbonyl allylation by allylic alcohols with palladium–tin dichloride in nonpolar solvents: Inversed regiocontrol of carbonyl allylation in polar solvents. *J. Chem. Soc. Chem. Commun.* 1102.

Matthews, R.W. 1998. Kinetics of photocatalytic oxidation of organic solutes over titanium dioxide. *J. Catal.* 111: 264–278.

Mazellier, P., J. Jirkovsky, and M. Bolte, *Pestic. Sci.* 1997. Degradation of diuron photoinduced by iron(III) in aqueous solution. 49: 259–267.

Mehrab, M. and V.S. Haffield. 2005. Photocatalytic treatment of linear alkylbenzene sulfonate (LAS) in water. *J. Environ. Sci. Health A* 40: 1003–1012.

Merabet, S., D. Robert, J.V. Weber, M. Bouhelassa, and S.M. Benkhanouche. 2009. Photocatalytic degradation of indole in UV/TiO_2: Optimizations and modeling using the response surface methodology (RSM). *Environ. Chem. Lett.* 7: 45–49.

Metcalf & Eddy Inc. 2003. *Wastewater Engineering Treatment and Reuse*, 4th edn. New York: McGraw-Hill.

Mills, A. and G. Williams. 1987. Methyl orange as a probe of the semiconductor–electrolytic interfaces in CdS suspensions. *J. Chem. Soc. Faraday Trans. I* 83: 2647–2661.

Modrishahla, N., M.A. Behnajady, and F. Ghanbary. 2007. Decolorization and mineralization of C.I. acid yellow 23 by Fenton and photo-Fenton processes. *Dyes Pigments* 73: 305–310.

Mogra, D., R. Agarwal, P.B. Punjabi, and S.C. Ameta. 2003. Photochemical degradation of o-chlorobenzoic acid by photo-Fenton's reagent. *Chem. Environ. Res.* 12 (3&4): 227–235.

Mokrini, A., D. Oussi, and S. Esplugas. 1997. Oxidation of aromatic compounds with UV radiaton/Ozone/hydrogen peroxide, *Water Sci. Technol.* 35: 95–102.

Mortazavi, S.B., A. Khavanin, G. Moussavi, and A. Azhdarpoor. 2008. Removal of sodium dodecyl sulfate in an intermittent cycle extended aeration system. *Pak. J. Biol. Sci.* 11: 290–293.

Movahedyan, H., A.M. Seid Mohammadi, and A. Assadi. 2009. Comparison of different advanced oxidation processes degrading p-chlorophenol in aqueous solution. *Iran. J. Environ. Health. Sci. Eng.* 6: 153–160.

Mrowetz, M., C. Pirola, and E. Selli. 2003. Degradation of aqueous water pollutants through sonophotocatalysis in the presence of TiO_2. *Ultrason. Sonochem.* 10: 247–250.

Mrowtez, M. and E. Selli. 2004. Effects of iron species in the photocatalytic degradation of an azo dye. *J. Photochem. Photobiol. A. Chem.* 162: 89–95.

Mu, Y., H.Q. Yu, J.C. Zheng, and S.J. Zheng. 2004. TiO_2-mediated photocatalytic degradation of orange II with the presence of Mn^{2+} in solution. *J. Photochem. Photobiol. A. Chem.* 163: 311–316.

Mukhtar-Al-Hassan, J. Anwar, and M. J. Saif. 2008. Photocatalytic degradation of hexadecyltrimethylammonium chloride $C_{19}H_{42}NCl$. *J. Sci. Res.* 38: 29–34.

Munter, R., J. Kallas, S. Preis, S. Kamenev, M. Trapido, and Y. Veressinina. 1995. Comparative studies of AOP for aromatic and PAH destruction. In *Proceedings of the 12th World Ozone Congress*, May 15–18, Lille, France, 1, pp. 395–406.

Murphy, A.P., W.J. Boegli, M.K. Price, and C.D. Moody. 1989. A Fenton like reaction to mineralize formaldehyde waste solutions. *Environ. Sci. Technol.* 23: 166–169.

Muszkat, L., L. Bir, and L. Feigelson. 1995. Solar photocatalytic mineralization of pesticides in polluted water. *J. Photochem. Photobiol. A. Chem.* 87: 85–88.

Naddeo, V., V. Belgiorno, D. Ricco, and D. Kassinos. 2009. Effect of sonolysis on waste activated sludge solubilisation and anerobic biodegradability. *Ultrason. Sonochem.* 16: 790–794.

Nadtochenko, V. and J. Kiwi. 1998. Photoinduced mineralization of xylidine by the Fenton reagent. 2: Implications of the precursors formed in the dark. *Environ. Sci. Technol.* 32: 3282–3285.

Natts, R.J., A.P. Jones, P.H. Chen, and A. Kenny. 1997. Mineral-catalyzed Fenton like oxidation of sorbed chlorobenzene. *Water Environ. Res.* 69: 369–275.

Neyens, E. and J. Baeyens. 2003. A review of classic Fenton's peroxidation as an advanced technique. *J. Hazard. Mater. B* 98: 33–50.

Nogueira, R.F.P., A.G. Trovó, and D.F. Mode. 2002. Solar photodegradation of dichloroacetic acid and 2,4-dichlorophenol using an enhanced photo-Fenton process. *Chemosphere* 48: 385–391.

Nosaka, Y., M. Kishimoto, and J. Nishino. 1998. Factors governing the initial process of TiO_2 photocatalysis studied by means of in situ electron spin resonance measurements. *J. Phys. Chem. B* 102: 10279–10283.

Obee, T.N. and S. Satyapal. 1998. Photocatalytic decomposition of DMMP on titania. *J. Photochem. Photobiol. A. Chem.* 118: 45–54.

Ollis, D.F., C. Hasiao, L. Budirman, and C. Lee. 1984. Heterogeneous photoassisted catalysis: Conversions of perchloroethylenes, dichloroethylene, chloroacetic acids and chlorobenzenes. *J. Catal.* 88: 89–96.

Osawa, Z., T. Sunakani, and Y. Fukude. 1994. Photodegradation of blends of poly (vinyl chloride) and polyurethane. *Polymer Degrad. Stability* 150: 61–66.

Oturan, M.A. 2000. An ecologically effective water treatment technique using electrochemically generated hydroxyl radicals for *in situ* destruction of organic pollutants: Application to herbicide 2,4-D. *J. Appl. Electrochem.* 30: 475–482.

Oturan, M.A., J.J. Aaron, N. Oturan, and J. Pikson. 1999. Degradation of chlorophenoxy acids herbicides in aqueous media, using a novel electrochemical method. *Pestic. Sci.* 55: 558–562.

Panwar, C., A. Kumar, M. Paliwal, P.B. Punjabi, and S.C. Ameta. 2007. Sonochemical degradation of some triphenylmethane dyes. *Polish J. Chem.* 81: 1777–1784.

Panwar, C., A. Kumar, M. Paliwal, P.B. Punjabi, and S.C. Ameta. 2008. Sonochemical degradation of brilliant cresyl blue. An ecofriendly approach. *J. Indian Chem. Soc.* 85: 654–657.

Parra, S., S. Malato, J. Blanco, P. Peringer, and C. Pulgarin. 2001. Concentrating and non-concentrating reactors for solar photocatalytic degradation of p-nitrotoluene-o-sulphonic acids. *Water Sci. Technol.* 44: 219–227.

Patel, S.G., S. Bhatnagar, J. Vardia, and S.C. Ameta. 2006. Use of photocatalysts in solar desalination. *Desalination* 189: 287–291.

Pelizzetti, E., C. Minero, V. Maurino, A. Sclafani, H. Hidaka, and N. Serpone. 1989. Photocatalytic degradation of nonylphenol ethoxylated surfactants. *Environ. Sci. Technol.* 23: 1380–1385.

Peller, J., O. Wiest, and P.V. Kamat. 2001. Sonolysis of 2,4-dichlorophenoxyacetic acid in aqueous solutions. Evidence for •OH-radical-mediated degradation. *J. Phys. Chem. A.* 105: 3176–3181.

Perez, M., F. Torrades, X. Domenech, and J. Peral. 2002. Fenton and photo-Fenton oxidation of textile effluents. *Water Res.* 36: 2703–2710.

Perez, M., F. Torrades, X. Domenech, and J. Peral. 1983. Removal of organic contaminants in paper pulp effluents by AOP's: An economic study. *J. Chem. Technol. Biotechnol.* 72: 525–532.

Peyton, G.R., F.Y. Huang, J.L. Burleson, and W.H. Glaze. 1982. Destruction of pollutants in water with ozone in combination with UV radiation. *Environ. Sci. Technol.* 16: 448–453.

Philippopoulos, C.J. and S.G. Poulopoulos. 2003. Photo-assisted oxidation of an oily wastewater using hydrogen peroxide *J. Hazard. Mater. B.* 98: 201–210.

Pignatello, J.J. and L.Q. Huang. 1993. Degradation of polychlorinated dibenzo-p-dioxin and dibenzofuran contaminants in 2,4,5-T by photoassisted iron catalyzed hydrogen peroxide. *Water Res.* 27: 1731–1736.

Pignatello, J.J., E. Oliveros, and A. Mackay. 2006. Advanced oxidation processes for organic contaminant destruction based on the Fenton reaction and related chemistry. *Crit. Rev. Environ. Sci. Technol.* 36: 1–84.

Pinto, L.D.S., L.M.F. dos Santos, B. Al-Duri, and R.C.D. Santos. 2006. Supercritical water oxidation of quinoline in a continuous plug flow reactor-Part 1: Effect of key operating parameters. *J. Chem. Technol. Biotechnol.* 81: 912–918.

Poulios, I., E. Micropoulou, R. Panou, and E. Kostopolou. 2003. Photooxidation of eosin Y in the presence of semiconducting oxides. *Appl. Catal.* 41: 345–355.

Pouretedal, H.R., H. Eskandari, M.H. Keshavarz, and A. Semnami. 2009. Photodegradation of organic dyes using nanoparticles of cadmium sulphide doped with manganese, nickel and copper as nanophotocatalyst. *Acta Chim. Slov.* 56: 353–361.

Pratap, K. and A.T. Lemley. 1998. Fenton electrochemical treatment of aqueous atrazine and metolachlor. *J. Agric. Food Chem.* 46: 3285–3291.

Pratt, E.F. and T.P. McGovern. 1964. Oxidation by Solids. III: Benzalanilines from N-benzylanilines and related oxidations by manganese dioxide. *J. Org. Chem.* 29: 1540–1545.

Priya, S.S., M. Premalatha, and N. Anantharaman. 2008. Solar photocatalytic treatment of phenolic waste water – Potential, challenges and opportunities. *ARPN J. Eng. Appl. Sci.* 3(6): 36–41.

Rajeshwari, R. and S. Kanmani. 2009. TiO_2-based heterogeneous photocatalytic treatment combined with ozonation for carbendazim degradation. *Iran J. Environ. Health Sci. Eng.* 6: 61–66.

Rao, K.V.S., B. Lavedrine, and P. Boule. 2003. Influence of metallic species on TiO_2 for the photocatalytic degradation of dyes and dye intermediates. *J. Photochem. Photobiol. A. Chem.* 154: 189–193.

Reddy, E.P., L. Davydov, and P.G. Smirniotis. 2003. TiO_2-loaded zeolites and mesoporous materials in the sonophotocatalytic decomposition of aqueous organic pollutants: The role of the support. *Appl. Catal. B. Environ.* 42: 1–11.

Rein, M. 2001. Advanced oxidation processes—Current status and prospects. *Proc. Estonian Acad, Sci. Chem.* 50: 59–80.

Rodriguez, M., V. Sarria, S. Esplugas, and C. Pulgarin. 2002. Photo-Fenton treatment of a biorecalcitrant wastewater generated in textile activities: Biodegradability of the photo-treated solution. *J. Photochem. Photobiol. A. Chem.* 151: 129–135.

Safarzadeh-Amiri, A., J. Bolton, and S. Cater 1996. Aniline mineralization by AOP's: Anodic oxidation, photocatalysis, electro-Fenton and photoelectro-Fenton processes. *Solar Energy* 56: 439–443.

Safarzadeh-Amiri, A., J.R. Bolton, and S.R. Cater. 1997. Ferrioxalate-mediated photodegradation of organic pollutants in contaminated water. *Water Res.* 31: 787–798.

Safarzadeh-Amiri, A., J.R. Bolton, S.R. Cater. 1996. The use of iron in advanced oxidation processes. *J. Adv. Oxid. Technol.* 1: 18–26.

Saritha, P., C. Aparna, V. Himabindu, and Y. Anjaneyulu. 2007. Comparison of various advanced oxidation processes for the degradation of 4-chloro-2-nitrophenol. *J. Hazard. Mater.* 149: 609–614.

Savall, A. 1995. Electrochemical treatment of industrial organic effluents. *Chimia.* 49: 23–27.

Scheeren, C.W., T.M. Paniz, and A.F. Martins. 2002. Comparison of advanced processes on the oxidation of acid orange 7 dye. *J. Environ. Sci. Health A.* 37: 1253–1261.

Segura, Y., R. Molina, F. Martínez, and J.A. Melero. 2009. Integrated heterogeneous sono-photo-Fenton processes for the degradation of phenolic aqueous solution. *Ultrason. Sonochem.* 16(3): 417–424.

Sehgal, C.M. and S.Y. Wang. 1981. Threshold intensities and kinetics of sonoreaction of thymine in aqueous solutions at low ultrasonic intensities. *J. Am. Chem. Soc.* 103: 6606–6611.

Selli, E., C.L. Bianchi, C. Pirola, and M. Bertelli. 2005. Degradaton of methyl tert-butyl ether in water: Effects of the combined use of sonolysis and photocatalysis. *Ultrason. Sonochem.* 12: 395–400.

Selli, E., C.L. Blanchi, C. Pirola, G. Cappelletti, and V. Ragaini. 2008. Efficiency of 1,4-dichlorobenzene degradation in water under photolysis, photocatalysis on TiO_2 and sonolysis. *J. Hazard. Mater.* 153: 1136–1141.

Serpone, N., P. Muruthamuthu, P. Pichat, E. Pelizzetti, and H. Hidaka. 1995. Exploiting the interparticle electron transfer process in the photocatalysed oxidation of phenol, 2-chlorophenol and pentachlorophenol: Chemical evidence for electron and hole transfer between coupled semiconductors. *J. Photochem. Photobiol. A. Chem.* 85: 247–255.

Shanhu, F., S. Yong, and C. Liuping. 2002. Photocatalytic degradation of cefradine in water over immobilized TiO_2 catalyst in continuous flow reactor. *Chin. J. Catal.* 23: 100–112.

Sharma, A., P. Rao, R.P. Mathur, and S.C. Ameta. 1998. Phtocatalytic reactions of bromopyrogallol red on zinc oxide particles. *Res. J. Chem. Environ.* 2: 17–20.

Sherrard, K.B., P.J. Marriott, M.J. Mc Cormick, R. Cotton, and G. Smith. 1994. Electrospray mass spectrometric analysis and photocatalytic degradation of polyethoxylate surfactants used in wool scorning. *Anal. Chem.* 66: 3394–3399.

Shimookawa, A., N. Kometni, and Y. Yonewaza. 2010. Degradation of chlorobenzene by the hybrid process of supercritical water oxidation and TiO_2 photocatalysis, *Sep. Sci. Technol.* 45(11): 1538–1545.

Shu, H.Y., C.P. Huang, and M.C. Chang. 1994. Photooxidative degradation of azo dyes in water using hydrogen peroxide and UV radiation. In *Proceedings of the 26th Mid-Atlantic Industrial Waste Conference*, University of Delaware, Newyork, pp. 186–193.

Silva, A.M.T., E. Nouli, Â.C. Carmo-Apolinário, N.P. Xekoukoulotakis, and D. Mantzavinos. 2007. Sonophotocatalytic/H_2O_2 degradation of phenolic compounds in agro-industrial effluents. *Catal. Today* 124: 232–239.

Singhal, B., A. Porwal, A. Sharma, R. Ameta, and S.C. Ameta. 1997. Photocatalytic degradation of cetylpyridinium chloride over titanium dioxide powder. *J. Photochem. Photobiol. A. Chem.* 108: 85–88.

Sir'es, J., C. Arias, P.L. Cabot, et al. 2007. Degradation of clofibric acid in acidic aqueous medium by electro-Fenton and photoelectron-Fenton processes. *Chemosphere* 66: 1660–1669.

Sir'es, I., J.A. Garrido, R.M. Rodriguez, et al. 2006. Electrochemical degradation of paracetamol from water by catalytic action of Fe^{2+}, Cu^{2+} and UVA light on electrogenerated hydrogen peroxide. *J. Electrochem. Soc.* 153: D1–D9.

Soana, F., M. Sturini, L. Cermenti, and A. Albini. 2000. Titanium dioxide photocatalyzed oxygenation of naphthalene and some of its derivatives. *J. Chem. Soc. Perkin Trans.* 2: 699–704.

Sorensen, M. and F.H. Frimmel. 1997. Photochemical degradation of hydrophilic xenobiotic in the UV/H_2O_2 process: Influence of nitrate on the degradation rate of EDTA, 2-amino-1-napthalene sulfonate, diphenyl-4-sulfonate and 4,4′-diaminostilbene-2,2′-disulfonate. *Water Res.* 31: 2885–2891.

Stepnowski, P., E.M. Siedlicka, P. Behrend, and B. Jastooff. 2002. Enhanced photodegradation of contaminants in petroleum refinery wastewater. *Water Res.* 36: 2167–2172.

Stock, N.L., J. Peller, K. Vinodgopal, and P.V. Kamat. 2000. Combinative sonolysis and photocatalysis for textile dye degradation. *Environ. Sci. Technol.* 34: 1747–1750.

Subramani, A.K., K. Byrappa, S. Anaunde, et al. 2007. Photocatalytic degradation of indigo carmine dye using TiO_2 impregnated activated carbon, *Bull. Mater. Sci.* 30(1): 37–41.

Sun, Y. and J.J. Pignatello. 1993. Photochemical reactions involved in the total mineralization of 2,4-D by iron(III)/hydrogen peroxide/UV. *Environ Sci. Technol.* 27: 304–310.

Sunder, M. and D.C. Hempel. 1997. Oxidation of tri- and perchloroethene in aqueous solution with ozone and hydrogen peroxide in a tube reactor. *Water Res.* 31: 33–40.

Suslick, K.S. (ed.). 1988. *Ultrasound: Its Chemical, Physical and Biological Effects*. VCH: New York.

Suslick, K.S., S.B. Choe, A.A. Chichowlas, and M.W. Grinstaff. 1991. Sonochemical synthesis of amorphous iron. *Nature* 253: 414–416.

Suslick, K.S., D.A. Hammerton, and D.E. Cline Jr. 1986. The sonochemical hot spot. *J. Am. Chem. Soc.* 100: 5641–5642.

Taicheng, A., C. Jiaxin, L. Guiying, et al. 2008. Characterization and the photocatalytic activity of TiO_2 immobilized hydrophobic montmorillonite photocatalysts: Degradation of decabromodiphenyl ether (BDE 209). *Catal. Today* 139: 69–76.

Takahashi, N. 1990. Ozonation of several organic compounds having low molecular weight under UV irradiation. *Ozone Sci. Eng.* 12: 1–18.

Takizawa, T., T. Watanabe, and K. Honda. 1978. Photocatalysis through excitation of adsorbents. 2: A comparative study of rhodamine B and methylene blue cadmium sulphide. *J. Phys. Chem.* 82: 1391–1396.

Tanaka, K., K. Harda, and S. Murata. 1986. Photocatalytic deposition of metal ions on to TiO_2 powder. *Sol. Energy* 36: 159–161.

Teel, A.L. and R.J. Watts. 2002. Degradation of carbon tetrachloride by modified Fenton's reagent. *J. Hazard. Mater.* 94: 179–189.

Terzian, R., N. Serpone, R. Barton Draper, M.A. Fox, and E. Pelizzetti. 1991. Pulse radiolytic studies of the reaction of pentahalophenols with OH radicals: Formation of pentahalophenoxyl, dihydroxypentahalocyclohexadienyl, and semiquinone radicals. *Langmuir* 7: 3081–3089.

Theron, P., P. Pichat, C. Petrier, and C. Guillard. 2001. Water treatment by TiO_2 photocatalysis and/or ultrasound: Degradation of phenyl trifluoromethyl ketone, a trifluoroacetic acid forming pollutant and octan-1-o1, a very hydrophobic pollutant. *Water Sci. Technol.* 44: 263–270.

Trapido, M. and J. Kallas. 2000. Advanced oxidation processes for the degradation and detoxification of 4-nitrophenol. *Environ. Technol.* 21: 799–808.

Trapido, M., Y. Veressinina, and J. Kallas. 2001. Degradation of aqueous nitrophenols by ozone combined with UV radiation and hydrogen peroxide. *Ozone Sci. Eng.* 23: 333–342.

Turchi, C.S. and D.F. Ollis. 1990. Photocatalytic degradation of organic water contaminants: Mechanisms involving hydroxyl radical attack. *J. Catal.* 122: 178–192.

Uchida, H., S. Itoh, and H. Youneyama. 1993. Photocatalytic decomposition of propyzamide using TiO_2 supported on activated carbon. *Chem. Lett.* 1995–1998.

Uchida, H., S. Katoh, and M. Watarloe. 1995. Photocatalytic decomposition of trichlorobenzene using TiO_2 supported on nickel-poly(tetrafluoroethylene) composite plate. *Chem. Lett.* 4: 261–262.

Unkroth, A., V. Wagner, and R. Sauerbry. 1997. Laser-assisted photochemical wastewater treatment. *Water Sci. Technol.* 35: 181–188.

Vaishnav, P., G. Ameta, A. Kumar, S. Sharma, and S.C. Ameta. 2010. Sonophoto-Fenton and photo-Fenton degradation of methylene blue: A comparative study. *J. Indian Chem. Soc.* 87 (in press).

Villa, R.D. and R.F.P. Nogueira. 2006. Oxidation of p,p′-DDT and p,p′-DDE in highly and long term contaminated soil using Fenton reaction in a slurry system. *Sci. Total Environ.* 371: 11–18.

Wahi, R.K., W.W. Yu, Y.P. Liu, et al. 2005. Photodegradation of congo red catalysed by nanosized TiO_2. *J. Catal.* 242A: 45–56.

Walling, C. and T. Weil. 1994. The ferric ion catalyzed decomposition of hydrogen peroxide in perchloric acid solution. *Int. J. Chem. Kinet.* 6: 507–516.

Wang, Q. and A.T. Lemley. 2003. Competitive degradation and detoxification of corbonate insecticides by membranes anodic Fenton treatment. *J. Agric. Food Chem.* 51: 5382–5390.

Wang, S. 2008. A comparative study of Fenton and Fenton like reaction kinetics in decolourisation of wastewater. *Dyes Pigments* 76: 714–720.

Wang, S.S., Z.H. Wang, and Q.X. Zhuang. 1992. Photocatalytic reduction of the environmental pollutant Cr^{VI} over a cadmium sulphide powder under visible light illumination. *Appl. Catal.* 1: 257–270.

Wentworth, P., L.H. Jones, A.D. Wentworth, et al. 2001. Antibody catalysis of the oxidation of water. *Science* 293: 1806–1811.

Wood, R.W. and A.L. Loomis. 1927. The physical and biological effects of high frequency sound waves of great intensity. *Philos. Mag.* 4: 414.

Yang, X. and N. Tamai. 2001. How fast is interfacial hole transfer? In situ monitoring of carrier dynamics in anatase TiO_2 nanoparticles by femtosecond laser spectroscopy. *Phys. Chem. Chem. Phys.* 3: 3393–3398.

Ye, C., Y. Bando, G. Shen, and D. Golberg. 2006. Thickness dependent photocatalytic performance on ZnO nanoplatelets. *J. Phys. Chem.* 110: 15146–15151.

Yeber, M.C., J. Rodriquez, J. Freer, N. Duran, and H.D. Mansilla. 2000. Photocatalytic degradation of cellulose bleaching effluent by supported TiO_2 and ZnO. *Chemosphere* 41: 1193–1197.

Yeshodhran, S. 2002. Supercritical water oxidation: An environmentally safe method for the disposal of organic wastes. *Curr. Sci.* 82(9): 1112–1122.

Zepp, L., B. Faust, and J. Hoigne. 1992. Hydroxyl radical formation in aqueous solution (pH 3–8) of iron(II) with hydrogen peroxide: The photo-Fenton reaction. *Environ. Sci. Technol.* 26: 313–319.

Zhang, G. and I. Hua. 2003. Supercritical water oxidation of nitrobenzene. *Ind. Eng. Chem. Res.* 42(2): 285–289.

Zhang, L. and T. Shen. 1996. Photocatalytic reduction of methyl yellow by CdS nanoparticles mediated in reverse micelles: Microheterogeneous electron transfer across a water–oil boundary. *J. Chem. Soc. Chem. Commun.* 473.

Zhang, S.J., H.Q. Yu, X.W. Ge, and R.F. Zhu. 2005. Optimization of radiolytic degradation of poly(vinyl alcohol). *Ind. Eng. Chem. Res.* 44(7): 1995–2001.

Zhihui, A., Y. Peng, and L. Xiaohua. 2005. Degradation of 4-chlorophenol by a microwave assisted photocatalysis method. *J. Hazard. Mater. B.* 124: 147–152.

Zhiyong, Y., E. Mielezarski, J. Mielezarski, et al. 2003. Preparation, stabilization and characterization of TiO_2 on thin polyethylene films (LDPE). Photocatalytic applications. *Water Res.* 41: 862–874.

Zhou, W., K. Pan, M. Qu, et al. 2010. Photodegradation of organic contamination in wastewaters by bonding TiO_2/single-walled carbon nanotube composites with enhanced photocatalytic activity. *Chemosphere* 81(5): 555–561.

Zhou, Y. and J. Hoigne. 1992. Formation of hydrogen peroxide and depletion of oxalic acid in atmospheric water by photolysis of iron(III)-oxalato complexes. *Environ. Sci. Technol.* 26: 1014–1022.

Zhu, W., H. Hurang, H. Hu, M. Yang, and Y. Wu. 2010. *International Conference on Biomformatics and Biomedical Engineering (ICBBE)*, Chengdu, China, pp. 1–4.

5

Impinging-Jet Ozone Bubble Column Reactors

Mahad S. Baawain

CONTENTS

5.1 Introduction

Advanced oxidation processes (AOPs) are advanced processes of wastewater treatment used for oxidizing complex organic constituents present in wastewater that are difficult to degrade biologically into simpler end products (Metcalf & Eddy 2003). The ozonation process is an AOP that can be utilized to degrade complex organic compounds and can also be used as a disinfection process. Ozone is also used in wastewater treatment for odor control purposes (Metcalf & Eddy 2003). The ozonation process can be conducted in different chambers among which bubble columns have gained much popularity in the last decade (Gamal El-Din and Smith 2001a).

Bubble columns, simple and inexpensive reactors, are perhaps the most common type of ozone contactor that can be operated in both cocurrent and countercurrent flow modes, as well as in semibatch mode. These contactors are capable of operating successfully when efficient and fast ozone dissolution is required, or when ozonation is required in situations where chemical reactions control the rate of the ozonation process. The high mass transfer efficiency in these contactors is influenced by the ozone concentration in the gas phase, the liquid and gas flow rates, the column water depth, and the size of the bubbles created. Furthermore, bubble column contactors offer several advantages, including minimal maintenance, high liquid-phase content for reactions to occur, superior heat transfer properties and temperature control, good interphase mass transfer, minimal space requirements, and relatively cheap construction costs (Gamal El-Din and Smith 2001d).

The main disadvantage of a conventional bubble column is the occurrence of a large degree of liquid-phase backmixing. Therefore, proper modeling is essential for accurately estimating its treatment capability. Other disadvantages include high gas pressure drop due to the high static head of the liquid, decrease in the specific interfacial area for length/diameter ratios because of coalescence, and channeling of gas bubbles. In addition, inadequate contact between the ozone and the liquid can occur under low gas flow rates. However, bubble columns can be modified to overcome their disadvantages and to enhance their efficiencies in treating and disinfecting water and wastewater. Impinging-jet ozone bubble columns are an excellent alternative technique for improving the performance of bubble columns (Gamal El-Din and Smith 2001c).

This chapter will discuss the use of an impinging-jet bubble column as an ozone contactor. It will cover the models used to predict dissolved and gaseous ozone concentrations in the column, the ways of predicting the overall ozone mass transfer coefficient, and the ozone decay rate for different waters and wastewaters under different operating conditions. A comparison of the impinging-jet bubble column with other contactors will be illustrated by utilizing laser measurement techniques. Also, the applications of artificial neural networks (ANNs) in characterizing this ozone contactor will be introduced at the end of the chapter.

5.2 Impinging-Jet Ozone Bubble Column Modeling

Accurate modeling of contactor hydrodynamics should take into account the geometry of the reactor, the retention time of the liquid phase, the operational conditions of the reactor, and the mixing characteristics of the gas and liquid phases, with the latter being the most difficult parameter to model, as the actual flow pattern of the liquid phase typically deviates from the assumed ideal flow regimes (Bellamy 1995; Deckwer et al. 1983). Previous researchers have determined that the intensity of the liquid-phase backmixing occurring at the inlet of a reactor is quite high and must, therefore, be modeled appropriately (Deckwer et al. 1983; Salazar et al. 1993). In general, bubble column reactors are modeled on the basis of the variations of the axial dispersion model (ADM) and the cell-based models.

The basis of the ADM assumes that the flow inside the reactor resembles an ideal plug flow with some degree of backmixing superimposed on top (Zhou et al. 1994). The ADM assumes that the dispersion is represented by diffusion laws, the concentration profile is uniform in the radial direction, and the axial dispersion is uniform throughout the water column (Marinas et al. 1993). Kawagoe et al. (1989) and Shetty et al. (1992) applied the ADM model to describe the flow of gas bubbles through their bubble column reactors. Other researchers have successfully applied the model in characterizing the flow of liquids in bubble column reactors (Deckwer et al. 1974, 1983; Gamal El-Din and Smith 2001a; Houzelot et al. 1985; Kantak et al. 1994; Marinas et al. 1993). An example of the performance of one-phase ADM for predicting the overall mass transfer coefficient in the ozone contactor was discussed by Gamal El-Din and Smith (2001d).

Cell-based models generally assume four types of liquid flow zones: the zone of plug flow, the zone of axial dispersed flow, the zone of perfect mixing, and the dead zones (Kastánek et al. 1993). These zones of liquid flow are interconnected by the main flow of the liquid, the cross-flow stream, the circulation-flow stream (backmixing and exchange flows), or a by-pass flow stream (Kastánek et al. 1993). Several researchers have studied the backmixing in the liquid phase to understand and explain the mixing behavior of the bubble columns (Deckwer et al. 1973; Hikita and Kikukawa 1974; Houzelot et al. 1985; Lesauze et al. 1992; Ohki and Inoue 1970). Moreover, several researchers have studied gas-phase backmixing in bubble columns (Deckwer 1976; Kawagoe et al. 1989; Reith et al. 1968; Shetty et al. 1992; Wachi and Nojima 1990; Zahradnik and Fialova 1996).

5.2.1 Bubble Column Ozone Contactors

In bubble columns with diameters larger than 140 mm, the gas-phase dispersion coefficient is usually high due to the occurrence of large-scale eddies. In bubble columns with diameters less than 140 mm and large aspect ratios (length/diameter ratio $(L/D) \gg 1$), the gas-phase dispersion coefficient is usually less than the liquid-phase dispersion coefficient by an order of magnitude (Reith et al. 1968). Based on the theoretical calculations of Deckwer (1976), gas-phase backmixing can be neglected in bubble columns with diameters less than 500 mm. Therefore, the gas-phase flow can be considered in the plug flow regime.

The effect of the liquid-phase backmixing can be studied by using the dimensionless back flow ratio (r), which can be represented by the following Equation 5.1:

$$r = \frac{N_{BFCM}}{Pe_L} - 0.5 = \frac{D_L \varepsilon_L N_{BFCM}}{u_L L} - 0.5, \tag{5.1}$$

where N_{BFCM} is the number of cells in series, Pe_L is the dimensionless liquid-phase Péclet number, D_L is the liquid-phase axial dispersion coefficient (m^2/s), ε_L is the dimensionless liquid-phase holdup, u_L is the superficial liquid velocity (m/s), and L is the length (or height) of the reactor (m). Researchers have applied cell-based models in an attempt to describe the hydrodynamics of bubble column reactors (Gamal El-Din and Smith 2001b; Roustan et al. 1996; Zahradnik and Fialova 1996; Zhou and Smith 1995).

Depending on the degree of mixing, the gas properties in a gas–liquid reactor will be affected. For example, in a cocurrent bubble column reactor, an increase in the liquid-phase turbulence leads to higher shear stresses affecting the gas bubbles' sizes (Gamal El-Din and Smith 2003a). Consequently, the large gas bubbles are sheared into smaller bubbles. The rise in velocity for the small bubble sizes is slower in an upflow reactor and faster (less drag) in a downflow reactor. In turn, the smaller bubbles display a smaller rise in velocity in the upflow bubble column and are thus retained for longer periods of time than the larger bubbles, resulting in a larger amount of gas being retained in the reactor. The ratio of the volumetric fraction of gas to the total volume of the reactor, the gas holdup (ε_G), is a parameter that is essential for achieving an efficient reactor design. This parameter reflects the amount of gas available for transfer into the liquid phase and hence for the treatment of water and wastewater (Kastánek et al. 1993).

The diameter of a gas bubble is also affected by the bubble's coalescence and breakup rate, which are influenced by the superficial liquid and gas velocities and the properties of the liquid (Akita and Yoshida 1974). Smaller bubbles are more desirable, as they increase the surface area available for the transfer of the ozone gas into the liquid phase. By knowing the bubble diameter and the gas holdup, an important parameter recognized as the specific bubble interfacial area (a) can be determined on the basis of the following relationship (Equation 5.2):

$$a = \frac{6\varepsilon_G}{d_S}, \tag{5.2}$$

where a = mean specific interfacial area (m^2/m^3), ε_G = dimensionless gas holdup, and d_S = Sauter mean bubble diameter (m). An increase in the interfacial area can result in an increase in the overall mass transfer rate ($k_L a$). The absorption of gas via diffusion at the gas–liquid interface is governed by diffusion through the liquid film, as the diffusivity of the ozone gas in the liquid phase is much smaller than that in the gas phase. The surface renewal theory proposed by Danckwerts (1970) indicates that an increase in turbulence can cause an increase in the rate of mass transfer. This increase occurs because, with increased turbulence, the elements of liquid at the interface are replaced more frequently, allowing those fluid elements not saturated with the ozone gas to absorb more gas at an increased rate. This increase in turbulence also causes the liquid film thickness to decrease, allowing the gas to diffuse through the film at an increased rate (Danckwerts 1970; Gamal El-Din and Smith 2003b).

Some of the factors influencing the mass transfer of the ozone gas into the liquid phase in bubble columns include the ability of the ozone to diffuse into the liquid solution, the rate of the autodecomposition of the ozone (influenced by liquid temperature and pH), and the occurrence of any chemical reactions between the ozone and other dissolved and/or undissolved constituents in the water or wastewater matrix (Sotelo et al. 1989). In the case of deep ozone contactors (e.g., deep U-tubes), the static water pressure can significantly impact the mass transfer of the ozone (Roustan et al. 1992). The type of gas distributor also plays a significant role. In trying to improve the performance of the reactors, researchers have attempted to maximize the rate of gas absorption by changing the type of distributor used. Huynh et al. (1991) found an increased rate of mass transfer when using a Venturi injector compared with that obtained from a porous plate distributor. Zhou and Smith (2000) found that Venturi injectors produced significantly smaller bubbles than those generated when the bubble column was fitted with glass disks or a crystalline alumina diffusing stone. These researchers speculated that these smaller bubble sizes would subsequently increase the rate of mass transfer.

Gas injectors are gaining popularity as an effective means of dissolving ozone gas into liquids because of the high mixing intensity they generate (Zhou and Smith 2000). Gamal El-Din and Smith (2001a, 2003a) also used Venturi injectors in their bubble column and noticed a marked increase in the mass transfer of the system when compared with that of other bubble column designs.

5.2.2 Transient Back Flow Cell Model of Impinging-Jet Bubble Column

It is difficult to characterize the liquid-phase hydrodynamics of bubble column reactors; however, a model that can accurately represent the hydrodynamic behavior of the flow of the liquid phase inside an impinging-jet bubble column reactor has been developed. One primary assumption is that the liquid-phase flow follows the axially dispersed flow regime, which is characterized by a back flow ratio (r) and an exchange flow ratio (r'), with the gas phase flow following the plug flow regime.

In essence, the transient back flow cell model (TBFCM) consists of two series of completely mixed cells: one series representing the gas phase and the other representing the liquid phase. With respect to the cells describing the liquid phase, there exists a main flow stream (exchange flow) of the liquid and a backmixed flow (i.e., back flow) of the liquid. The gas-phase flow is assumed to follow the plug flow regime due to the buoyancy of the gas bubbles. Zhou (1995) developed a version of TBFCM that was applied to fine-diffuser bubble columns. Gamal El-Din and Smith (2001b) later modified and expanded this model to account for variable backmixing, cross-sectional area, and cell volume along the bubble columns and/or the ozone contactors.

The TBFCM's ability to predict a variable backmixing parameter or back flow ratio (r) along the height of the reactor is unique. This feature allows an accurate representation of the mixing occurring along the height of the reactor, as it is known that there is a larger degree of mixing occurring at the inlet of the reactor than at its outlet. Therefore, it is prudent, when modeling a bubble column reactor, to include a variable degree of mixing along its height, especially for tall or long bubble columns.

Another unique feature of TBFCM is its ability to accommodate a varying cell volume along the height of the reactor. This feature allows easy use of the model when dealing with the ozone reactors (or contactors) with multiple chambers of variable dimensions. The governing assumptions and details regarding the development of TBFCM and the equations used to describe it can be found in Gamal El-Din and Smith (2001b).

Baawain et al. (2007b) used a steady-state BFCM to model the performance of the impinging-jet bubble column. Figure 5.1 plots the regression-fitted k_La

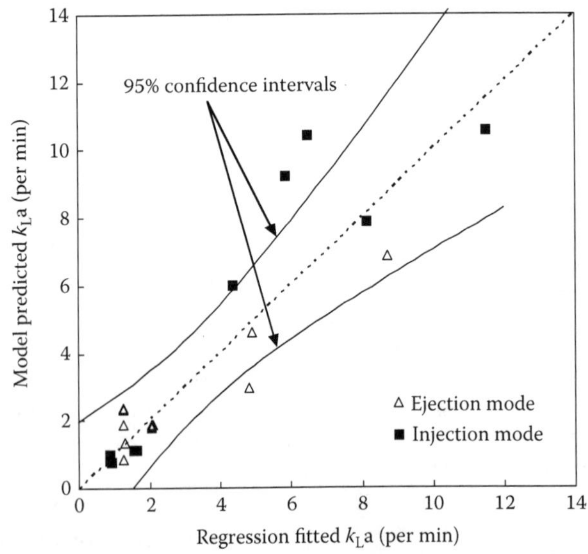

FIGURE 5.1
Comparison between the BFCM-predicted and the regression-fitted k_La. (Adapted from Baawain, M.S., Gamal El-Din, M., and Smith, D.W., *Ozone Sci. Eng.*, 29(4), 245–259, 2007.)

against the steady-state BFCM-fitted k_La. The data lie within close proximity to the 1:1 line, and the majority of the data points are within the 95% confidence interval. Furthermore, the k_La values predicted for the impinging-jet bubble column were compared with those predicted for other bubble column designs. The impinging-jet bubble column reactor displays greater mass transfer capabilities, within comparable ranges of superficial gas velocities (Baawain et al. 2007b).

5.3 Particle Image Velocimetry/Planar Laser-Induced Fluorescence Study of Impinging-Jet Bubble Column with Mixing Nozzles

As mentioned earlier, Gamal El-Din and Smith (2001b) introduced a novel design for the bubble column by utilizing two impinging jets through which gas and liquid enter the contactor under a turbulent flow condition. These researchers showed that this new design is very effective compared with that of conventional bubble columns. They also introduced a one-phase ADM and a TBFCM to predict the dissolved ozone concentration along the column (Gamal El-Din and Smith 2001a,b; Baawain et al. 2007a).

Baawain et al. (2010) aimed at characterizing the hydrodynamics of an impinging-jet bubble column with mixing nozzles by evaluating the concentration distributions of the tracer dye, the liquid velocity components, and the gas velocity components in the bubble column under different operating conditions. Then they attempted to predict the longitudinal dispersion coefficients of the bubble column under each operating condition by analyzing the residence time distribution (RTD) curves. They utilized nonintrusive laser-based techniques in their study. A brief description of these techniques is provided next.

5.3.1 Laser Measurement Techniques

Laser-based measuring techniques are popular because they are nonintrusive, direction-sensitive, highly spatial and temporal in resolution, and highly accurate (Albrecht et al. 2002). Some of these are pointwise measuring techniques, such as those using a laser Doppler anemometry and a particle dynamics analyzer (PDA), while others such as those using particle image velocimetry (PIV) and planar laser-induced fluorescence (PLIF) are simultaneous flow field measuring techniques. PIV and PLIF techniques are considered to be effective experimental tools for characterizing the hydrodynamics of different reactors.

The laser Doppler anemometer (LDA) is an optical method for flow measurement utilizing the Doppler principle, which states that the coherent light (laser) reflected from a moving particle exhibits a frequency relative to a fixed observer that depends on the known laser light source wavelength and the velocity of the moving particle (Bernard and Wallace 2002). LDA has been the single most important instrumental technique that has contributed to the investigation of complex flows of fundamental and practical interest (Tropea 1995). This importance is due to its nonintrusiveness (except for the presence of the seeding particles), directional sensitivity, and high accuracy (Bernard and Wallace 2002). LDA uses monochromatic laser light as a light source. The idea is to have two laser beams crossing in one volume (measuring volume). The interference of the two beams in the measuring volume or the interference of the two scattering waves (due to the moving particle) on the detector creates a fringe pattern. The velocity information is contained in the scattered field due to the Doppler effect.

The PDA is considered a well-established measuring technique for simultaneous measurements of particle velocity and size. The most accurate PDA configuration is the planar PDA systems (Durst et al. 1997). The basic configurations of the PDA systems are similar to those of the LDA systems, with difference in the number and alignment of detectors, usually two, and the laser beam intersection angles (Albrecht et al. 2002). The phase difference of the two signals received at the same time for both the detectors is employed to determine the size of the particle (Albrecht et al. 2002; Durst et al. 1997).

The PIV system aims at measuring the displacement of seeded particles over a short time in order to determine the velocity components in an image plane. The particles' displacement is determined through a pulsed light velocimetry in which the positions of the marker particles in a plane are noted at some time step by illuminating through successive pulses of laser light and capturing the images either on film or via a charge-coupled device (CCD) camera (Bernard and Wallace 2002; Raffel et al. 1998). The experimental setup of a PIV system normally consists of several subsystems. Most applications add tracer particles to the flow. These particles have to be illuminated in a plane of the flow at least twice within a short time interval, as mentioned earlier. The scattered light has to be recorded either on a single frame or on a sequence of frames. The displacement of the particle images between pulses can then be determined by evaluating the PIV recordings. A sophisticated postprocessing system is required to handle the large amount of data that can be collected by using the PIV technique (Raffel et al. 1998; Stanislas et al. 2000).

The PLIF system is based on obtaining the planar measurements of a scalar concentration field in water. In such a case, a fluorescent dye is mixed and carried as a passive scalar in the flow. During the illumination process of the laser, the dye absorbs incident light at one wavelength and reemits it at a different wavelength. According to the Beer–Lambert law, the reemitted wavelength of the light intensity is proportional to the dye concentration at the measuring point (Bernard and Wallace 2002).

Additional details about the basic principles behind different planar laser measurement techniques can be found in Vancruyningen et al. (1990) and Willert and Gharib (1991).

5.3.2 Experimental Setup

The experimental setup used by Baawain et al. (2010) is similar to the one shown in Figure 5.2, which was introduced originally by Gamal El-Din and Smith (2001d). A slight modification to the bubble column introduced by Gamal El-Din and Smith (2001d) was implemented in the study of Baawain et al. (2010) by adding mixing nozzles to the outlet of the injectors. The distance between the centers of the two nozzles was measured to be 40 mm. This modification was expected to enhance the momentum of the entering jets and to provide better mixing at the bottom of the column.

The PIV/PLIF setup used in the hydrodynamic analysis of the ozone contactor included a laser source, CCD cameras, and processing units. An Nd:Yag dual cavity laser was utilized in this study for both the PIV and the PLIF experiments. The emitted wavelength of the utilized Nd:Yag laser was 532 nm, with a pulse duration of 10 ns. The period between the pulses was set to 1000 μs during the PIV measurements and 100 μs during the PLIF measurements, with a maximum repetition rate of 8.0 Hz. The measurements were obtained at a time interval of 1000 ms during the PIV measurements and at 125 ms during the PLIF measurements. The CCD cameras were configured to use double frames for the PIV measurements (velocity measurements) and a single frame for the PLIF measurements (concentration measurements). A FlowMap System Hub produced by Dantec Dynamics was used to transfer the data to a PC, and the FlowMap Software was used for further analysis of the collected data.

The PLIF system was first calibrated by measuring the intensity of five different concentrations of Rhodamine 6G (Rh6G) solutions ranging from zero to 400 μg/L at a power level ranging from 50 to 150 mJ. The concentration versus the intensity was plotted to determine the most appropriate calibration curve for this study. The calibration curve obtained for the 150 mJ power gave the highest correlation coefficient (0.88). Therefore, this power level was used during all the PLIF measurements.

The PIV measurements were performed by using two CCD cameras with a double-frame mode for measuring the velocity of both phases (liquid and gas) simultaneously by utilizing special filters. Melamine–formaldehyde (MF) spheres, coated with Rhodium B (RhB), were used as seeding particles to obtain the liquid velocity measurements, while gas bubbles represented the seeding particles for the gas velocity measurements. The 2-D velocity vectors were then obtained by employing an interrogation cell of 64 × 64 pixels, which was a subdomain of a 1344 × 1024 pixel viewing area. The interrogation cell was then shifted with 25% overlap and, thus, 21 × 21 velocity vectors

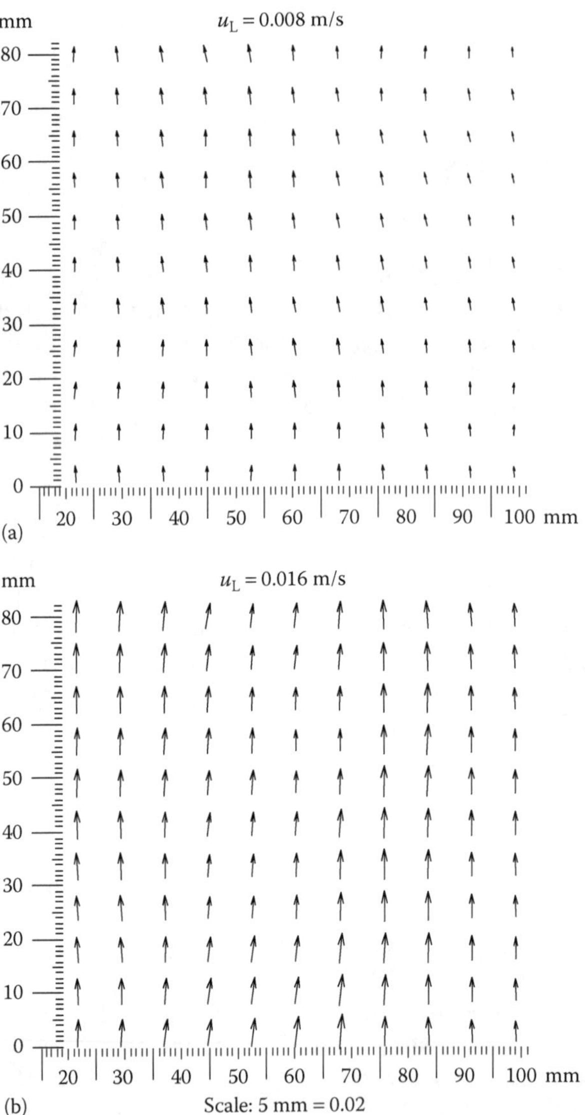

FIGURE 5.2
(a,b) Velocity vectors for different liquid flow rates without gas injection. (Adapted from Baawain, M.S., Gamal El-Din, M., and Smith D.W., *Ozone Sci. Eng.*, 32(2), 99–112, 2010.)

were obtained in each instantaneous PIV sample. However, for illustration purposes, 11 × 11 velocity vector maps were produced.

The PDA setup included a PDA system and the same bubble column as used in the PIV/PLIF experiments. The PDA system utilized during the study, produced by Dantec Dynamics, consisted of a 60X FiberFlow transmitting probe, a FiberPDA 58N70 detector unit, and a BSA P60 flow and particle processor. An Argon-ion laser with a beam wavelength of 514.5 nm was utilized in this study. This system was used to provide the bubble size measurements for the same operational conditions conducted during the PIV/PLIF study.

5.3.3 Measurement Description

For the purposes of comparison, Baawain et al. (2010) used operational conditions (for liquid and gas flow rates) in their study similar to those used by Gamal El-Din and Smith (2001d). Three liquid flow rates (Q_L) were used: 6.3×10^{-5}, 1.26×10^{-4}, and 1.89×10^{-4} m^3/s. The gas flow rate (Q_G) ranged from 8.3×10^{-6} to 1.0×10^{-4} m^3/s. The resulting gas-to-liquid ratio was from 4% to 53%. After a steady-state flow condition was reached, a continuous injection of the Rh6G tracer at about 1.4×10^{-5} m^3/s was introduced into the system through two injection points at the entrance of each injector. Three concentrations of the Rh6G were used to yield a 45 μg/L average concentration when fully mixed with the entering flow (250, 425, and 625 μg/L for Q_L of 6.3×10^{-5}, 1.26×10^{-4}, and 1.89×10^{-4} m^3/s, respectively). The continuous (step) input of the tracer was chosen over the slug input for this contactor, due to the difficulties associated with injecting the tracer manually at the two inlets. The PLIF process of image capturing covered about two times the detention time required for the tracer to pass through the system under each operating condition (300 images at 1 s intervals were captured for Q_L of 6.3×10^{-5} m^3/s, 130 images at 1 s intervals were captured for Q_L of 1.26×10^{-4} m^3/s, and 100 images at 1 s intervals were captured for Q_L of 1.89×10^{-4} m^3/s). (Duplicate measurements were applied to reduce the uncertainty associated with the measurements.) The average water temperature was (17 ± 1)°C.

During the PIV experiments, the flow patterns of the system were studied for 1-phase and 2-phase flow conditions for the same experimental conditions used in the PLIF experiments. The PIV measurements were taken at the same location where the PLIF measurements were taken. The PIV process of image capturing was taken in 30 duplicated measurements for each experimental condition. The average water temperature was 20 ± 1°C.

The bubble-sizing measurements were also conducted for the same range of operating conditions mentioned earlier. The measurements were conducted around the middle height of the column by using the PDA system. Different acquisition times were used for the bubble size measurements, depending on the rate of the bubbles formed in the column. The acquisition time ranged from 100 to 600 s.

Since the bubble size measurements were conducted, it was thought that estimating the gas holdup (ε_G) would provide a basis for determining the specific interfacial area (a, per meter), which in turn would provide useful information for characterizing the gas mass transfer efficiency of the contactor. The value of ε_G for each operating condition was calculated by inserting two tubes through the sides of the bubble column (one was fixed in the mixing zone, which is the zone where the two jets impinged into each other, and the other one was fixed just above the measurement location used for the PIV, PLIF, and bubble sizing). A change in the water level in the tubes was observed, and the following equation (Equation 5.3) was applied to determine ε_G:

$$\varepsilon_G = \frac{\Delta h}{\Delta x}, \tag{5.3}$$

where Δh (m) is the difference between the water level in the tubes and Δx (m) is the axial distance between the measuring points.

5.3.4 Results and Discussion

All images captured during the PLIF experiments were converted to 2D concentration fields through the obtained calibration relation by using the FlowMap Software®. A resampling of these concentration fields yielded contour maps that show the concentration distribution of the tracer along the cross section of the contactor parallel to the flow direction. Further analysis of the mixing and dispersion in the contactor was achieved through velocity measurements obtained for the studied operating conditions by using the PIV system. Figure 5.2 shows the averaged velocity vectors for two different liquid flow rates (Q_L of 6.3×10^{-5} and 1.3×10^{-4} m^3/s) without gas injection. This figure shows that the axial flow is dominating and, hence, the flow with the higher axial velocity will exhibit a higher dispersion. By contrast, for the same liquid flow rate, as the gas flow rate increased, the radial mixing increased in both phases (liquid and gas) and, hence, the axial dispersion decreased.

In order to evaluate the mixing in the contactor numerically, the following differential equation (Equation 5.4), introduced by Levenspiel (1999) and representing the dispersion of a conservative tracer (C, μg/L), was considered:

$$\frac{\partial C}{\partial \theta} = \left(\frac{D_L}{uL}\right)\frac{\partial^2 C}{\partial z^2} - \frac{\partial C}{\partial z}, \tag{5.4}$$

where θ is the dimensionless time ($\theta = t/\tau = tu/L$, t is time (s)), (D_L/uL) is the dispersion number (the inverse of the Peclet number, *Pe*), D_L is the liquid axial dispersion coefficient (m^2/s), u is the pipe flow average velocity (m/s), L is the axial distance between the tracer input point and the measurement point (m), and z is the dimensionless axial distance ($z = (ut + x)/L$, x is the axial distance along the pipe (m)). For a step input of a tracer, the shape of the tracer at the measurement point is S-shaped and referred to as the F-curve. The F-curves normally represent the dimensionless concentration (F), the ratio between the tracer concentration at the measurement point (C, μg/L) and the initial mixed tracer concentration (C_o, μg/L), as a function of time.

The shape of the F-curve depends on the boundary conditions of the contactor and the dispersion number (D_L/uL). The value (D_L/uL) can be obtained directly by plotting the experimental data on a probability graph paper or by differentiating the S-shaped response curve and considering the boundary conditions (Levenspiel and Smith 1957).

The numerical concentration of the Rh6G values obtained by the PLIF system can be extracted for all images under each operating condition and the step response curves (F-curves) at any position of interest can then be plotted. Figure 5.3 shows an F-curve obtained from the PLIF experiments for the tracer concentration at the center of the images captured for $Q_L = 1.26 \times 10^{-4}$ m^3/s and $Q_G = 8.3 \times 10^{-6}$ m^3/s. In order to obtain the dispersion number of the dye under each operating condition, the F-curves were differentiated, and the open vessel condition was assumed, since the flow patterns at the boundaries were not disturbed (Levenspiel 1999). Therefore, the RTD curves at any position of interest were obtained for the dimensionless concentration (E_θ) as a function of the dimensionless time (θ). The value of $\theta(t_i/\tau)$ is the ratio of the data acquisition time (t_i) to the mean residence time (τ). For discrete tracer data, τ can be determined as follows:

$$\tau = \frac{\sum t_i C_i \Delta t_i}{\sum C_i \Delta t_i}, \tag{5.5}$$

where t_i is the data acquisition time (s), Δt_i is the time step (s), and C_i is the instantaneous (recorded) concentration (μg/L).

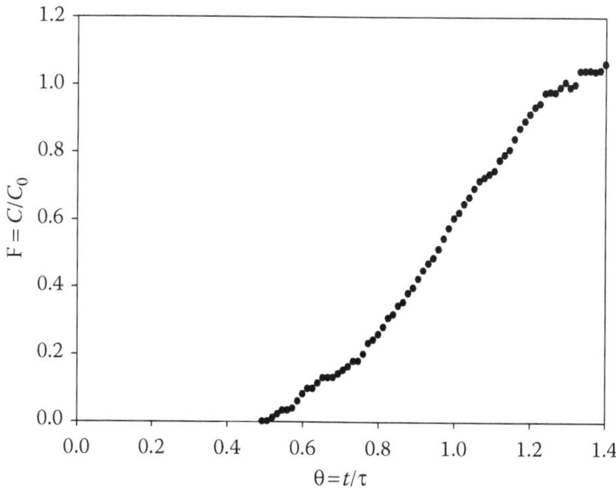

FIGURE 5.3
Step response curve from PLIF images ($u_L = 0.016$ m/s, $u_G = 0.001$ m/s). (Adapted from Baawain, M.S., Gamal El-Din, M., and Smith D.W., *Ozone Sci. Eng.*, 32(2), 99–112, 2010.)

The dispersion number can be obtained from the RTD curves, which are characterized by τ (Equation 5.5) and the variance (σ_t^2), which can be obtained for discrete tracer test data as follows (Equation 5.6):

$$\sigma_t^2 = \frac{\sum t_i^2 C_i \Delta t_i}{\sum C_i \Delta t_i} - \tau^2. \tag{5.6}$$

The value of the dispersion number can be obtained from the dimensionless variance (σ_θ^2), which can be determined as follows (Equation 5.7):

$$\sigma_\theta^2 = \frac{\sigma_t^2}{\tau^2} = 2\frac{D_L}{uL} + 8\left(\frac{D_L}{uL}\right)^2. \tag{5.7}$$

The value of D_L for all operating conditions was determined and correlated with the superficial liquid velocity (u_L) and the superficial gas velocity (u_G) by using a nonlinear regression as follows:

$$D_L = 8.52 \times 10^{-3} u_G^{0.06} u_L^{0.29}. \tag{5.8}$$

A plot of the experimentally determined versus the predicted D_L, obtained by using Equation 5.8, is shown in Figure 5.4. The obtained correlation of the

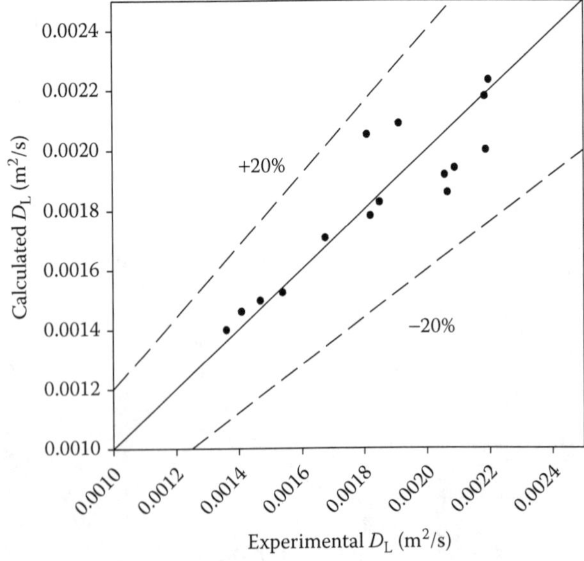

FIGURE 5.4
Comparison between the predicted and the experimental D_L. (Adapted from Baawain, M.S., Gamal El-Din, M., and Smith, D.W., *Ozone Sci. Eng.*, 32(2), 99–112, 2010.)

multiple determination coefficient (R^2) is 0.90, indicating an excellent agreement. According to Equation 5.8, as u_G and u_L increased, D_L also increased. This result agrees with the observations reported by Zhou (1995) and Gamal El-Din and Smith (2001c, 2003b). Furthermore, Equation 5.8 yielded much lower D_L values (about 80% lower) compared with those of the impinging-jet bubble column studied by Gamal El-Din and Smith (2001b), indicating that a major improvement in the mixing regime resulted from the use of the mixing nozzles.

However, the results obtained from the PDA measurements of the bubble sizing showed that most of the bubble sizes ranged from 0.0003 to 0.003 m. It was observed that for the same gas flow rate, as the liquid flow rate increased, the bubble sizes decreased. This result can be related to the higher shearing rates exerted by the higher flow rates. Furthermore, when the gas flow rates increased, at a fixed liquid flow rate, higher bubble sizes could be observed. The value of the Sauter mean bubble diameter (d_S, m) was correlated with u_L and u_G as follows:

$$d_S = 3.6 \times 10^{-3} u_G^{0.32} u_L^{-0.19}. \tag{5.9}$$

Figure 5.5 shows a comparison between the measured and the calculated d_S. A very good agreement was obtained as the R^2 value was found to be 0.91. The value of d_S obtained by using the nozzles was found to be about 50% to

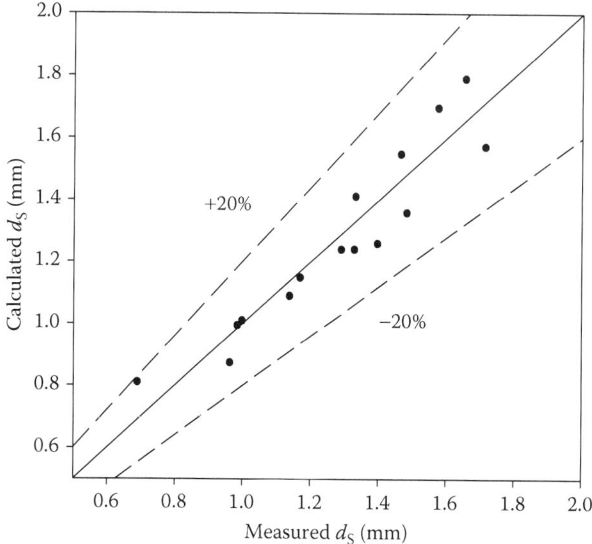

FIGURE 5.5
Comparison between the predicted and the experimental d_S. (Adapted from Baawain, M.S., Gamal El-Din, M., and Smith D.W., *Ozone Sci. Eng.*, 32(2), 99–112, 2010.)

60% of the d_S for the column without nozzles. Therefore, the mass transfer efficiency is expected to be higher for the bubble column with the mixing nozzle, as the available bubble surface area is larger.

The results obtained from the gas holdup (ε_G) measurements were collected and correlated with u_L and u_G as follows ($R^2 = 0.95$):

$$\varepsilon_G = 3.25 u_G^{1.54} u_L^{-0.67}. \tag{5.10}$$

The values of ε_G obtained from Equation 5.10 were found to be comparable to the values of ε_G obtained for the same bubble column without nozzles. This result is mainly due to the high dependence of ε_G on u_G compared with u_L. Since ε_G and d_S have been determined, one can estimate the value of the specific bubble interfacial area (a, per meter) by using the following equation:

$$a = \frac{6\varepsilon_G}{d_S}. \tag{5.11}$$

Substituting Equations 5.9 and 5.10 into Equation 5.11 will yield the following (Equation 5.12):

$$a = 5.42 \times 10^3 u_G^{1.22} u_L^{-0.48}. \tag{5.12}$$

The nozzles have enhanced the values of a by a factor of 1.5–2. This result indicates that the mass transfer efficiency was consequently enhanced. If the value of a local mass transfer coefficient (k_L) of the contactor determined by Baawain et al. (2007b) remains the same for both the cases of the bubble column (i.e., with and without nozzles), then the overall mass transfer coefficient (k_La, per second) can be expressed as follows:

$$k_La = 88.8 u_G^{1.57} u_L^{-0.36}. \tag{5.13}$$

Equation 5.13 indicates that k_La will range from 0.01 to 0.33/s for the range of operational conditions covered by this study. The corresponding values for the bubble column without nozzles are from 0.006 to 0.16/s.

5.4 ANNs Modeling of Ozone Bubble

Bubble columns are commonly used in chemical, biochemical, and environmental industries as gas–liquid mass transfer systems, due to their simple construction, operation, and maintenance. More details on the applications and advantages of bubble columns have been reviewed recently by Kantarci et al. (2005). It is evident that a precise knowledge of the hydrodynamics is required for an accurate design of such contactors in order to evaluate their performance. The overall performance of bubble columns is often limited by the gas–liquid mass transfer (Cramers et al. 1992), which is controlled by the gas holdup, specific interfacial area, and bubble-size distribution in the contactors (Patel et al. 1989; Weisweiler and Rosch 1978). Therefore, many empirical correlations have been introduced to estimate these key design parameters of bubble columns (Akita and Yoshida 1973, 1974; Deckwer and Schumpe 1993; Hughmark 1967; Kawase et al. 1992; Koide 1996; Kumar et al. 1997; Kundu et al. 2003; Lau et al. 2004; Nakao et al. 1983; Patel et al. 1989; Poulsen and Iversen 1998; Shah et al. 1982; Vasconcelos et al. 2003; Winterton 1994; Yamashita et al. 1979). However, all these correlations were obtained for specific experimental conditions and cannot be used for a wide range of operating conditions, physical properties, and reactor geometries. Hence, a need for new empirical correlations arises whenever a new bubble column is to be designed and run under new operating conditions.

5.4.1 Artificial Neural Network Modeling

Good experimental techniques are costly and require qualified personnel. Therefore, a general modeling approach that can be implemented easily with satisfactory efficiency is necessary to obtain reliable designs of bubble columns. ANN models appear to be a good choice as they have been recognized to perform exceptionally well in capturing complex interactions within the used input parameters without prior knowledge about the nature of the problem. They can tolerate imprecise data and approximate results with less vulnerability to outliers compared with that of deterministic models (Haykin 1999). They are based on the decomposition of input–output relationship into a series of linearly separable steps by using hidden layer(s) neurons (Haykin 1999). Generally, ANN models are developed in four steps: (1) data transforming (scaling); (2) network architecture defining (setting the number of hidden layers, the number of neurons in each layer, and the connectivity between the neurons); (3) training (calibrating) the network to respond adequately to a given set of inputs; and (4) validating the network to a set of inputs to ensure the generality of the ANN model's prediction ability. Although several architectures of neural networks are available, the feedforward multilayer perceptron (MLP) ANNs trained with a backpropagation algorithm (BP) are considered among the most commonly used networks. MLP-BP ANNs with only one hidden layer have been reported to be the universal approximators of all nonlinear functions and can be sufficient for most important applications (Hornik et al. 1989).

A review of the available literature illustrated that ANNs have been applied successfully to predict the values of the overall mass transfer coefficient (k_La, per second), the gas holdup (ε_G, dimensionless), the specific interfacial area (a, per meter) of a gas bubble, and the bubble diameter (d_b, m) in a number of industrial reactors (see Table 5.1). The presented models cover a wide range of gas–liquid systems. These developed ANN models are based on the reactors' geometry, operating conditions, and/or physicochemical properties (viscosity, density, surface tension, and diffusivity). According to the provided statistical measures, ANNs can provide a promising tool for predicting the important design parameters of the gas–liquid systems.

TABLE 5.1
Summary of ANN Applications in Reactor Modeling

References	Reactor Type	Flow Mode	Data Points	Predicted Parameter	Statistical Measures		
					MSE	*AARE*	R^2
Al-Masry and Abdennour (2006)	BCs	a	200	ε_G	2%	—	—
Alvarez et al. (2000)	BCs	a	—	k_La	—	1%	—
Baawain et al. (2005)	BCs	b	300	k_La	—	—	0.85
Behkish et al. (2005)	BCs	a	3880	ε_G	—	16%	0.90
Djebbar and Narbaitz (2002)	ASPTs	b	1078	k_La	—	24%	—
Fonseca et al. (2000)	SPECs	a	60	k_La	—	7%	—
Garcia-Ochoa and Castro (2001)	STRs	a	450	k_La	—	10%	—
Iliuta et al. (1999)	TBRs	b	3200	k_La, a	—	29%	—
Jamialahmadi et al. (2001)	BCs	a	200	d_b	—	2%	—
Lemoine et al. (2003)	SGRs	a	4435	k_La	—	—	0.91
Shaikh and Al-Dahhan (2003)	BCs	a	3500	ε_G	—	15%	—
Supardan et al. (2004)	BCs	a	178	k_La, ε_G	—	15%	—
Utomo et al. (2001)	BCs	a	102	ε_G, d_b	—	1%	—
Wu et al. (2003)	BCs	a	3000	ε_G	—	6%	0.89
Yang et al. (1999)	STRs	a	824	k_La	—	25%	—

Source: Adapted from Baawain, M.S., Gamal El-Din, M., and Smith, D.W., *Ozone Sci. Eng.*, 29(5), 343–352, 2007.

Notes: BCs, bubble columns; ASPTs, air stripping packed towers; SPECs, spray and packed extraction columns; STRs, stirred tank reactors; TBRs, trickle-bed reactors; SGRs, surface aeration and gas-induced reactors.

[a] Semibatch flow mode (injection of gas into a constant liquid volume).

[b] Continuous flow mode (both gas and liquid are injected continuously into the reactor).

Baawain et al. (2007a) conducted a study to develop general, simple, and relatively accurate ANN models that can be used to design and evaluate the ozone bubble columns operating under different experimental conditions. The aim of the developed ANN models was to adequately predict k_La, ε_G, and the Sauter mean bubble diameter (d_S, m) by using simple inputs such as the bubble columns' geometry and operational parameters. Once these parameters are successfully predicted, the local mass transfer coefficient (k_L, m/s) can be estimated.

5.4.2 Preparation of ANN Models

As Table 5.1 shows, the majority of ANN models were developed for semi-batch systems, which sparge a gas into a stagnant liquid, because of the relative ease in controlling the experimental conditions and in measuring the parameter(s) of interest. However, practical conditions require the use of continuous liquid and gas flow conditions. Only three studies have considered systems with a continuous flow mode, two of which are not relevant to this work. Baawain et al. (2005) provided a preliminary study of the applicability of the ANN modeling technique in predicting k_La in continuous-flow bubble columns. Furthermore, Baawain et al. (2007a) modified the previous k_La ANN model and developed two more ANN models to predict ε_G and d_S.

The data used for developing the three ANN models were obtained from 18 research studies performed on different bubble columns operating under continuous flow conditions (Table 5.2). Almost every study used a unique bubble column with a different geometry and gas-injection method. Also, the operating conditions (liquid and gas flow rates) varied from one column to the other. Among the 18 research studies summarized in Table 5.2, only five examined the ozone–water bubble columns under continuous flow conditions (Beltran et al. 1995; Chen et al. 2002; Gamal El-Din and Smith 2003a; Roustan et al. 1996; Xu and Liu 1990). The other studies considered the oxygen–water bubble columns operating under continuous flow conditions. Baawain et al. (2007a) assumed that the gas type has a negligible effect on the values of ε_G and d_S. However, according to the surface renewal theory, the values of oxygen-based k_La have to be converted to that of ozone-based k_La (Danckwerts 1970).

The k_La values were first corrected for the water temperature using the following modified Arrhenius equation (Roustan et al. 1996) (Equation 5.14):

$$(k_La) = (k_La)_T\, 1.024^{(20-T)}, \tag{5.14}$$

where $(k_La)_{20}$ is the overall mass transfer coefficient at 20°C (per second) and $(k_La)_T$ is the overall mass transfer coefficient (per second) at the water temperature in the bubble column (T, °C). Then, the surface renewal theory

TABLE 5.2
Sources of Data Used in Training and Validating the Developed ANN Models

References	Data Points	Gas Diffuser	Flow Direction[a]	Measured Parameter(s)
Akita and Yoshida (1973)	28	SN	1, −1	k_La, ε_G
Akita and Yoshida (1974)	28	SN	1, −1	d_S
Akosman et al. (2004)	10	SN	1	k_La, ε_G, d_S
Alvarezcuenca et al. (1980)	32	SSN	1	k_La
Alvarezcuenca and Nerenberg (1981)	32	SSN	1	k_La
Beltran et al. (1995)	4	DP	1	k_La, ε_G
Briens et al. (1992)	13	NVT	1	k_La, ε_G
Chen et al. (2002b)	7	PCP	−1	k_La, ε_G
Deckwer et al. (1974)	48	CwO, SP	1, −1	ε_G
Deckwer et al. (1983)	48	CwO, SP	1, −1	k_La
Gamal El-Din and Smith (2003a)[b]	12	VI	1	k_La, ε_G, d_S
Hikita et al. (1981)	12	SN	1	k_La, ε_G
Huynh et al. (1991)	31	NVT	1	k_La, ε_G
Kulkarni and Shah (1984)[c]	11	RD	1	k_La, ε_G, d_S
Kulkarni et al. (1983)	25	RD	1	k_La, ε_G, d_S
Roustan et al. (1996)	52	PCP	1, −1	k_La, ε_G, d_S
Wang and Fan (1978)	25	SN	1	k_La, ε_G
Xu and Liu (1990)	36	MPTP	−1	k_La

Source: Adapted from Baawain, M.S., Gamal El-Din, M., and Smith, D.W., *Ozone Sci. Eng.*, 29(5), 343–352, 2007.

Notes: SN, single nozzle; SSN, stainless-steel nozzle, DP, diffuser plate; NVT, nozzle with Venturi throat; PCP, perforated ceramic plate; CwO, cross with orifices; SP, sintered plate; VI, Venturi injector; RD, ring-type distributor; MPTP, microporous titanium plate.

[a] 1 = cocurrent flow, −1 = countercurrent flow.

[b] Values of ε_G were obtained from Gamal El-Din and Smith (2003b).

[c] Values of d_S were obtained from Kulkarni et al. (1983).

introduced by Danckwerts (1970) and validated by Sherwood et al. (1975) and Beltran et al. (1998) was used to convert the oxygen-based k_La to an ozone-based k_La as follows (Equation 5.15):

$$\frac{(k_La)_{O_3}}{(k_La)_{O_2}} = \sqrt{\frac{D_{O_3}}{D_{O_2}}}, \tag{5.15}$$

where D_{O_3} is the molecular diffusivity of the ozone gas in water (=1.74×10^{-9} m^2/s) and D_{O_2} is the molecular diffusivity of the oxygen gas in water (=2.50×10^{-9} m^2/s).

After the parameters of interest (k_La, ε_G, and d_S) were collected, three MLP-BP ANN models were developed by utilizing NeuroShell 2 Software from Ward Systems Group Inc. The parameters' cross-sectional area (A_{cs}, m^2),

TABLE 5.3

Ranges of Input and Output Parameters of ANN Models

Variable	Minimum	Maximum	Mean	Standard Deviation
A_{cs} (m^2)[a]	0.002	0.031	0.013	0.009
H_{bc} (m)[a]	0.112	7.2	2.4	1.86
d_{dif} (m)[a]	0.003	0.18	0.052	0.047
d_{pore} (m)[a]	0	0.019	0.005	0.006
α[a]	−1	1	0.523	0.853
u_L (m/s)[a]	0	0.399	0.095	0.114
u_G (m/s)[a]	0.0003	0.4	0.063	0.086
k_La (per s)[b]	0.001	1.038	0.11	0.175
ε_G[b]	0.004	0.348	0.084	0.073
d_S (m)[b]	0.001	0.028	0.006	0.006

Source: Adapted from Baawain, M.S., Gamal El-Din, M., and Smith, D.W., *Ozone Sci. Eng.*, 29(5), 343–352, 2007.

[a] Input variables.

[b] Actual outputs.

effective water height in the bubble column (H_{bc}, m), ozone diffuser diameter (d_{dif}, m), diffuser pore size (d_{pore}, m), flow direction (α; +1 for the cocurrent flow condition and −1 for the countercurrent flow conditions), u_L (m/s), and u_G (m/s) were used as inputs for the three MLP-BP ANN models, while the k_La, ε_G, and d_S were set to be the models' outputs.

Table 5.3 shows the ranges of input and output parameters used in the developed ANN models. The collected data for each ANN model (378 data points for the k_La network, 278 data points for the ε_G network, and 138 data points for the d_S network) were divided randomly into training and validation data sets at a ratio of 3:1 (training to validation). The input parameters for all networks were scaled with a linear scaling function over an open interval from 0 to 1 to overcome the large variation in the input data, which can slow down if not prevent the calibration of the network. One hidden layer with a number of hidden neurons was employed in developing the networks. The output layer of each ANN model consisted of one neuron to resemble the value of the predicted parameter. The neurons in the input layer were related to the hidden layer neurons, which were in turn related to the output layer by using connection weights. The input data were propagated, during the training process of a network, by using a feedforward method to produce output data according to the connection weights and activation functions. The outputs from the neurons in the preceding layer were multiplied by the connecting weights to the neuron in the next layer. An activation function was then applied to the sum of these products, and the result was introduced in the neuron of the succeeding layer.

Different activation functions were explored in both the hidden and the output layers in order to obtain the best model network. The training process was propagated by using a number of iterations (epochs) and automatic adjustment of the connection weights until the minimum mean square error (*MSE*) was achieved. Further details regarding the MLP-BP ANN's structure and activation functions can be found elsewhere (Haykin 1999; Hornik et al. 1989).

5.4.3 Results and Discussion

The MLP-BP ANN architecture that yielded the best training and validation results for k_La consisted of one hidden layer with 20 hidden neurons, with logistic activation functions in both the hidden and the output layers. This network was trained with 4000 epochs with TurboProbe weight updates (by updating the weight size independently for each different weight with an adaptive adjustment of the step size as the learning progressed) and rotation pattern selection (by selecting the training pattern as they appeared in the provided file according to the training-to-validation ratio mentioned earlier). The developed MLP-BP ANN model simulated the k_La values very well, as the coefficient of multiple determination (R^2) was 00.98 for the validation data set. Furthermore, the performance of the developed model was evaluated by determining the average absolute relative error (*AARE*), which was found to be 9% for the training data set and 13% for the validation data set. The high performance of the developed ANN model is strongly supported by Figure 5.6, which shows an excellent agreement

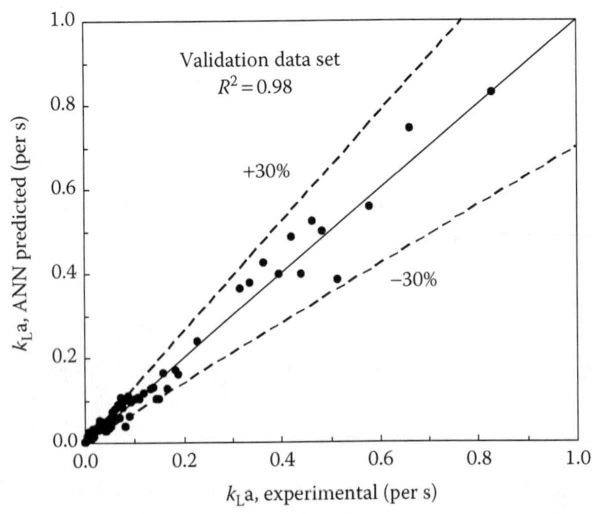

FIGURE 5.6
ANN-predicted versus experimental results for k_La. (Adapted from Baawain, M.S., Gamal El-Din, M., and Smith, D.W., *Ozone Sci. Eng.*, 29(5), 343–352, 2007.)

between the experimentally measured k_La values and the ANN predicted ones in both the training and validation data sets. The values of the *AARE* are relatively small (compared with most *AARE* values shown in Table 5.1; only the ANN models developed for specific experiments yielded lower *AARE* than those obtained in this study). The small difference between the ANN model's predictions and the measured k_La values is probably due to the wide range of some operating conditions such as the range of H_{bc} (7.2 m used by Deckwer et al. (1983) and 0.11 m used by Alvarezcuenca et al. (1980)).

The MLP-BP ANN architecture that resulted in the minimum *MSE* of ε_G simulations for both the training and the validation of data sets consisted of 12 hidden neurons and used logistic activation functions in both the hidden and output layers. The ε_G ANN model was trained with 5000 epochs with TurboProbe weight updates and rotation pattern selection. The model predicted ε_G values in validation data set excellently, as R^2 and *AARE* were 0.99 and 6%, respectively. Figure 5.7 illustrates the high performance of the developed ε_G ANN model through the excellent agreement shown between the measured and the ANN-predicted ε_G values in both the training and validation data sets. Although previous ANN models that aimed at predicting ε_G were developed for semibatch systems (with simpler operating conditions than those of the continuous flow systems considered in this study), the current ε_G ANN model outperformed most of them.

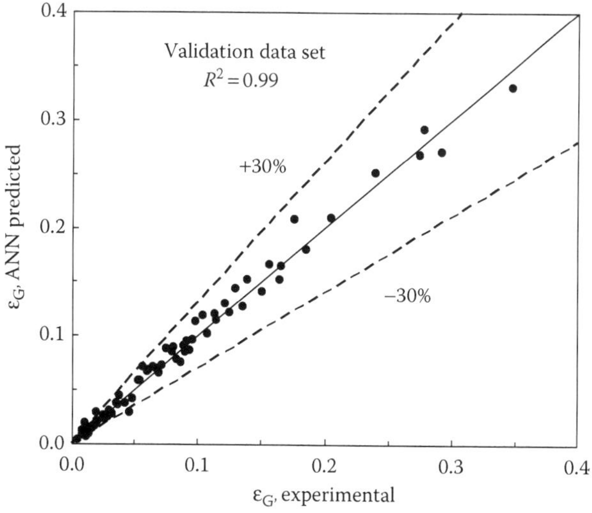

FIGURE 5.7
ANN-predicted versus experimental results for ε_G. (Adapted from Baawain, M.S., Gamal El-Din, M., and Smith, D.W., *Ozone Sci. Eng.*, 29(5), 343–352, 2007.)

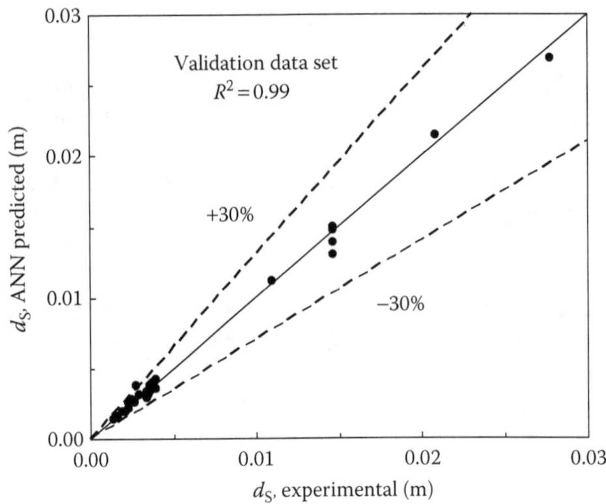

FIGURE 5.8
ANN-predicted versus experimental results for d_S. (Adapted from Baawain, M.S., Gamal El-Din, M., and Smith, D.W., *Ozone Sci. Eng.*, 29(5), 343–352, 2007.)

The MLP-BP ANN architecture developed for modeling d_S values consisted of three hidden neurons and used a symmetric logistic activation function in the hidden layer and a logistic activation function in the output layer. This model was trained with 3500 epochs with TurboProbe weight updates and rotation pattern selection. Figure 5.8 depicts an excellent agreement between the measured and the ANN-predicted d_S values in the validation data set. The obtained R^2 and *AARE* values were 0.99 and 5%, respectively, which demonstrate the validity of the developed model in predicting d_S in bubble columns. The slightly higher *AARE* associated with d_S compared with that of the ANN models of d_S in semibatch systems (Table 5.1) can be related to the uncertainty associated with the turbulence intensity induced by u_L in the bubble columns.

As Table 5.2 shows, most researchers explored the use of only one gas diffuser in their bubble columns. Therefore, their experimental procedures were based on relating k_La, ε_G, and/or d_S to the operating conditions (i.e., u_G and/or u_L) and/or the physiochemical properties through nonlinear regressions for the bubble column design under investigation. As a result, each regression model (developed for k_La, ε_G, and/or d_S) was case-sensitive as it excluded the effects of several other factors, including the bubble column's geometry and the gas-sparging technique. The ANN models showed that u_G had a higher effect than u_L on k_La and ε_G values. This finding agrees with the findings in most of the available literature. Moreover, the variables d_{pore}, d_{dif}, and H_{bc} have a relatively high effect on k_La and ε_G, and this finding also agrees with the observations of several researchers (Alvarezcuenca et al.

1980; Alvarezcuenca and Nerenberg 1981; Huynh et al. 1991). However, the effect of u_G was found to be lower than that of u_L for the d_S model. This finding also concurs with some previous studies (Akita and Yoshida 1974; Unno and Inoue 1980; Varley 1995). The variables d_{pore} and d_{dif} have a relatively high effect on d_S, as Unno and Inoue (1980) also found. The flow direction was found to have minimal effect on the predicted parameters in bubble columns. This finding agrees with those of Gamal El-Din and Smith (2001c) and Roustan et al. (1996).

After the determination of k_La, ε_G, and d_S by utilizing the developed ANN models, the values of k_L were calculated as follows:

$$k_L = \frac{k_La \cdot d_S}{6\varepsilon_G}. \tag{5.16}$$

The k_L values predicted by using the results obtained from the three developed ANN models and Equation 5.16 were then plotted against the experimental values of k_L, as shown in Figure 5.9. The overall prediction for k_L is very good as the obtained R^2 and *AARE* values were found to be 0.85 and 20%, respectively. The relatively high error in k_L values is due to the propagation of the errors in predicting k_La, ε_G, and d_S. Nevertheless, the combination of the three ANN models and Equation 5.16 provided an adequate tool to predict k_L when compared with the cumbersome experimental methods.

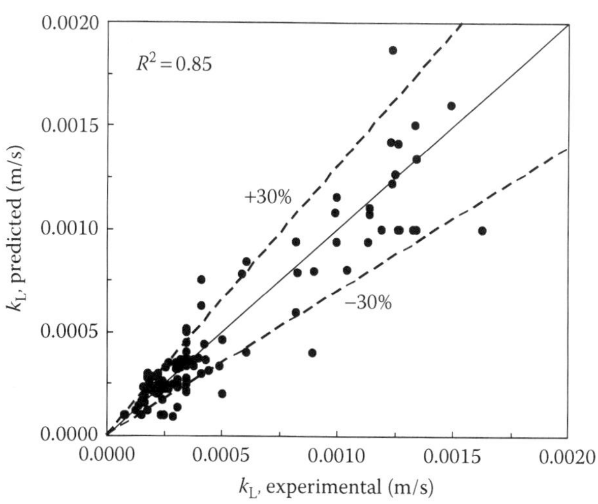

FIGURE 5.9
Predicted (using developed ANN models) versus experimental results for k_L. (Adapted from Baawain, M.S., Gamal El-Din, M., and Smith, D.W., *Ozone Sci. Eng.*, 29(5), 343–352, 2007.)

References

Akita, K. and F. Yoshida. 1974. Bubble size, interfacial area, and liquid-phase mass transfer coefficient in bubble columns. *Ind. Eng. Chem. Process Des. Dev.* 13(1): 84–91.

Akita, K. and F. Yoshida. 1973. Gas holdup and volumetric mass-transfer coefficient in bubble columns – Effects of liquid properties. *Ind. Eng. Chem. Process Des. Dev.* 12(1): 76–80.

Akosman, C., R. Orhan, and G. Dursun. 2004. Effects of liquid property on gas holdup and mass transfer in co-current downflow contacting column. *Chem. Eng. Process.* 43(4): 503–509.

Albrecht, H.E., M. Borys, N. Damaschke, and C. Tropea. 2002. *Laser Doppler and Phase Doppler Measurement Techniques*. New York: Springer-Velag Berlin Heidelberg.

Al-Masry, W.A. and A. Abdennour. 2006. Gas hold-up estimation in bubble columns using passive acoustic waveforms with neural networks. *J. Chem. Technol. Biotech.* 81(6): 951–957.

Alvarez, E., J.M. Correa, C. Riverol, and J.M. Navaza. 2000. Model based in neural networks for the prediction of the mass transfer coefficients in bubble columns. Study in Newtonian and non-Newtonian fluids. *Int. Commun. Heat Mass Transfer* 27(1): 93–98.

Alvarezcuenca, M. and M.A. Nerenberg. 1981. Oxygen mass-transfer in bubble-columns working at large gas and liquid flow-rates. *AIChE J.* 27(1): 66–73.

Alvarezcuenca, M., C.G.J. Baker, and M.A. Bergougnou. 1980. Oxygen mass-transfer in bubble-columns. *Chem. Eng. Sci.* 35(5): 1121–1127.

Baawain, M.S., M. Gamal El-Din, and D.W. Smith. 2007a. Artificial neural networks modeling of ozone bubble columns: Mass transfer coefficient, gas hold-up, and bubble size. *Ozone Sci. Eng.* 29(5): 343–352.

Baawain, M.S., M. Gamal El-Din, and D.W. Smith. 2007b. Impinging-jet ozone bubble column modeling: Hydrodynamics, gas hold-up, bubble characteristics, and ozone mass transfer. *Ozone Sci. Eng.* 29(4): 245–259.

Baawain, M.S., M. Gamal El-Din, and D.W. Smith. 2005. Mass transfer analysis in ozone bubble columns using artificial neural netwroks. In *Proceedings of the Eighth International Conference on the Application of Artificial Intelligence to Civil, Structural and Environmental Engineering,* Rome. (30 August–2 September 2005).

Baawain, M.S., M. Gamal El-Din, and D.W. Smith. 2010. PIV/PLIF study of impinging jet ozone bubble column with mixing nozzles. *Ozone Sci. Eng.* 32(2): 99–112.

Behkish, A., R. Lemoine, L. Sehabiague, R. Oukaci, and B. Morsi. 2005. Prediction of the gas holdup in industrial-scale bubble columns and slurry bubble column reactors using back-propagation neural networks. *Int. J. Chem. Reactor Eng.* 3: 1–35.

Bellamy, W.D. 1995. *Full-Scale Ozone Contactor Study*. Denver: A.W.W.A. Research Foundation.

Beltran, F.J., J.M. Encinar, and J.F. Garciaaraya. 1995. Modeling industrial waste-water ozonation in bubble contactors. 2: Scale-up from bench to pilot-plant. *Ozone Sci. Eng.* 17(4): 379–398.

Beltran, F.J., L.A. Fernandez, P. Alvarez, and E. Rodriguez. 1998. Comparison of ozonation kinetic data from film and Danckwerts theories. *Ozone Sci. Eng.* 20(5): 403–420.

Bernard, P.S. and J.M. Wallace. 2002. *Turbulent Flow: Analysis, Measurement and Prediction*. Hoboken, NJ: Wiley.

Briens, C.L., L.X. Huynh, J.F. Large, A. Catros, J.R. Bernard, and M.A. Bergougnou. 1992. Hydrodynamics and gas–liquid mass-transfer in a downward venturi-bubble column combination. *Chem. Eng. Sci.* 47(13–14): 3549–3556.

Chen, Y.H., C.Y. Chang, Y.H. Yu, P.C. Chiang, C.Y. Chiu, Y. Ku, and J.N. Chen. 2002. A dynamic model of ozone disinfection in a bubble column with oxygen mass transfer. *J. Chinese Inst. Chem. Eng.* 33(3): 253–265.

Cramers, P., A. Beenackers, and L.L. Vandierendonck. 1992. Hydrodynamics and mass-transfer characteristics of a loop-venturi reactor with a downflow liquid jet ejector. *Chem. Eng. Sci.* 47(13–14): 3557–3564.

Danckwerts, P.V. 1970. *Gas–Liquid Reactions*. New York: McGraw-Hill.

Deckwer, W., U. Graeser, Langeman. H, and Y. Serpemen. 1973. Zones of different mixing in liquid-phase of bubble columns. *Chem. Eng. Sci.* 28(5): 1223–1225.

Deckwer, W.D. 1976. Non-isobaric bubble-columns with variable gas velocity. *Chem. Eng. Sci.* 31(4): 309–317.

Deckwer, W.D., R. Burckhar, and G. Zoll. 1974. Mixing and mass-transfer in tall bubble columns. *Chem. Eng. Sci.* 29(11): 2177–2188.

Deckwer, W.D., K. Nguyentien, B.G. Kelkar, and Y.T. Shah. 1983. Applicability of axial-dispersion model to analyze mass-transfer measurements in bubble-columns. *AIChE J.* 29(6): 915–922.

Deckwer, W.D. and A. Schumpe. 1993. Improved tools for bubble column reactor design and scale-up. *Chem. Eng. Sci.* 48(5): 889–911.

Djebbar, Y. and R.M. Narbaitz. 2002. Neural network prediction of air stripping k_La. *J. Environ. Eng. ASCE* 128(5): 451–460.

Durst, F., G. Brenn, and T.H. Xu. 1997. A review of the development and characteristics of planner phase-Doppler anemometry. *Meas. Sci. Technol.* 8: 1203–1221.

Fonseca, A.P., J.V. Oliveira, and E.L. Lima. 2000. Neural networks for predicting mass transfer parameters in supercritical extraction. *Braz. J. Chem. Eng.* 17(4–7): 517–524.

Gamal El-Din, M. and D.W. Smith. 2001a. Designing ozone bubble columns: A spreadsheet approach to axial dispersion model. *Ozone Sci. Eng.* 23(5): 369–384.

Gamal El-Din, M. and D.W. Smith. 2001b. Development of transient back flow cell model (Bfcm) for bubble columns. *Ozone Sci. Eng.* 23(4): 313–326.

Gamal El-Din, M. and D.W. Smith. 2003a. Mass transfer analysis in ozone bubble columns. *J. Environ. Eng. Sci.* 2(1): 63–76.

Gamal El-Din, M. and D.W. Smith. 2001c. Maximizing the enhanced ozone oxidation of kraft pulp mill effluents in an impinging-jet bubble column. *Ozone Sci. Eng.* 23(6): 479–493.

Gamal El-Din, M. and D.W. Smith. 2003b. Measurements of the size, rise velocity, and specific interfacial area of bubbles in an impinging-jet bubble column. *J. Environ. Eng. Sci.* 2(2): 127–138.

Gamal El-Din, M. and D.W. Smith. 2001d. Ozone mass transfer in water treatment: Hydrodynamics and mass transfer modelling of ozone bubble columns. *Water Sci. Technol. Water Supply* 1(2): 123–130.

Garcia-Ochoa, F. and E.G. Castro. 2001. Estimation of oxygen mass transfer coefficient in stirred tank reactors using artificial neural networks. *Enzyme Microb. Technol.* 28(6): 560–569.

Haykin, S. 1999. *Neural Networks a Comprehensive Foundation*. New Jersey: Prentice Hall.

Hikita, H., S. Asai, K. Tanigawa, K. Segawa, and M. Kitao. 1981. The volumetric liquid-phase mass-transfer coefficient in bubble-columns. *Chem. Eng. J. Biochem. Eng. J.* 22(1): 61–69.

Hikita, H. and H. Kikukawa. 1974. Liquid-phase mixing in bubble columns: Effect of liquid properties. *Chem. Eng. J.* 8(3): 191–197.

Hornik, K., M. Stinchcombe, and H. White 1989. Multilayer feedforward networks are universal approximators. *Neural Netw.* 2(5): 359–366.

Houzelot, J.L., M.F. Thiebaut, J.C. Charpentier, and J. Schiber. 1985. Contribution to the hydrodynamic study of bubble columns. *Int. Chem. Eng.* 25(4): 645–650.

Hughmark, G.A. 1967. Holdup and mass transfer in bubble columns. *Ind. Eng. Chem. Process Des. Dev.* 6(2): 218.

Huynh, L.X., C.L. Briens, J.F. Large, A. Catros, J.R. Bernard, and M.A. Bergougnou. 1991. Hydrodynamics and mass-transfer in an upward venturi bubble column combination. *Can. J. Chem. Eng.* 69(3): 711–722.

Iliuta, I., F. Larachi, B.P.A. Grandjean, and G. Wild 1999. Gas–liquid interfacial mass transfer in trickle-bed reactors: State-of-the-art correlations. *Chem. Eng. Sci.* 54(23): 5633–5645.

Jamialahmadi, M., M.R. Zehtaban, H. Muller-Steinhagen, A. Sarrafi, and J.M. Smith. 2001. Study of bubble formation under constant flow conditions. *Chem. Eng. Res. Des.* 79(A5): 523–532.

Kantak, M.V., S.A. Shetty, and B.G. Kelkar. 1994. Liquid-phase backmixing in bubble-column reactors – a new correlation. *Chem. Eng. Commun.* 127: 23–34.

Kantarci, N., F. Borak, and K.O. Ulgen. 2005. Bubble column reactors. *Process Biochem.* 40(7): 2263–2283.

Kastánek, F., J. Zahradník, J. Kratochvíl, and J. Čermák. 1993. *Chemical Reactors for Gas–Liquid Systems*. Chichester, West Sussex, England: Ellis Horwood.

Kawagoe, M., T. Otake, and C.W. Robinson. 1989. Gas-phase mixing in bubble-columns. *J. Chem. Eng. Jpn* 22(2): 136–142.

Kawase, Y., S. Umeno, and T. Kumagai. 1992. The prediction of gas hold-up in bubble column reactors – Newtonian and non-Newtonian fluids. *Chem. Eng. J. Biochem. Eng. J.* 50(1): 1–7.

Koide, K. 1996. Design parameters of bubble column reactors with and without solid suspensions. *J. Chem. Eng. Jpn* 29(5): 745–759.

Kulkarni, A. and Y.T. Shah. 1984. Gas phase dispersion in a downflow bubble column. *Chem. Eng. Commun.* 28: 311–326.

Kulkarni, A., Y.T. Shah, and A. Schumpe. 1983. Hydrodynamics and mass transfer in downflow bubble column. *Chem. Eng. Commun.* 24: 307–337.

Kumar, S.B., D. Moslemian, and M.P. Dudukovic. 1997. Gas-holdup measurements in bubble columns using computed tomography. *AIChE J.* 43(6): 1414–1425.

Kundu, A., E. Dumont, A.M. Duquenne, and H. Delmas. 2003. Mass transfer characteristics in gas–liquid–liquid system. *Can. J. Chem. Eng.* 81(3–4): 640–646.

Lau, R., W. Peng, L.G. Velazquez-Vargas, G.Q. Yang, and L.S. Fan. 2004. Gas–liquid mass transfer in high-pressure bubble columns. *Ind. Eng. Chem. Res.* 43(5): 1302–1311.

Lemoine, R., B. Fillion, A. Behkish, A.E. Smith, and B.I. Morsi. 2003. Prediction of the gas–liquid volumetric mass transfer coefficients in surface-aeration and gas-inducing reactors using neural networks. *Chem. Eng. Process.* 42(8–9): 621–643.

Lesauze, N., A. Laplanche, M.T.O. Develasquez, G. Martin, B. Langlais, and N. Martin. 1992. The residence time distribution of the liquid-phase in a bubble column and its effect on ozone transfer. *Ozone Sci. Eng.* 14(3): 245–262.

Levenspiel, O. 1999. *Chemical Reaction Engineering*, 3rd edn. Hoboken, NJ: Wiley.

Levenspiel, O. and W.K. Smith. 1957. Notes on the diffusion-type model for the longitudinal mixing of fluids in flow. *Chem. Eng. Sci.* 6(4–5): 227–233.

Marinas, B.J., S. Liang, and E.M. Aieta. 1993. Modeling hydrodynamics and ozone residual distribution in a pilot-scale ozone bubble-diffuser contactor. *J. Am. Water Works Assoc.* 85(3): 90–99.

Metcalf & Eddy. 2003. *Wastewater Engineering: Treatment and Reuse*. New York: MacGraw-Hill.

Nakao, K., H. Takeuchi, H. Kataoka, H. Kaji, T. Otake, and T. Miyauchi. 1983. Mass-transfer characteristics of bubble-columns in recirculation flow regime. *Ind. Eng. Chem. Process Des. Dev.* 22(4): 577–582.

Ohki, Y. and H. Inoue. 1970. Longitudinal mixing of liquid phase in bubble columns. *Chem. Eng. Sci.* 25(1): 1.

Patel, S.A., J.G. Daly, and D.B. Bukur. 1989. Holdup and interfacial area measurements using dynamic gas disengagement. *AIChE J.* 35(6): 931–942.

Poulsen, B.R. and J.J.L. Iversen. 1998. Characterization of gas transfer and mixing in a bubble column equipped with a rubber membrane diffuser. *Biotechnol. Bioeng.* 58(6): 633–641.

Raffel, M., C. Willert, and J. Kompenhans. 1998. *Particle Image Velocimetry: A Practical Guide*. New York: Springer-Velag Berlin Heidelberg.

Reith, T., S. Renken, and B.A. Israel. 1968. Gas hold-up and axial mixing in fluid phase of bubble columns. *Chem. Eng. Sci.* 23(6): 619.

Roustan, M., A. Line, J.P. Duguet, J. Mallevialle, and O. Wable. 1992. Practical design of a new ozone contactor – The deep U-tube. *Ozone Sci. Eng.* 14(5): 427–438.

Roustan, M., R.Y. Wang, and D. Wolbert. 1996. Modeling hydrodynamics and mass transfer parameters in a continuous ozone bubble column. *Ozone Sci. Eng.* 18(2): 99–115.

Salazar, J.A., K.D. Wisecarver, Y.T. Shah, and B. Solari. 1993. Gas–liquid mass-transfer in jet bubble-column. *Chem. Eng. Commun.* 124: 177–188.

Shah, Y.T., B.G. Kelkar, S.P. Godbole, and W.D. Deckwer. 1982. Design parameters estimations for bubble column reactors. *AIChE J.* 28(3): 353–379.

Shaikh, A. and M. Al-Dahhan. 2003. Development of an artificial neural network correlation for prediction of overall gas holdup in bubble column reactors. *Chem. Eng. Process.* 42(8–9): 599–610.

Sherwood, T.K., L.P. Robert, and R.W. Charles. 1975. *Mass Transfer*. New York: McGraw-Hill.

Shetty, S.A., M.V. Kantak, and B.G. Kelkar. 1992. Gas-phase backmixing in bubble-column reactors. *AIChE J.* 38(7): 1013–1026.

Sotelo, J.L., F.J. Beltran, F.J. Benitez, and J. Beltranheredia. 1989. Henrys law constant for the ozone water-system. *Water Res.* 23(10): 1239–1246.

Stanislas, M., J. Kompenhans, and J. Westerweel. 2000. *Particle Image Velocimetry: Progress Towards Industrial Application*. Dordrecht, The Netherlands: Kluwer Academic Publishers.

Supardan, M.D., Y. Masuda, A. Maezawa, and S. Uchida. 2004. Local gas holdup and mass transfer in a bubble column using an ultrasonic technique and a neural network. *J. Chem. Eng. Jpn* 37(8): 927–932.

Tropea, C. 1995. Laser-Doppler anemometry – Recent developments and future challenges. *Meas. Sci. Technol.* 6(6): 605–619.

Unno, H. and I. Inoue. 1980. Size reduction of bubbles by orifice mixer. *Chem. Eng. Sci.* 35: 1571–1579.

Utomo, M.B., T. Sakai, S. Uchida, and A. Maezawa. 2001. Simultaneous measurement of mean bubble diameter and local gas holdup using ultrasonic method with neural network. *Chem. Eng. Technol.* 24(5): 493–500.

Vancruyningen, I., A. Lozano, and R.K. Hanson. 1990. Quantitative imaging of concentration by planar laser-induced fluorescence. *Exp. Fluids* 10(1): 41–49.

Varley, J. 1995. Submerged gas–liquid jets: Bubble size prediction. *Chem. Eng. Sci.* 50(5): 901–905.

Vasconcelos, J.M.T., J.M.L. Rodrigues, S.C.P. Orvalho, S.S. Alves, R.L. Mendes, and A. Reis. 2003. Effect of contaminants on mass transfer coefficients in bubble column and airlift contactors. *Chem. Eng. Sci.* 58(8): 1431–1440.

Wachi, S. and Y. Nojima. 1990. Gas-phase dispersion in bubble columns. *Chem. Eng. Sci.* 45(4): 901–905.

Wang, K.B. and L.T. Fan. 1978. Mass transfer in bubble columns packed with motionless mixers. *Chem. Eng. Sci.* 33: 945–952.

Weisweiler, W. and S. Rosch. 1978. Interfacial area and bubble-size distribution in jet reactors. *German Chem. Eng.* 1: 212–218.

Willert, C.E. and M. Gharib. 1991. Digital particle image velocimetry. *Exp. Fluids* 10(4): 181–193.

Winterton, R.H.S. 1994. A simple method of predicting bubble-size in bubble-columns. *Chem. Eng. Process.* 33(1): 1–5.

Wu, Y.X., X.H. Luo, Q.M. Chen, D.H. Li, S.R. Li, M.H. Al-Dahhan, and M.P. Dudukovic. 2003. Prediction of gas holdup in bubble columns using artificial neural network. *Chinese J. Chem. Eng.* 11(2): 162–165.

Xu, F.C. and C.L. Liu. 1990. Mass balance analysis of ozone in a conventional bubble column. *Ozone Sci. Eng.* 12(3): 269–279.

Yamashita, F., Y. Mori, and S. Fujita. 1979. Sizes and size distributions of bubbles in a bubble column – Comparison between the 2 point electric probe method and the photographic method. *J. Chem. Eng. Jpn* 12(1): 5–9.

Yang, H., B.S. Fang, and M. Reuss. 1999. $k_L a$ correlation established on the basis of a neural network model. *Can. J. Chem. Eng.* 77(5): 838–843.

Zahradnik, J. and M. Fialova. 1996. The effect of bubbling regime on gas and liquid phase mixing in bubble column reactors. *Chem. Eng. Sci.* 51(10): 2491–2500.

Zhou, H. 1995. Investigation of ozone disinfection kinetics and contactor performance modeling. Ph.D., University of Alberta, Edmonton, Alberta, Canada.

Zhou, H. and D.W. Smith. 1995. Modelling of mass transfer and ozone decomposition in a bubble column: Experimental verification. In *12th World Congress of the International Ozone Association*, Lille, France.

Zhou, H.D. and D.W. Smith. 2000. Ozone mass transfer in water and wastewater treatment: Experimental observations using a 2d laser particle dynamics analyzer. *Water Res.* 34(3): 909–921.

Zhou, H.D., D.W. Smith, and S.J. Stanley. 1994. Modeling of dissolved ozone concentration profiles in bubble-columns. *J. Environ. Eng. ASCE*, 120(4): 821–840.

6

Biological Treatment of Wastewaters: Recent Trends and Advancements

K. Vijayaraghavan

CONTENTS

6.1 Introduction

Water constitutes over 70% of the earth's surface and it is undeniably the most valuable natural resource that exists on our planet. Without this invaluable compound, which is composed of hydrogen and oxygen, life on earth would be nonexistent. Although we as humans recognize this fact, we sometimes disregard it by polluting our water resources. Consequently, we are slowly but surely harming our planet to the point where organisms are dying at an alarming rate. In addition, our drinking water has been greatly affected and, in many instances, has lost its original purpose. In order to combat water pollution, we must understand the problems associated with it and become part of the solution.

There are many sources of water pollution, but two general categories exist: direct and indirect contaminant sources. Direct sources include effluent outfalls from industries, refineries, waste treatment plants, etc. Indirect sources include contaminants that enter the water supply from soils and groundwater systems and from the atmosphere via rainwater. The soils and groundwaters contain the residues of human agricultural practices (fertilizers, pesticides, etc.) and improperly discharged industrial effluents. Atmospheric contaminants are mainly derived from human practices, such as gaseous emissions from automobiles and factories.

The pollutants in water include a wide spectrum of chemicals and pathogens, with different physical chemistry or sensory changes. In general, contaminants fall under two broad classes, namely, organic and inorganic. Some organic water pollutants include industrial solvents, volatile organic compounds (VOC), insecticides, pesticides, dyes, and food processing wastes. Inorganic water pollutants include metals, fertilizers, and acidity caused by industrial discharges. Among these pollutants, metals are a cause of concern in many cases as they are nonbiodegradable. In some cases, metals play an integral role in the life processes of living organisms. Several metals and metalloids, such as Zn, Cu, Fe, and Mg, which are essential for biochemical and cellular processes, are taken up to different extents. They may be essential for normal functions, but excessive doses can become toxic (Marmiroli and Maestri 2008). Other metallic elements with no biological role can enter the system and damage the normal processes (Maestri et al. 2010). Toxicity occurs through the displacement of essential metals from their native binding sites or through ligand interactions (Bruins et al. 2000).

Considering the toxic nature of these pollutants, it is desirable to treat the wastewaters prior to utilization. Currently, various physical, chemical, and biological processes, either stand-alone or in combination, are available to treat and manage these wastewaters. However, some of the uneconomical physicochemical methods are slowly losing ground due to the inherent problems of secondary contamination and nonsustainable control of contaminants.

In recent years, research attention has been focused on the biological methods of wastewater treatment, which are in the process of commercialization. Biological treatment involves the utilization of biomaterials to remove contaminants. In the case of organic pollutants, the biological treatment involves enhanced degradation by transforming organic compounds into innocuous substances. However, inorganic compounds undergo sorption or accumulation onto biological materials. There are three principal advantages of biological technologies for the removal of pollutants. First, biological processes can be carried out in situ at the contaminated site. Second, bioprocess technologies are usually environmentally benign (no secondary pollution).

Third, they are cost-effective. Several biological methods are available to treat organic and inorganic pollutants–bearing wastewaters, including biodegradation, biosorption, bioaccumulation, phytoremediation, biosparging, bioleaching, bioaugmentation, biostimulation, and bioventing. Of these methods, several researchers focused on biosorption, bioaccumulation, and phytoremediation because they proved efficient for the removal of metal ions from contaminated waters (Volesky and Holan 1995; Prasad and Freitas 2003; Malik 2004). Hence, this chapter will account these important biological treatment techniques for metal-bearing wastewaters. The recent trends and advancements in the biological treatment of wastewaters will be included and discussed.

6.2 Biosorption

During the last several decades, heavy metal removal by biosorption has been extensively investigated (Volesky 1987; Volesky and Holan 1995; Vijayaraghavan and Yun 2008). Biosorption can be defined as the passive uptake of toxicants by dead/inactive biological materials or by materials derived from biological sources. Biosorption is caused by a number of metabolism-independent processes that essentially take place in the cell wall, where the mechanisms that are responsible for the pollutant uptake will differ according to the types of biomass. The mechanisms of metal removal usually come under physical adsorption, ion exchange, complexation, chelation, and microprecipitation (Vegliò and Beolchini 1997). The biosorbents for metal ions broadly come under the following divisions: bacteria, fungi, algae, industrial wastes, agricultural wastes, and other polysaccharide materials (Vijayaraghavan and Yun 2008). On the basis of their biosorption capacities, all these biosorbents showed good removal efficiency toward different metal ions.

Table 6.1 lists some of the important results of metal biosorption using different biosorbents. Several bacterial and fungal biomasses showed good biosorption potential toward different metal ions. Since these microorganisms are widely used in different food and pharmaceutical industries, they are generated as wastes, which can be obtained for free or at a low cost from these industries. Another important biosorbent that has gained momentum in recent years is seaweed. Marine algae, popularly known as seaweeds, are biological resources that are available in many parts of the world. Algal divisions include red, green, and brown seaweeds; of these, brown seaweeds have been found to be excellent biosorbents (Davis et al. 2003). Also, their macroscopic structure offers a convenient basis for the production of biosorbent particles that are suitable for sorption process applications (Vieira and Volesky 2000). Owing to their cell wall chemical

TABLE 6.1

Important Results from the Literature on Heavy Metal Biosorption by Different Biosorbents

Metal	Biosorbent	pH	Biosorption Capacity (mg/g)	Reference
Cadmium	*Pseudomonas* sp. (B)	7.0	278.0	Ziagova et al. (2007)
	Staphylococcus xylosus (B)	6.0	250.0	Ziagova et al. (2007)
	Mucor rouxii (F)	6.0	20.3	Yan and Viraraghavan (2003)
	Ascophyllum nodosum (S)	4.9	215.0	Holan et al. (1993)
	Crab shell (IW)	5.0	199.0	An et al. (2001)
	Black gram husk (AW)	5.0	40.0	Saeed et al. (2005)
	Activated sludge (IW)	5.0	69.2	Al-Qodah (2006)
Chromium (VI)	*Aeromonas caviae* (B)	2.5	284.4	Loukidou et al. (2004)
	Staphylococcus xylosus (B)	1.0	143.0	Ziagova et al. (2007)
	Rhizopus nigricans (F)	2.0	43.5	Bai and Abraham (2001)
	Sargassum sp. (S)	4.0	68.9	Cossich et al. (2002)
	Crab shell (IW)	2.0	28.1	Niu and Volesky (2003)
Cobalt	Crab shell (IW)	5.0	510.0	Lee et al. (2004)
	Ascophyllum nodosum (S)	—	156.0	Kuyucak and Volesky (1989a)
Copper	*Pseudomonas putida* (B)	5.5	96.9	Uslu and Tanyol (2006)
	Thiobacillus ferrooxidans (B)	6.0	198.5	Ruiz-Manriquez et al. (1997)
	Aspergillus niger (F)	5.0	28.7	Dursun (2006)
	Turbinaria ornata (S)	6.0	147.1	Vijayaraghavan et al. (2005a)
	Crab shell (IW)	6.0	243.9	Vijayaraghavan et al. (2006a)
	Black gram husk (AW)	5.0	27.7	Saeed et al. (2005)
	Activated sludge (IW)	5.0	80.6	Al-Qodah (2006)
Lead	*Corynebacterium glutamicum* (B)	5.0	567.7	Choi and Yun (2004)
	Pseudomonas putida (B)	5.5	270.4	Uslu and Tanyol (2006)
	Aspergillus niger (F)	4.0	32.6	Dursun (2006)
	Penicillium chrysogenum (F)	4.5	116.0	Niu et al. (1993)
	Fucus vesiculosus (B)	6.0	600.0	Holan and Volesky (1994)
	Crab shell (IW)	3.0	870.0	Lee et al. (1998)
	Black gram husk (AW)	5.0	50.2	Saeed et al. (2005)

TABLE 6.1 (Continued)

Important Results from the Literature on Heavy Metal Biosorption by Different Biosorbents

Metal	Biosorbent	pH	Biosorption Capacity (mg/g)	Reference
Nickel	*Bacillus thuringiensis* (B)	6.0	45.9	Öztürk (2007)
	Corynebacterium glutamicum (B)	6.0	111.2	Vijayaraghavan et al. (2008)
	Aspergillus niger (F)	6.3	4.8	Amini et al. (2009)
	Ascophyllum nodosum (S)	6.0	135.9	Holan and Volesky (1994)
	Crab shell (IW)	4.5	169.5	Vijayaraghavan et al. (2005b)
	Activated sludge (IW)	5.0	69.2	Al-Qodah (2006)
	Black gram husk (AW)	5.0	20.2	Saeed et al. (2005)
Zinc	*Thiobacillus ferrooxidans* (B)	6.0	172.4	Liu et al. (2004)
	Aphanothece halophytica (B)	6.5	133.0	Incharoensakdi and Kitjaharn (2002)
	Mucor rouxii (F)	6.0	53.9	Yan and Viraraghavan (2003)
	Sargassum uitans (S)	5.0	65.0	Leusch and Volesky (1995)
	Crab shell (IW)	6.0	123.7	Vijayaraghavan et al. (2011)
	Black gram husk (AW)	5.0	34.3	Saeed et al. (2005)

Notes: B, bacterial biomass; F, fungal biomass; S, seaweed; IW, industrial waste; AW, agricultural waste.

composition, many seaweeds excel at metal biosorption (Table 6.1). In addition to the aforementioned biosorbents, recently, numerous approaches have been made to develop low-cost sorbents from industrial and agricultural wastes. Table 6.1 highlights the performance of some of the waste materials in metal biosorption.

6.2.1 Biosorption Mechanism

The mechanism of biosorption depends on the type of sorbent. Different types of biomass exhibit different mechanisms, including physical adsorption, ion exchange, complexation, chelation, and microprecipitation. Since the mode of solute uptake by dead or inactive cells is extracellular, the chemical functional groups of the cell wall play vital roles in biosorption. For microbial biosorption, the cell wall is the first component that comes into contact with the metal ions, where the solutes can be deposited on the surface or within the cell wall structure. Owing to the nature of the cellular components, several functional groups are present on the microbial cell wall, including carboxyl, phosphonate, amine, and hydroxyl groups.

As they are negatively charged and abundantly available, the carboxyl groups actively participate in the binding of heavy metal cations. Golab and Breitenbach (1995) indicated that the carboxyl groups of the cell wall, the peptidoglycans of *Streptomyces pilosus*, were responsible for the binding of copper. Yee and Fein (2001) confirmed that the carboxyl groups were responsible for the binding of cadmium via the Cd–carboxyl complex on the bacterial surface. Conversely, the amine groups are very effective at removing heavy metal ions, as they not only chelate the cationic metal ions, but also adsorb the anionic metal species via electrostatic interaction or hydrogen bonding. Kang et al. (2007) observed that the amine groups, protonated at a pH of 3, attracted negatively charged chromate ions via electrostatic interaction. In general, increasing the pH increases the overall negative charge on the surface of the cells until all the relevant functional groups are deprotonated, which favors the electrochemical attraction and adsorption of the cations. Anions would be expected to interact more strongly with cells with increasing concentrations of positive charges, due to the protonation of the functional groups at lower pH values.

In marine algal biosorption, the functional groups, such as the carboxyl, phosphonate, and amine groups, play a vital role in metal biosorption. Brown algae are mainly composed of alginates (alginic acids), which constitute 10%–40% of the dry weight of the algae (Davis et al. 2003). The alginic acids are linear carboxylated copolymers constituted by different proportions of 1,4-linked β-D-mannuronic acid (M-block) and α-L-guluronic acid (G-block) (Haug et al. 1966). The M-block and G-block sequences display significantly different structures and their proportions in the alginates determine the physical properties and reactivity of the polysaccharides (Haug et al. 1967). The most abundant carboxyl groups of alginates, the second abundant sulfonate groups of fucoidans, and the hydroxyl groups in other polysaccharides are found to play an important role in metal binding at different pH conditions.

6.2.2 Biosorption Process

The biosorption process proceeds via several modes, of which batch and continuous modes of operation are frequently employed to conduct laboratory-scale biosorption processes. Although most industrial applications prefer a continuous mode of operation, in this chapter, batch experiments will be used to evaluate the required fundamental information, such as the biosorbent efficiency, optimum experimental conditions, biosorption rate, and possibility of biomass regeneration.

6.2.2.1 Batch Biosorption

Batch experiments usually focus on the study of the factors influencing biosorption, which are important in the evaluation of the full biosorption potential of any biomaterial. The important factors include

- Solution pH
- Temperature
- Biosorbent dosage
- Biosorbent size
- Initial solute concentration
- Agitation rate

Of these, the solution pH usually plays a major role in biosorption and seems to affect the solution chemistry of metals and the activity of the functional groups of the biomass. The pH strongly influences the speciation and biosorption availability of the heavy metals. At a higher solution pH, the solubility of the metal complexes decreases sufficiently for precipitation to occur, which may complicate the sorption process. The activity of the binding sites can also be altered by adjusting the pH. For instance, during the biosorption of a heavy metal by bacterial biomass, a pH in the range of 3–6 favors biosorption, due to the negatively charged carboxyl groups (pKa = 3–5), which are responsible for the binding of metal cations via the ion-exchange mechanism. The solution pH affects not only the biomass surface chemistry, but also the metal speciation. Metal ions in solution undergo hydrolysis with an increasing pH, the extent of which differs at different pH values and with each metal; however, the usual sequence of hydrolysis is the formation of hydroxylated monomeric species, followed by the formation of polymeric species, and then the formation of crystalline oxide precipitates after aging (Baes and Mesmer 1976). For example, in the case of a nickel solution, López et al. (2000) indicated that within the pH range of 1–7, nickel existed in solution as Ni^{2+} ions (90%), whereas at a pH of 9, Ni^{2+} (68%), $Ni_4OH_4^{4+}$ (10%), and $Ni(OH)^+$ (8.6%) coexisted. The different chemical species of a metal occurring at varying pH values will have variable charges and adsorbabilities at the solid–liquid interfaces. In many instances, biosorption experiments conducted at high alkaline pH values have been reported to complicate the evaluation of the biosorbent potential as a result of metal precipitation (Selatnia et al. 2004; Iqbal and Saeed 2007).

The temperature only seems to affect biosorption to a less extent within the range of 20°C–35°C (Vegliò and Beolchini 1997). Higher temperatures usually enhance sorption due to the increased surface activity and the kinetic energy of the solute (Sağ and Kutsal 2000); however, physical damage to a biosorbent can be expected at higher temperatures. Owing to the exothermic nature of some adsorption processes, an increase in temperature will lead to a reduction in the biosorption capacity of the biomass. It is always desirable to conduct and evaluate/maximize biosorption at room temperature, as this condition is easy to replicate.

The dosage of a biosorbent strongly influences the extent of biosorption. In many instances, lower biosorbent dosages yield higher uptakes and lower the percentage of removal efficiencies (Aksu and Çağatay 2006; Vijayaraghavan et al. 2006a). An increase in the biomass concentration generally increases the amount of solute biosorbed, due to the increased surface area of the biosorbent, which in turn increases the number of binding sites. Conversely, the quantity of the biosorbed solute per unit weight of biosorbent is decreased with an increasing biosorbent dosage, which may be due to the complex interaction of several factors. An important factor at high sorbent dosages is that the available solute is insufficient to completely cover the available exchangeable sites on the biosorbent, usually resulting in a low solute uptake. Also, as suggested by Gadd et al. (1988), interference between the binding sites due to increased biosorbent dosages cannot be overruled, as this will result in a low specific uptake.

The size of the biosorbent also plays a vital role in biosorption. Smaller-sized particles have a higher surface area, which in turn favors biosorption and results in a shorter equilibration time. Simultaneously, a particle for biosorption should be sufficiently resilient to withstand the applicable pressures and the extreme conditions applied during the regeneration cycles. Therefore, preliminary experiments are mandatory to decide on the suitable size of a biosorbent. If a biosorbent is available in powdered form, such as industrial waste, efforts should be made to improve its mechanical strength, such as granulation, for its effective use in biosorption columns.

The initial solute concentration seems to have an impact on biosorption, with a higher concentration resulting in a high solute uptake, because at lower initial solute concentrations, the ratio of the initial moles of solute to the available surface area is low; subsequently, the fractional sorption becomes independent of the initial concentration. However, at higher concentrations, the sites available for sorption become fewer than the moles of the solute present; therefore, the removal of the solute is strongly dependent on the initial solute concentration. It is always necessary to identify the maximum saturation potential of a biosorbent, where experiments at the highest possible initial solute concentration should always be conducted.

In some instances, external film diffusion can influence the rate of a biosorption process. With the appropriate agitation, this mass transfer resistance can be minimized. On increasing the agitation rate, the diffusion rate of a solute from the bulk liquid to the liquid boundary layer surrounding the particles becomes higher due to the enhanced turbulence and the decrease in the thickness of the liquid boundary layer. Under these conditions, the value of the external diffusion coefficient becomes larger. Finally, at higher agitation rates, the boundary layer becomes very thin, which usually enhances the rate at which a solute will diffuse through the boundary layer.

The quality of a biosorbent is judged by how much sorbate it can attract and retain in an immobilized form. The determination of the solute uptake by a biosorbent is most often based on the material balance of the sorption system: the solute that has disappeared from the solution must be attached to the solid. To be precise, the amount of solute that is biosorbed can be calculated from the differences between the initial quantities of the solute added to that contained in the supernatant, which is achieved using the following Equation 6.1:

$$Q = \left(V_0 C_0 - V_f C_f\right)/M, \tag{6.1}$$

where Q is the solute uptake (mg/g); C_0 and C_f are the initial and equilibrium solute concentrations in the solution (mg/L), respectively; V_0 and V_f are the initial and final solution volumes (L), respectively; and M is the mass of biosorbent (g).

A biosorption isotherm, the plot of the uptake (Q) versus the equilibrium solute concentration in the solution (C_f), is often used to evaluate the sorption performance. The isotherm curves can be evaluated by varying the initial solute concentrations while fixing the environmental parameters, such as pH, temperature, and ionic strength. In general, the uptake increases with increasing concentrations and reaches saturation at higher concentrations. In most biosorption studies, pH seems to be an important parameter for the evaluation of an isotherm.

Biosorption kinetics, the plot of the solute concentration (C_f) versus time, is very important because it provides valuable insights into the reaction pathways and the mechanism of a sorption reaction. Also, the kinetics describes the solute uptake, which in turn controls the residence time of a sorbate at the solid–solution interface. Since biosorption is metabolism-independent, it would be expected to be a rapid process. Usually, free cell microbial biosorption comprises two phases: a very fast initial uptake for 30–60 min, followed by a slow attainment of equilibrium within 2–3 h. However, when the microbial biomass is immobilized, a delay in the attainment of equilibrium would be expected as a result of mass transfer resistances (Vegliò and Beolchini 1997; Wu and Yu 2007).

Sorption is a multistep process, comprising four consecutive elementary steps in the case of immobilized beads (Guo et al. 2003): (1) transfer of the solute from the bulk of the solution to the liquid film surrounding the beads; (2) transport of the solute from the boundary liquid film to the surface of the bead (external diffusion); (3) transfer of the solute from the surface to the internal active binding sites (intraparticle diffusion); and (4) interaction of the solute with the active binding sites. In general, the first two steps (external diffusion) are usually fast, as long as sufficient agitation is provided to avoid the formation of a concentration gradient within the solution. If the fourth step is assumed to be rapid, the subsequent intraparticle diffusion becomes the rate-limiting step. Frequently, intraparticle diffusion has been shown to be an important factor in deciding the attainment of equilibrium with the use of immobilized beads (Yahaya et al. 2009; Mata et al. 2009).

6.2.2.2 Column Biosorption

Continuous biosorption studies are of utmost importance to evaluate the technical feasibility of a process for real applications. Among the different column configurations, packed bed columns have been established as effective, economical, and very convenient for biosorption processes (Zhao et al. 1999; Saeed and Iqbal 2003; Volesky et al. 2003). They make the best use of the concentration difference, which is known to be the driving force for sorption, and allow a more efficient utilization of the sorbent capacity, resulting in better effluent quality. Also, packed bed sorption has a number of process engineering merits, including a high operational yield and the relative ease of scaling-up procedures. Other column contactors, such as fluidized and continuous stirred tank reactors, are very rarely used for the purpose of biosorption (Prakasham et al. 1999; Solisio et al. 2000). Continuous stirred tank reactors are useful when the biosorbent is in the form of a powder; however, they suffer from high capital and operating costs (Volesky 1987). Fluidized bed systems, which operate continuously, require high flow rates to keep the biosorbent particles in suspension.

A packed bed configuration basically consists of a cylindrical column packed firmly with a sorbent, through which wastewater is allowed to flow. Initially, most of the solute will be sorbed as it is exposed to the fresh biosorbent bed; therefore, almost zero concentration would be expected at the column outlet. Theoretically, this is where the highest mass transfer occurs. However, as time is required (and column length) for a stabilized performance, the initial column behavior cannot be considered because this will only represent a transient and unsteady-state regime (Naja and Volesky 2006). With increasing time, the biosorbent bed will become saturated with the solute, the concentration of which will gradually increase at the column outlet. Here, the breakthrough/service concentration can be fixed depending on the toxicity of the solute. For most solutes, 0.01–1 mg/L is considered the breakthrough concentration. When the solute concentration exceeds this limit in real industrial applications, the column has to be removed from active operation, with the column regenerated or the flow switched to another column. However, for laboratory trials, the operation of the column should be terminated only when the inlet solute concentration approximately equals that at the outlet. This is due to complete column saturation, which results in an S-shaped breakthrough curve; therefore, it is important to evaluate the characteristics and the dynamic response of a biosorption column (Aksu 2005). Recording the concentration profile at the column exit usually results in a typical S-shaped curve, whose shape and slope are the result of the equilibrium sorption isotherm relationships, mass transfer to and throughout the sorbent in the column, and operation macroscopic fluid flow parameters.

Some microorganisms show high biosorption capacities in batch tests, but fail when applied to continuous-flow processes. This is because the performance in a column mode strongly depends on the mechanical strength of the biosorbent and the kinetics of the process. This is the reason why some free biomass cannot be used effectively in a column. Vijayaraghavan and Yun (2007) indicated that it was not possible to use the free biomass of *C. glutamicum* in a packed column, as it tended to swell and form a dense slurry, blocking the liquid flow; nevertheless, they suggested immobilization as a potential remedy to this limitation. However, a column operation provides only a short contact time for the solute and the sorbent, and the mass transfer resistances prevailing in the immobilized beads strongly affect the column biosorption performance. Therefore, utmost care must be taken in, specifically, preparing a biosorbent for use in the column mode. In addition, a systematic evaluation of the biosorbent and the parameters affecting the biosorption should be performed. Some important parameters affecting the biosorption in a packed column include the bed depth, the flow rate, and the initial solute concentration.

The accumulation of a solute in a fixed column is largely dependent on the amount of biosorbent loaded into the column. Zulfadhly et al. (2001) reported that the metal uptake increased on increasing the bed height in a *Pycnoporus sanguineus*–loaded fixed column. Similarly, Vijayaraghavan et al. (2004) observed an increase in the nickel biosorption capacity when the bed height was increased from 15 to 25 cm in a crab shell–loaded packed column. The increase in the uptake capacity with the increasing bed depth was due to the increased surface area of the sorbent, providing a greater amount of available binding sites for biosorption.

The flow rate is a crucial characteristic in the evaluation of sorbents for the continuous treatment of effluents on an industrial scale. In general, a low flow rate favors biosorption, which can be explained as follows: (1) when the flow rate increases, the residence time of the solute in the column decreases, which causes the effluent to leave the column prior to the attainment of equilibrium; and (2) when the process is controlled by intraparticle mass transfer, a slower flow rate enhances sorption, but if it is controlled by external mass transfer, a higher flow rate will decrease the film resistance.

The driving force for a sorption process is the concentration difference between the solute in the sorbent and that in the solution. Thus, an increased inlet solute concentration increases the concentration difference, which favors biosorption.

Various parameters can be used to characterize the performance of a packed bed biosorption, including the length of the sorption zone, the uptake, the removal efficiency, and the slope of the breakthrough curve. A mass transfer zone will develop between the gradually saturated section of the column and the fresh biosorbent section. The length of this zone is important practically, which can be calculated from (Equation 6.2)

$$Z_m = Z\left(1 - \frac{t_b}{t_e}\right), \tag{6.2}$$

where Z denotes the bed depth (cm) and t_b and t_e are the column breakthrough and exhaustion times (h), respectively.

The uptake is an important parameter, often used to characterize the performance of a biosorbent in a packed column. The column uptake (Q_{col}) can be calculated by dividing the total mass of the biosorbed sorbate (m_{ad}) by that of the biosorbent (M). The mass of the biosorbed sorbate is calculated from the area above the breakthrough curve (C vs t) multiplied by the flow rate.

The removal efficiency (%) can be calculated from the ratio of the sorbate mass biosorbed to the total mass of the sorbate sent to the column as follows (Equation 6.3):

$$\text{Removal efficiency (\%)} = \frac{m_{ad}}{C_0 F t_e} \times 100, \tag{6.3}$$

where C_0 and F are the inlet solute concentration (mg/L) and the flow rate (L/h), respectively. It is important to note that the removal efficiency is independent of the biosorbent mass, but it is solely dependent on the flow volume. Therefore, it is necessary to consider both the uptake and the removal efficiency when evaluating the biosorbent potential.

The slope of the breakthrough curve from t_b to t_e (dC/dt) is often used to characterize the shape of the curve (Volesky et al. 2003). It is always preferential to have an extended breakthrough curve with a steep slope, as a steep slope is usually the result of a shorter mass transfer zone, which implies a longer column service time and a greater utilization of the sorbent portion inside the column.

Thus, for good biosorbents, a delayed breakthrough, an earlier exhaustion, a shortened mass transfer zone, a high uptake, a steep breakthrough curve, and a high removal efficiency would be expected.

6.2.3 Desorption and Regeneration

Biosorption is a process of treating a pollutant-bearing solution to make it contaminant-free. However, it is also necessary to be able to regenerate the biosorbent. This is possible only with the aid of an appropriate elutant, which usually results in a concentrated pollutant solution. Therefore, the overall achievement of a biosorption process is to concentrate the solute, that is, sorption followed by desorption. Desorption is of utmost importance when the biomass preparation or generation is costly, as it is possible to decrease the process cost and also the dependency of the process on a continuous supply of biosorbent. A successful desorption process requires the proper selection of elutants, which strongly depends on the type of biosorbent and the mechanism of biosorption. Also, the elutant must be (i) nondamaging to the biomass, (ii) less costly, (iii) environmentally friendly, and (iv) effective. Several researchers have conducted exhaustive screening experiments to identify the appropriate elutants for this process. Of these, the work of Kuyucak and Volesky (1989b) is noteworthy; they examined several chemical agents to desorb Co^{2+} from cobalt-laden *Ascophyllum nodosum* and identified $CaCl_2$ in the presence of HCl as a suitable elutant. Vijayaraghavan et al. (2004) conducted screening experiments to identify a potent elutant for nickel-loaded crab shell particles and identified EDTA in the presence of NH_4OH to be a practical elutant.

The purpose of desorption is to unbind a contaminant from a biosorbent so that both the recovered solute and the biosorbent can be reused. After desorption, the biosorbent should be close to its original form, both morphologically and effectually. Also, during the desorption process, the removal of all the bound sorbate from the biosorbent should be ensured. If this does not happen, an undiminished uptake cannot be expected in the next cycle. Therefore, careful and uniform desorption processes are necessary to ensure the regeneration of the biomass. In the field of biosorption, Puranik and Paknikar (1999) regenerated and reused a polysulfone-immobilized *Citrobacter* strain over three cycles for the biosorption of lead, cadmium, and zinc, using 0.1 *M* HCl and 0.1 *M* EDTA as elutants; however, they had only limited success and emphasized the need for further screening work. Beolchini et al. (2003) immobilized *Sphaerotilus natans* into a polysulfone matrix for the biosorption of copper and, with the aid of 0.05 *M* $CaCl_2$, regenerated and reused the beads over 10 cycles, with satisfactory results. Vijayaraghavan et al. (2007) employed polysulfone-immobilized *C. glutamicum* for the biosorption of Reactive black 5 and were able to successfully regenerate the beads, using 0.1 *M* NaOH as the elutant, over 20 sorption–desorption cycles.

One of the main attractions of biosorption is its potential ability to regenerate the biomass; however, very little research has focused specifically on desorption. Most published works have aimed to evaluate the binding ability of the biomass and the parameters affecting the process. Less attention has been paid to the regeneration ability of the biosorbent, which often decides the industrial applicability of a process. Thus, biosorption studies should emphasize the possibility of biomass regeneration to improve the process viability.

6.2.4 Scope and Future Directions of Biosorption

Biosorption is an efficient and cheap technique for metal removal, proven by hundreds of research reports published in the last few decades. However, most of these research reports portray a particular biomaterial that can efficiently bind a particular metal ion from a single-solute system.

However, in reality, contaminated solutions are composed of several organic and inorganic ions, dissolved, colloidal, or suspended matters (Volesky 2001). Hence, severe competition between the solutes is expected and will eventually affect the performance of a biosorbent. Considering this, few researchers have attempted to examine the performance of a biosorbent in real industrial effluents (Vijayaraghavan et al. 2006b; Prigione et al. 2009). An in-depth analysis of these research reports revealed that most of the effluent studies are performed to highlight the possibility of a biosorbent to remove a metal ion of interest. High biosorptional uptake was not considered as a main criterion in effluent studies, and for some, the mere removal efficiency is sufficient to recommend the biosorbent for industrial applications. Not much effort was taken to optimize the process so that improved efficiency or specificity could be obtained. Other important factors, such as the physical characteristics, the availability, and the cost of the biosorbent, have always been overlooked. Atkinson et al. (1998) highlighted the questions to be considered concerning the feasibility of a potential biosorbent for metal removal from industrial effluents. These include the effluent characteristics, such as volume, type of contaminant and competitive ions, solution chemistry, pH, and temperature adjustment; the biomass characteristics, such as availability, mechanical stability, regeneration ability, contaminant specificity, and reaction kinetics; and the process characteristics, such as capital and operating costs and batch/continuous and land space requirements.

Hence, in order to compete with the current treatment technologies, more research should be focused on the industrial feasibility of the biosorption process. Conversely, it is no small feat to replace well-established conventional techniques. However, in addition to being cost-effective, biosorption has a huge potential, as many biosorbents are known to perform as well as, if not better than, most conventional methods. As well as being aware of the hundreds of biosorbents that are able to bind various pollutants, sufficient research has been performed on various biomaterials to understand the mechanism responsible for biosorption. Therefore, through continued research, especially on pilot- and full-scale biosorption processes, the situation is likely to change in the near future, with biosorption technology becoming more beneficial and attractive than currently used technologies.

6.3 Bioaccumulation

Another fascinating biological technique for the removal of metal ions is bioaccumulation. Bioaccumulation can be defined as the uptake of metal ions by living cells. Biosorption and bioaccumulation differ in that in the first process, the pollutants are bound to the surface of the cell wall, and in the second, they also accumulate inside the cell. To be precise, bioaccumulation comprises two steps, namely, the rapid sorption of the metal ions onto the cellular components of the biomass, followed by a slower metabolism-dependent active uptake of the metals (Dönmez and Aksu 1999). Most of the studies dealing with microbial metal remediation via growing cells describe the biphasic uptake of the metals, that is, the initial rapid phase of biosorption followed by the slower, metabolism-dependent active uptake of the metals (Garnham et al. 1992; Dönmez and Aksu 1999). Recent reports employing growing cultures of marine microalgae indicate that the intracellular Cd levels are often higher than the biosorbed Cd levels (Perez-Rama et al. 2002).

When considering the operational aspects, bioremoval by growing cells is usually performed in batch systems (Aksu and Dönmez 2000). The advantage of the process of bioaccumulation is that it is not necessary to include a separate biomass cultivation mode or a harvesting biomass from the environment. Also, additional unit processes are reduced: harvesting, drying, processing, and storing (Aksu and Dönmez 2005). The process of bioremoval of pollutants is highly affected by the operational conditions, in particular by the presence of pollutants in the growth medium, which can inhibit the growth of cells and also the bioaccumulation itself. This is a severe limitation of the process, because it makes it impossible to treat a solution with a high load of pollutants. Moreover, it is necessary to supply an external source of energy to the growing cells (e.g., sucrose from molasses) (Aksu and Dönmez 2005). However, if proper strains are selected, it is possible to propose a self-replenishing system (Aksu and Dönmez 2005), whereby a biological material that accumulates pollutants (either inorganic ions or organic compounds) is generated in the bioaccumulation unit. The hope for the practical application of bioaccumulation is that the majority of conventional municipal wastewater treatment plants will be based on living organisms, with a significant contribution from bioaccumulation itself (Aksu and Dönmez 2000).

In bioaccumulation, the pollutants are transported across the cell wall and the membrane. Inside, the cells are bound to the intracellular structures (Kujan et al. 1995). The literature lists the following processes contributing to the mechanism of bioaccumulation, including intracellular accumulation and oxidation or reduction reactions (Yilmazer and Saracoglu 2009). This process is very complex and depends on several factors (which are almost identical to the factors influencing the cultivation of an organism): the composition of the growth medium or, in other words, in this case, wastewater, pH, temperature, the presence of other pollutants, which are growth inhibitors as well, or other inhibitors, surfactants, etc. (Kujan et al. 1995).

The bioaccumulation literature revealed that several organisms are efficient in the accumulation of metal ions. Dönmez and Aksu (1999) reported the bioaccumulation of Cu(II) by various strains of yeast: *Saccharomyces cerevisiae, Kluyveromyces marxianus, Schizosaccharomyces pombe,* and *Candida* sp.

It was found that the bioaccumulation was dependent on the initial concentration of Cu(II) and also on the pH. The best bioaccumulators were found to be *Candida* sp. and *K. marxianus.* Yilmazer and Saracoglu (2009) compared the bioaccumulation capacities of the adapted and nonadapted *Pichia stipitis* yeasts toward Cu(II) and Cr(III) ions. The authors identified that the adapted cells performed well and exhibited a maximum specific capacity of 15.85 and 9.10 mg/g, respectively, at 100 mg/L of the initial Cu(II) and Cr(III) concentrations, respectively. Kujan et al. (1995) identified that the cadmium bioaccumulation potential of *Candida utilis* was dependent on the carbon source (xylose and glucose). In the presence of glucose, *C. utilis* exhibited a cadmium uptake capacity of 0.18 mg/g, whereas in the presence of xylose, *C. utilis* exhibited a cadmium uptake capacity of 0.26 mg/g. Uslu et al. (2003) compared the bioaccumulation property of *Rhizopus arrhizus* toward Cd(II), Pb(II), and Cu(II) and evaluated that there was greater accumulation of both Pb(II) and Cd(II) than Cu(II). Dursun et al. (2003) evaluated the bioaccumulation capacity of *Aspergillus niger* toward Cu(II), Pb(II), and Cr(VI) and identified that the strain is capable of accumulating 15.6, 34.4, and 6.6 mg/g of Cu(II), Pb(II), and Cr(VI), respectively.

6.3.1 Scope and Future Directions of Bioaccumulation

The process of bioaccumulation has the potential to find practical applications in the future separation technologies and become part of the hybrid or integrated installations for wastewater treatment. Even less advantageous when compared with biosorption, the process of bioaccumulation has been well researched in recent years to suit commercial applications. To an extent, the isolation of super-resistant strains from contaminated sites has eliminated the primary hurdle for the application of growing cells. Genetic engineering may further enhance the potential of robust environmental strains. The exploitation of locally available support material and cheap carbon/nutrition sources for the cells appears to promise an economically favorable process. However, the choice of consortia and carbon/nutrition source must depend on the nature of the effluents because of the varied complexing properties of the metals and the nutritional content of the effluent itself. Thus, continued research on the successful reproduction of the process on a commercial scale, as well as identification of novel bioaccumulators, is required for the success of bioaccumulation technology.

6.4 Phytoremediation

Phytoremediation refers to a diverse collection of plant-based technologies that use either naturally occurring or genetically engineered plants for cleaning contaminated environments (Cunningham and Ow 1996; Prasad and Freitas 2003). It is an efficient technique for the remediation of several organic compounds (Campanella et al. 2002; Susarla et al. 2002) and inorganics (Prasad and Freitas 2003). Phytoremediation consists of four different plant-based technologies, each having a different mechanism of action for the remediation of the metal-polluted soil, sediment, or water. These include: rhizofiltration, which involves the use of hydroponically cultivated plant roots to remediate contaminated water through the adsorption, concentration, and precipitation of the pollutants; phytostabilization, where plants are used to stabilize rather than clean the contaminated soil; phytovolatilization, which refers to the uptake and transpiration of contaminants by plants; and phytoextraction, which refers to the uptake of contaminants by the plant roots and the movement of the contaminants from the roots to the aboveground parts of the plant. These approaches of plant-based technologies for environmental restoration allow the treatment of many sites that cannot be addressed with currently available methods. However, phytoremediation should be viewed as a long-term remediation solution because many cropping cycles may be needed over several years to reduce the contaminant to acceptable regulatory levels. This new remediation technology is competitive and may be superior to the existing conventional technologies at sites where phytoremediation is applicable.

6.4.1 Rhizofiltration

Rhizofiltration involves raising plants hydroponically and transplanting them into metal-polluted waters where the plants absorb and concentrate the metals in their roots and shoots. Either the contaminated water is collected from a waste site where plants are cultivated, or the plants are planted in the contaminated area, where the plant roots then take up the water and the contaminants dissolved in it. Many plant species naturally take up heavy metals and excess nutrients for a variety of reasons: sequestration, drought resistance, disposal by leaf abscission, interference with other plants, and defense against pathogens and herbivores. Additionally, root exudates and changes in the rhizosphere pH may cause the metals to precipitate onto the root surfaces. As they become saturated with the metal contaminants, the roots or the whole plants are harvested for disposal (Prasad and Freitas 2003). This process is very similar to phytoextraction in that it removes contaminants by trapping them into a harvestable plant biomass. Both phytoextraction and rhizofiltration follow the same basic path to remediation. First, the plants are put in contact with the contaminants. They absorb the contaminants through their root systems and store them in the root biomass and/or transport them up into the stems and/or leaves. The plants continue to absorb the contaminants until they are harvested. The plants are then replaced to continue the growth/harvest cycle until satisfactory levels of the contaminants are achieved. Both processes are also aimed more toward concentrating and precipitating the heavy metals than the organic contaminants. The major difference between rhizofiltration and phytoextraction is that rhizofiltration is used for treating aquatic environments, while phytoextraction deals with soil remediation. Rhizofiltration may be applicable to the treatment of surface water and groundwater, industrial and residential effluents, stormwaters, acid mine drainage, agricultural runoffs, and radionuclide-contaminated solutions. Dushenkov et al. (1995) described the characteristics of the ideal plants for rhizofiltration. The plants should be able to accumulate and tolerate significant amounts of the target metals in conjunction with easy handling, low maintenance cost, and a minimum of secondary waste requiring disposal. It is also desirable that the plants produce significant amounts of root biomass or root surface area. The roots of many hydroponically grown terrestrial plants, such as Indian mustard, sunflower, and various grasses, can be used to remove toxic metals such as Cu^{2+}, Cd^{2+}, Cr^{6+}, Ni^{2+}, Pb^{2+}, and Zn^{2+} from aqueous solutions (Raskin and Ensley 2000; Dushenkov et al. 1995). Several other plant species showed good rhizofiltration ability to remove metals and radionuclides, including *Helianthus annuus* (Lee and Yang 2010; Tomé et al. 2008), *Phaseolus vulgaris* (Lee and Yang 2010; Laroche et al. 2005), water hyacinth (Kay et al. 1984; Zhu et al. 1999), and duckweed (Mo et al. 1989).

Rhizofiltration is cost-effective for large volumes of water having low concentrations of contaminants that are subjected to stringent standards. It is relatively inexpensive, yet potentially more effective than comparable technologies. Rhizofiltration may be conducted in situ, with plants being grown directly in the contaminated water body. This allows for a relatively inexpensive procedure with low capital costs. The operation costs are also low, but depend on the types of contaminants. This treatment method is also aesthetically pleasing and results in a decrease in the water infiltration and leaching of the contaminants. After harvesting, the crop may be converted to a biofuel briquette, a substitute for fossil fuel.

6.4.2 Phytoextraction

Phytoextraction is the uptake of contaminants by the plant roots and the movement of the contaminants from the roots to the aboveground parts of the plants. Generally, the contaminants are removed from the site by harvesting the plants. Phytoextraction accumulates the contaminants in a much smaller amount of material to be disposed of (the contaminated plants) than does the excavation of the soil or the sediment. The technique is mostly applied to the heavy metals and the radionuclides in the soil, sediment, and sludge. It may use plants that naturally take up and accumulate extremely elevated levels of contaminants in their stems and leaves. It can also entail the use of plants that take up and accumulate aboveground significant amounts of contaminants only when special soil amendments are used. Another approach is the use of plants that trap the contaminants in their root systems and are then harvested whole (including the roots). Some researchers suggest that the incineration of the harvested plant tissues dramatically reduces the volume of the material requiring disposal (Kumar et al. 1995). In some cases, valuable metals can be extracted from the metal-rich ash and serve as a source of revenue, thereby offsetting the expense of remediation (Cunningham and Ow 1996).

6.4.3 Phytostabilization

Phytostabilization is another mechanism that can be used to minimize the migration of the contaminants in soils. This process takes advantage of the ability of plant roots to alter the soil environment conditions, such as pH and soil moisture content. Many root exudates cause metals to precipitate, thereby reducing the bioavailability. One advantage of this strategy over phytoextraction is that the disposal of metal-laden plant material is not required. By choosing and maintaining an appropriate cover of plant species, coupled with appropriate soil amendments, it may be possible to stabilize certain contaminants (particularly metals) in the soil and reduce the interaction of these contaminants with the associated biota (Susarla et al. 2002).

6.4.4 Phytovolatilization

Phytovolatilization involves the use of plants to take up the contaminants from the soil, transforming them into a volatile form and transpiring them into the atmosphere. Phytovolatilization occurs as growing trees and other plants take up water and organic and inorganic contaminants. Some of these contaminants can pass through the plants to the leaves and volatilize into the atmosphere at comparatively low concentrations (Ghosh and Singh 2005). Phytovolatilization has been primarily used for the removal of mercury; the mercuric ion is transformed into less toxic elemental mercury. The disadvantage of this is that the mercury released into the atmosphere is likely to be recycled by precipitation and then redeposited back into the ecosystem. Selenium phytovolatilization has also been given utmost attention (Lewis et al. 1966; Terry et al. 1992), because this element is a serious problem in many parts of the world where there are areas with Se-rich soil. Research on certain western US soils has led to proposed vegetation management systems that encourage Se volatilization through what appears to be a plant and/or a plant–microbe interaction (Banuelos et al. 1993; Zayed and Terry 1994). More recently, a bacterial mercuric ion reductase has been engineered into *Arabidopsis thaliana*, and the resulting transformant is capable of tolerating and volatilizing mercuric ions (Rugh et al. 1996). The toxic cation is absorbed by the root and is reduced to volatile Hg(0) by the introduced mercuric ion reductase. Heaton et al. (1998) suggest that the addition of Hg(0) into the atmosphere would not contribute significantly to the atmospheric pool. However, those who support this technique also agree that phytovolatilization would not be wise for sites near population centers or at places with unique meteorological conditions that promote the rapid deposition of volatile compounds (Heaton et al. 1998; Rugh et al. 2000). Unlike other remediation techniques, once the contaminants have been removed via volatilization, there is a loss of control over their migration to other areas. Despite the controversy surrounding phytovolatilization, this technique is a promising tool for the remediation of Se- and Hg-contaminated soils.

6.4.5 Scope and Future Directions of Phytoremediation

Phytoremediation is a fast-developing field. Over the last 10 years, many field applications have been initiated all over the world, including the phytoremediation of organic and inorganic contaminants and radionuclides. This sustainable and inexpensive process is fast emerging as a viable alternative to the conventional remediation methods. To date, commercial phytoextraction has been constrained by the expectation that site remediation should be achieved in a time comparable to other cleanup technologies. So far, most of the phytoremediation experiments have taken place on a laboratory scale, where the plants are grown in a hydroponic setting and are fed heavy metal diets. While these results are promising, scientists are ready to admit that a solution culture is quite different from that of a soil. In real soil, many metals are tied up in insoluble forms, making them less available, which is the biggest problem. The future of phytoremediation is still in the research and development phase, and many technical barriers need to be addressed. Both the agronomic management practices and the plant genetic abilities need to be optimized to develop commercially useful practices. Many hyperaccumulator plants remain to be discovered, and there is a need to know more about their physiology. The optimization of the process, the proper understanding of the plant heavy metal uptake, and the proper disposal of the biomass produced are still needed.

6.5 Conclusions

It is evident that biological treatment methods are an efficient alternative to the conventional methods, owing to their inherent advantages. However, as they are biological systems, their low predictability and sometimes longer reaction times act as a major drawback. Also, many biological treatment experiments are carried out in the laboratory and the same results cannot be expected on scale-up, owing to other external factors. Thus, for future advancement, more field data and pilot plot-scale experiments are essential in order to make biological techniques a reliable option for remediation activities. Also, research should continue to improve the knowledge about various systems and simultaneously decrease the overall process cost. All these efforts will ultimately lead to increasingly robust and reliable biological treatment methods that would be capable of replacing more invasive techniques.

References

Aksu, Z. 2005. Application of biosorption for the removal of organic pollutants: A review. *Proc. Biochem.* 40: 997–1026.

Aksu, Z. and Ş.Ş. Çağatay. 2006. Investigation of biosorption of Gemazol Turquoise Blue-G reactive dye by dried *Rhizopus arrhizus* in batch and continuous systems. *Sep. Purif. Technol.* 48: 24–35.

Aksu, Z. and G. Dönmez. 2000. The use of molasses in copper(II) containing wastewaters: Effects on growth and copper(II) bioaccumulation properties of *Kluyveromyces marxianus. Proc. Biochem.* 36: 451–458.

Aksu, Z. and G. Dönmez. 2005. Combined effects of molasses sucrose and reactive dye on the growth and dye bioaccumulation properties of *Candida tropicalis. Proc. Biochem.* 40: 2443–2454.

Al-Qodah, Z. 2006. Biosorption of heavy metal ions from aqueous solutions by activated sludge. *Desalination* 196: 164–176.

Amini, M., H. Younesi, and N. Bahramifar. 2009. Biosorption of nickel(II) from aqueous solution by *Aspergillus niger*: Response surface methodology and isotherm study. *Chemosphere* 75: 1483–1491.

An, H.K., B.Y. Park, and D.S. Kim. 2001. Crab shell for the removal of heavy metals from aqueous solution. *Water Res.* 35: 3551–3556.

Atkinson, B.W., F. Bux, and H.C. Kasan. 1998. Considerations for application of biosorption technology to remediate metal-contaminated industrial effluents. *Water SA* 24: 129–135.

Baes, C.F. and R.E. Mesmer. 1976. *The Hydrolysis of Cations.* New York: Wiley, pp. 241–310.

Bai, R.S. and E. Abraham. 2001. Biosorption of Cr(VI) from aqueous solution by *Rhizopus nigricans. Biores. Technol.* 79: 73–81.

Banuelos, G.S., G. Cardon, B. Mackey, I. Ben-Asherm, L. Wu, P. Beuselinck, S. Akohoue, and S. Zambrzuski. 1993. Boron and selenium removal in B-laden soils by four sprinkler irrigated plant species. *J. Environ. Qual.* 22: 786–797.

Beolchini, F., F. Pagnanelli, L. Toro, and F. Vegliò. 2003. Biosorption of copper by *Sphaerotilus natans* immobilised in polysulfone matrix: Equilibrium and kinetic analysis. *Hydrometallurgy* 70: 101–112.

Bruins, M.R., S. Kapil, and F.W. Oehme. 2000. Microbial resistance to metals in the environment. *Ecotoxicol. Environ. Safety* 45: 198–207.

Campanella, B.F., C. Bock, and P. Schröder. 2002. Phytoremediation to increase the degradation of PCBs and PCDD/Fs. *Environ. Sci. Poll. Res.* 9: 73–85.

Choi, S.B. and Y.S. Yun. 2004. Lead biosorption by waste biomass of *Corynebacterium glutamicum* generated from lysine fermentation process. *Biotechnol. Lett.* 26: 331–336.

Cossich, E.S., C.R.G. Tavares, and T.M.K. Ravagnani. 2002. Biosorption of chromium (III) by *Sargassum sp.* biomass. *Electron. J. Biotechnol.* 5: 133–140.

Cunningham, S.D. and D.W. Ow. 1996. Promises and prospects of phytoremediation. *Plant Physiol.* 110: 715–719.

Davis, T.A., B. Volesky, and A. Mucci. 2003. A review of the biochemistry of heavy metal biosorption by brown algae. *Water Res.* 37: 4311–4330.

Dönmez, G. and Z. Aksu. 1999. The effect of copper(II) ions on growth and bioaccumulation properties of some yeasts. *Proc. Biochem.* 35: 135–142.

Dursun, A.Y. 2006. A comparative study on determination of the equilibrium, kinetic and thermodynamic parameters of biosorption of copper(II) and lead(II) ions onto pretreated *Aspergillus niger*. *Biochem. Eng. J.* 28: 187–195.

Dursun, A.Y., G. Uslu, Y. Cuci, and Z. Aksu. 2003. Bioaccumulation of copper(II), lead(II) and chromium(VI) by growing *Aspergillus niger*. *Proc. Biochem.* 38: 1647–1651.

Dushenkov, V., P.B.A.N. Kumar, H. Motto, and I. Raskin. 1995. Rhizofiltration: The use of plants to remove heavy metals from aqueous streams. *Environ. Sci. Technol.* 29: 1239–1245.

Gadd, G.M., C. White, and L. DeRome. 1988. Heavy metal and radionuclide uptake by fungi and yeasts. In *Biohydrometallurgy*, eds. Norri, P.R. and D.P. Kelly. Chippenham, UK: A. Rowe, pp. 109–119.

Garnham, G.W., C.A. Codd, and G.M. Gadd. 1992. Kinetics of uptake and intracellular location of cobalt, manganese and zinc in the estuarine green alga *Chlorella salina*. *Appl. Microbiol. Biotechnol.* 37: 270–276.

Ghosh, M. and S.P. Singh. 2005. A review on phytoremediation of heavy metals and utilization of its byproducts. *Appl. Ecol. Environ. Res.* 3: 1–18.

Golab, Z. and M. Breitenbach. 1995. Sites of copper binding in *Streptomyces pilosus*. *Water Air Soil Poll.* 82: 713–721.

Guo, B., L. Hong, and H.X. Jiang. 2003. Macroporous poly(calcium acrylate-divinylbenzene) bead – a selective orthophosphate sorbent. *Ind. Eng. Chem. Res.* 42: 5559–5565.

Haug, A., B. Larsen, and O. Smidsrød. 1966. A study of the constitution of alginic acid by partial acid hydrolysis. *Acta Chem. Scand.* 20: 183–190.

Haug, A., S. Myklestad, B. Larsen, and O. Smidsrød. 1967. Correlation between chemical structure and physical properties of alginates. *Acta Chem. Scand.* 21: 768–778.

Heaton, A.C.P., C.L. Rugh, N. Wang, and R.B. Meagher. 1998. Phytoremediation of mercury and methylmercury – polluted soils using genetically engineered plants. *J. Soil Contam.* 7: 497–510.

Holan, Z.R. and B. Volesky. 1994. Biosorption of lead and nickel by biomass of marine algae. *Biotechnol. Bioeng.* 43: 1001–1009.

Holan, Z.R., B. Volesky, and I. Prasetyo. 1993. Biosorption of cadmium by biomass of marine algae. *Biotechnol. Bioeng.* 41: 819–825.

Incharoensakdi, A. and P. Kitjaharn. 2002. Zinc biosorption from aqueous solution by a halotolerant cyanobacterium *Aphanothece halophytica*. *Curr. Microbiol.* 45: 261–264.

Iqbal, M. and A. Saeed. 2007. Production of an immobilized hybrid biosorbent for the sorption of Ni(II) from aqueous solution. *Proc. Biochem.* 42: 148–157.

Kang, S.-Y., J.-U. Lee, and K.-W. Kim. 2007. Biosorption of Cr(III) and Cr(VI) onto the cell surface of *Pseudomonas aeruginosa*. *Biochem. Eng. J.* 36: 54–58.

Kay, S.H., W.T. Haller, and L.A. Garrard. 1984. Effect of heavy metals on water hyacinths [*Eichhornia crassipes* (Mart.) Solms]. *Aquat. Toxicol.* 5: 117–128.

Kujan, P., J. Votruba, and V. Kamenik. 1995. Substrate-dependent bioaccumulation of cadmium by growing yeast *Candida utilis*. *Folia Microbiol.* 40: 288–292.

Kumar, P.B.A.N., V. Dushenkov, H. Motto, and I. Raskin. 1995. Phytoextraction: The use of plants to remove heavy metals from soils. *Environ. Sci. Technol.* 29: 1232–1238.

Kuyucak, N. and B. Volesky. 1989a. Accumulation of cobalt by marine alga. *Biotechnol. Bioeng.* 33: 809–814.

Kuyucak, N. and B. Volesky. 1989b. Desorption of cobalt-laden algal biosorbent. *Biotechnol Bioeng.* 33: 815–822.

Laroche, L., P. Henner, V. Camilleri, M. Morello, and J. Garnier-Laplace. 2005. Root uptake of uranium by a higher plant model (*Phaseolus vulgaris*) –Bioavailability from soil solution. *Radioprotection* 40: 33–39.

Lee, M. and M. Yang. 2010. Rhizofiltration using sunflower (*Helianthus annuus* L.) and bean (*Phaseolus vulgaris* L. var. vulgaris) to remediate uranium contaminated groundwater. *J. Hazard. Mater.* 173: 589–596.

Lee, M.-Y., K.-J. Hong, T. Kajiuchi, and J.-W. Yang. 2004. Determination of the efficiency and removal mechanism of cobalt by crab shell particles. *J. Chem. Technol. Biotechnol.* 79: 1388–1394.

Lee, M.-Y., S.-H. Lee, H.-J. Shin, T. Kajiuchi, and J.-W. Yang. 1998. Characteristics of lead removal by crab shell particles. *Proc. Biochem.* 33: 749–753.

Leusch, H. and B. Volesky. 1995. Biosorption of heavy metals (Cd, Cu, Ni, Pb, Zn) by chemically reinforced biomass of marine algae. *J. Chem. Technol. Biotechnol.* 62: 279–288.

Lewis, B.G., C.M. Johnson, and C.C. Delwiche. 1966. Release of volatile selenium compounds by plants: Collection procedures and preliminary observations. *J. Agricult. Food Chem.* 14: 638–640.

Liu, H.-L., B.-Y. Chen, Y.-W. Lan, and Y.C. Cheng. 2004. Biosorption of Zn(II) and Cu(II) by the indigenous *Thiobacillus thiooxidans*. *Chem. Eng. J.* 97: 195–201.

López, A., N. Lázaro, J.M. Priego, and A.M. Marquès. 2000. Effect of pH on the biosorption of nickel and other heavy metals by *Pseudomonas fluorescens* 4F39. *J. Ind. Microbiol. Biotechnol.* 24: 146–151.

Loukidou, M.X., T.D. Karapantsios, A.I. Zouboulis, and K.A. Matis. 2004. Diffusion kinetic study of chromium(VI) biosorption by *Aeromonas caviae*. *Ind. Eng. Chem. Res.* 43: 1748–1755.

Maestri, E., M. Marmiroli, G. Visioli, and N. Marmiroli. 2010. Metal tolerance and hyperaccumulation: Costs and trade-offs between traits and environment. *Environ. Exp. Bot.* 68: 1–13.

Malik, A. 2004. Metal bioremediation through growing cells. *Environ. Int.* 30: 261–278.

Marmiroli, N. and E. Maestri. 2008. Trace elements contamination and availability: Human health implications of food chain and biofortification. In *Trace Elements as Contaminants and Nutrients: Consequences in Ecosystems and Human Health*, ed. Prasad, M.N.V. New York: Wiley, pp. 23–53.

Mata, Y.N., M.L. Blázquez, A. Ballester, F. González, and J.A. Muñoz. 2009. Biosorption of cadmium, lead and copper with calcium alginate xerogels and immobilized *Fucus vesiculosus*. *J. Hazard. Mater.* 163: 555–562.

Mo, S.C., D.S. Choi, and J.W. Robinson. 1989. Uptake of mercury from aqueous solution by duckweed: The effect of pH, copper, and humic acid. *J. Environ. Health* 24: 135–146.

Naja, G. and B. Volesky. 2006. Behavior of the mass transfer zone in a biosorption column. *Environ. Sci. Technol.* 40: 3996–4003.

Niu, H. and B. Volesky. 2003. Characteristics of anionic metal species biosorption with waste crab shells. *Hydrometallurgy* 71: 209–215.

Niu, H., X.S. Xu, J.H. Wang, and B. Volesky. 1993. Removal of lead from aqueous solutions by *Penicillium* biomass. *Biotechnol. Bioeng.* 42: 785–787.

Öztürk, A. 2007. Removal of nickel from aqueous solution by the bacterium *Bacillus thuringiensis*. *J. Hazard. Mater.* 147: 518–523.

Perez-Rama, M., J.A. Alonoso, C.H. Lopez, and E.T. Vaamonde. 2002. Cadmium removal by living cells of the marine microalgae *Tetraselmis suecica*. *Biores. Technol.* 84: 265–270.

Prakasham, R.S., J.S. Merrie, R. Sheela, N. Saswathi, and S.V. Ramakrishna. 1999. Biosorption of chromium VI by free and immobilized *Rhizopus arrhizus*. *Environ. Poll.* 104: 421–427.

Prasad, M.N.V. and H.M.O. Freitas. 2003. Metal hyperaccumulation in plants – Biodiversity prospecting for phytoremediation technology. *Electron. J. Biotechnol.* 6: 285–321.

Prigione, V., M. Zerlottin, D. Refosco, V. Tigini, A. Anastasi, and G.C. Varese. 2009. Chromium removal from a real tanning effluent by autochthonous and allochthonous fungi. *Biores. Technol.* 100: 2770–2776.

Puranik, P.R. and K.M. Paknikar. 1999. Biosorption of lead, cadmium, and zinc by *Citrobacter* strain MCM B-181: Characterization studies. *Biotechnol. Prog.* 15: 228–237.

Raskin, I. and B.D. Ensley. 2000. *Phytoremediation of Toxic Metals Using Plants to Clean Up the Environment*. New York: Wiley.

Rugh, C.L., S.P. Bizily, and R.B. Meagher. 2000. Phytoreduction of environmental mercury pollution. In *Phytoremediation of Toxic Metals: Using Plants to Clean-Up the Environment*, eds. Raskin, I. and B.D. Ensley. New York: Wiley, pp. 151–170.

Rugh, C.L., H.D. Wilde, N.M. Stack, D.M. Thompson, A.O. Summers, and R.B. Meagher. 1996. Mercuric ion reduction and resistance in transgenic *Arabidopsis thaliana* plants expressing a modified bacterial *merA* gene. *Proc. Natl. Acad. Sci. USA* 93: 3182–3187.

Ruiz-Manriquez, A., P.I. Magaña, V. López, and R. Guzmán. 1997. Biosorption of Cu by *Thiobacillus ferrooxidans*. *Bioproc. Eng.* 18: 113–118.

Saeed, A. and M. Iqbal. 2003. Bioremoval of cadmium from aqueous solution by black gram husk (*Cicer arientinum*). *Water Res.* 37: 3472–3480.

Saeed, A., M. Iqbal, and M.W. Akhtar. 2005. Removal and recovery of lead(II) from single and multimetal (Cd, Cu, Ni and Zn) solutions by crop milling waste (black gram husk). *J. Hazard. Mater.* 117: 65–73.

Sağ, Y. and T. Kutsal. 2000. Determination of the biosorption heats of heavy metal ions on *Zoogloea ramigera* and *Rhizopus arrhizus*. *Biochem. Eng. J.* 6: 145–151.

Selatnia, A., M.Z. Bakhti, A. Madani, L. Kertous, and Y. Mansouri. 2004. Biosorption of Cd^{2+} from aqueous solution by a NaOH-treated bacterial dead *Streptomyces rimosus* biomass. *Hydrometallurgy* 75: 11–24.

Solisio, C., A. Lodi, A. Converti, and M.D. Borghi. 2000. The effect of acid pretreatment on the biosorption of chromium(III) by *Sphaerotilus natans* from industrial wastewater. *Water Res.* 34: 3171–3178.

Susarla, S., V.F. Medina, and S.C. McCutcheon. 2002. Phytoremediation: An ecological solution to organic chemical contamination. *Ecol. Eng.* 18: 647–658.

Terry, N., C. Carlson, T.K. Raab, and A. Zayed. 1992. Rates of selenium volatilization among crop species. *J. Environ. Qual.* 21: 341–344.

Tomé, F.V., P.B. Rodríguez, and J.C. Lozano. 2008. Elimination of natural uranium and ^{226}Ra from contaminated waters by rhizofiltration using *Helianthus annuus* L. *Sci. Total Environ.* 393: 351–357.

Uslu, G., A.Y. Dursun, H.I. Ekiz, and Z. Aksu. 2003. The effect of Cd(II), Pb(II) and Cu(II) ions on the growth and bioaccumulation properties of *Rhizopus arrhizus*. *Proc. Biochem.* 30: 105–110.

Uslu, G. and M. Tanyol. 2006. Equilibrium and thermodynamic parameters of single and binary mixture biosorption of lead(II) and copper(II) ions onto *Pseudomonas putida*: Effect of temperature. *J. Hazard. Mater.* 135: 87–93.

Vegliò, F. and F. Beolchini. 1997. Removal of metals by biosorption: A review. *Hydrometallurgy* 44: 301–16.

Vieira, R.H.S.F. and B. Volesky. 2000. Biosorption: A solution to pollution. *Int. Microbiol.* 3: 17–24.

Vijayaraghavan, K., M.H. Han, S.B. Choi, and Y.-S. Yun. 2007. Biosorption of reactive black 5 by *Corynebacterium glutamicum* biomass immobilized in alginate and polysulfone matrices. *Chemosphere* 68: 1838–1845.

Vijayaraghavan, K., J. Jegan, K. Palanivelu, and M. Velan. 2004. Removal of nickel (II) ions from aqueous solution using crab shell particles in a packed bed up-flow column. *J. Hazard. Mater.* 113: 223–230.

Vijayaraghavan, K., J. Jegan, K. Palanivelu, and M. Velan. 2005a. Batch and column removal of copper from aqueous solution using a brown marine alga *Turbinaria ornata*. *Chem. Eng. J.* 106: 177–184.

Vijayaraghavan, K., J.R. Jegan, K. Palanivelu, and M. Velan. 2005b. Nickel recovery from aqueous solution using crab shell particles. *Adsorp. Sci. Technol.* 23: 303–311.

Vijayaraghavan, K., M.W. Lee, and Y.-S. Yun. 2008. Evaluation of fermentation waste (*Corynebacterium glutamicum*) as a biosorbent for the treatment of nickel(II)-bearing solutions. *Biochem. Eng. J.* 41: 228–233.

Vijayaraghavan, K., K. Palanivelu, and M. Velan. 2006a. Biosorption of copper(II) and cobalt(II) from aqueous solutions by crab shell particles. *Biores. Technol.* 97: 1411–1419.

Vijayaraghavan, K., K. Palanivelu, and M. Velan. 2006b. Treatment of nickel containing electroplating effluents with *Sargassum wightii* biomass. *Proc. Biochem.* 41: 853–859.

Vijayaraghavan, K., H.Y.N. Winnie, and R. Balasubramanian. 2011. Biosorption characteristics of crab shell particles for the removal of manganese(II) and zinc(II) from aqueous solutions. *Desalination* 266: 195–200.

Vijayaraghavan, K. and Y.-S. Yun. 2008. Bacterial biosorbents and biosorption. *Biotechnol. Adv.* 26: 266–291.

Volesky, B. 1987. Biosorbents for metal recovery. *TIBTECH* 5: 96–101.

Volesky, B. 2001. Detoxification of metal-bearing effluents: Biosorption for the next century. *Hydrometallurgy* 59: 203–216.

Volesky, B. and Z.R. Holan. 1995. Biosorption of heavy metals. *Biotechnol. Prog.* 11: 235–250.

Volesky, B., J. Weber, and J.M. Park. 2003. Continuous-flow metal biosorption in a regenerable *Sargassum* column. *Water Res.* 37: 297–306.

Wu, J. and H.-Q. Yu. 2007. Biosorption of 2,4-dichlorophenol by immobilized white-rot fungus *Phanerochaete chrysoporium* from aqueous solutions. *Biores. Technol.* 98: 253–259.

Yahaya, Y.A., M.M. Don, and S. Bhatia. 2009. Biosorption of copper(II) onto immobilized cells of *Pycnoporus sanguineus* from aqueous solution: Equilibrium and kinetic studies. *J. Hazard. Mater.* 161: 189–195.

Yan, G. and T. Viraraghavan. 2003. Heavy-metal removal from aqueous solution by fungus *Mucor rouxii*. *Water Res.* 37: 4486–4496.

Yee, N. and J. Fein. 2001. Cd adsorption onto bacterial surfaces: A universal adsorption edge? *Geochim. Cosmochim. Acta* 65: 2037–2042.

Yilmazer, P. and N. Saracoglu. 2009. Bioaccumulation and biosorption of copper(II) and chromium(III) from aqueous solutions by *Pichia stiptis* yeast. *J. Chem. Technol. Biotechnol.* 84: 604–610.

Zayed, A.M. and N. Terry. 1994. Selenium volatilization in roots and shoots: Effects of shoot removal and sulfate level. *J. Plant Physiol.* 143: 80–84.

Zhao, M., R. Duncan, and R.P. Van Hille. 1999. Removal and recovery of zinc from solution and electroplating effluent using *Azolla filiculoides*. *Water Res.* 33: 1516–1522.

Zhu, Y.L., A.M. Zayed, J.H. Quian, M. de Souza, and N. Terry. 1999. Phytoaccumulation of trace elements by wetland plants: II. water hyacinth. *J. Environ. Qual.* 28: 339–344.

Ziagova, M., G. Dimitriadis, D. Aslanidou, X. Papaioannou, E.L. Tzannetaki, and M. Liakopoulou-Kyriakides. 2007. Comparative study of Cd(II) and Cr(VI) biosorption on *Staphylococcus xylosus* and *Pseudomonas* sp. in single and binary mixtures. *Biores. Technol.* 98: 2859–2865.

Zulfadhly, Z., M.D. Mashitah, and S. Bhatia. 2001. Heavy metals removal in a fixed-bed column by the macro fungus *Pycnoporus sanguineus*. *Environ. Poll.* 112: 463–470.

7

Removal of Heavy Metals by Seaweeds in Wastewater Treatment

R. Senthilkumar, M. Velan, and S. Feroz

CONTENTS

7.1 Introduction

Metals are introduced into the aquatic systems as a result of the weathering of soils and rocks; from volcanic eruptions; and from a variety of human activities involving mining, processing, or the use of metals and/or substances. The most common metal pollution in freshwater comes from mining companies. Usually, mining companies use an acid mine drainage system to release heavy metals from ores, because metals are very soluble in an acid solution. After the drainage process, they disperse the acid solutions containing high levels of metals into the groundwater. When the pH in the water decreases, the metal solubility increases and the metal particles become more mobile. That is why metals are more toxic in soft waters.

Streams coming from mining areas are often very acidic and contain high concentrations of dissolved metals with little aquatic life. Both localized and dispersed metal pollutions cause environmental damage because metals are nonbiodegradable. Unlike some organic pesticides, metals cannot be broken down into less harmful components in the environment. As a result, the removal of these toxic metals from industrial effluents has become an important priority, which is reflected in the tightening and enforcement of the environmental regulations. Large volumes of industrial heavy metal–bearing wastewater require an efficient and a very cost-effective treatment. While conventional technologies cannot reliably remove trace metals or are too costly to implement, biosorption appears to offer a technically feasible and an economically attractive approach. Therefore, much effort has been directed at identifying a readily available biomass, which in its nonliving state is capable of effectively removing heavy metals. One of the most promising types of biosorbents are the marine algal biomasses, otherwise called seaweeds, in view of their high metal uptake capacity as well as the availability of the biomasses in many parts of the world's oceans.

The biosorption of metals is not based on only one type of mechanism. It consists of several mechanisms that quantitatively and qualitatively differ according to the type of biomass, its origin, and its processing. Metal sequestration may involve complex mechanisms, mainly ion exchange, chelation, adsorption by physical forces, and ion entrapment in the interfibrillar and intrafibrillar capillaries and spaces into the network of the structural polysaccharide cell wall. Thus, the uptake of heavy metals by seaweeds has gained increased credibility in recent years, as it offers a technically feasible and economical approach. It could be considered as an eco-friendly device compared with the existing high cost technologies. Generally, biosorptive processes can reduce the capital costs by 20%, the operational costs by 36%, and the total treatment costs by 28%, compared with the conventional systems (Silva et al. 2009).

The main objective of this chapter is to assess the potential of the different types of marine algae for the biosorption of strategic metals, namely, nickel, cobalt, and copper, as well as two toxic metals, namely, lead and zinc, both in batch and column modes. The equilibrium, isotherm, effect of pH, and temperature were studied. Desorption was also performed to recover metal ions from the biomass. These results would contribute a better understanding of the biosorption phenomena and aid in the development of locally derived potential biosorbents, which possess high capacities for heavy metal uptake from an aqueous solution.

7.2 Selection of Biosorbents

In the process of identifying suitable biosorbents, several important factors were taken into consideration. Availability and low cost are the major factors considered in the selection of biosorbents, while rigidity and mechanical strength are the other factors that are also taken into consideration. Seaweeds possess rigid physical shapes and structures that make their application in the biosorption processes particularly suitable (Vieira and Volesky 2000). Also, seaweeds are plentiful, especially in coastal areas, and can be obtained at a very low cost, as they are already a nuisance in some areas.

7.2.1 Seaweeds

Marine macroalgae, popularly known as seaweeds, have been extensively used in biosorption studies. They show an impressive metal uptake capacity and can be used for a wide variety of metals. Seaweeds offer advantages for biosorption because their macroscopic structures present a convenient basis for the production of biosorbent particles suitable for sorption process applications (Vieira and Volesky 2000). Seaweeds proliferate ubiquitously and abundantly in certain zones of the world's oceans, and they are rather stable and fast growing (Yang and Volesky 1999). At certain ocean locations, they threaten the tourism industry by spoiling the pristine environments and fouling the beaches (Volesky 2001). Alternative solutions, which utilize the potential of seaweeds, are significant and beneficial to the local communities. The biosorption by algae has mainly been attributed to the cell wall, which is composed of a fibrillar skeleton and an amorphous embedding matrix. Both the electrostatic attraction and the complexation of the metals in the biomaterial can play a role (Figueira et al. 2000).

A total of seven species of seaweeds were collected and distributed among the three major groups: green algae (*Ulva reticulate* and *U. lactuca*), brown algae (*Turbinaria conoides*, *T. Ornata*, and *Sargassum wightii*), and red algae (*Gracilaria edulis* and *Gelidium* sp.). The brown color of the Phaeophyta results from the dominance of the xanthophyll pigment fucoxanthin, which masks the other pigments, while the red color of the Rhodophyta is due to the presence of the pigment phycoerythrin, which reflects red light and absorbs blue light. The green color of the Chlorophyta comes from chlorophyll a and b (Hashim and Chu 2004). All the algal samples were washed with distilled water and sun-dried. The samples were ground, except *U. reticulata* and *U. Lactuca*, to the desired particle sizes and were subsequently used for biosorption experiments. In the case of *U. reticulata* and *U. lactuca*, the dry biomass was cut with a knife into irregular-shaped particles between 1 and 3 mm in size and the remaining procedure was the same as that of the other algal samples. Protonation of the algal biomass was carried out by soaking it in 0.1 *M* HCl for 3 h. The biomass was then washed with distilled water and dried at 60°C overnight. In general, protonation of the algal biomass was carried out unless or otherwise specifically stated.

7.2.2 Screening

Experiments were conducted to screen for the best performing algal species for the biosorption of copper, cobalt, nickel, lead, and zinc. The initial metal concentration was fixed at 100 mg/L and the initial solution pH was varied for all five metal ions. Due to the hydrogen ion competition at low solution pH values, the metal uptake was low (Volesky and Schiewer 1999). As the pH increased, the amount of metal uptake increased, and the sharpest increase was observed between the pH of 4 and 6 for copper, cobalt, nickel, lead, and zinc metal ions, for all the seaweed species. All the brown seaweed species exhibited a maximum uptake of copper at a pH of 6, cobalt at a pH of 4.5, nickel at a pH of 4, and lead at a pH of 4.5. In brown seaweeds, the alginate in the cell wall is the main component responsible for metal biosorption (Fourest and Volesky 1997; Davis et al. 2003). Alginate is present in gel form in the cell wall, which appears to be very porous and easily permeable to small ionic species (Dodge 1973).

Both the green seaweeds performed like the brown seaweeds in sequestering zinc ions. The green seaweeds are usually composed of xylans and mannans, which are responsible for metal biosorption. Among the seaweed species examined, *U. reticulata*, in particular, performed well in sequestering zinc and exhibited a maximum uptake at a pH of 5.5. This necessitated the use of *U. reticulata* in the present study of zinc. The red seaweed, *G. edulis*, exhibited the lowest metal uptake among the seaweed species examined. Since red seaweeds have more cationic sites than brown seaweeds, they have a relatively low affinity for the charged metal ions (Hashim and Chu 2004), and hence the use of red seaweed was insignificant for further studies.

Thus, *T. ornata* was selected for further studies on copper biosorption. Both *S. wightii* in the case of cobalt and nickel and *T. conoides* in the case of lead performed well among the seaweed species examined, and both were subsequently chosen for further study. *U. reticulata* was selected for further studies on zinc biosorption.

7.3 Batch Studies

Batch experiments were performed to optimize the particle size, biosorbent dosage, and pH for maximum metal biosorption. The influence of the pretreatment on the biosorption capacity of a given biosorbent was also examined. Batch desorption experiments were performed in an attempt to regenerate the biosorbent.

7.3.1 Sorption

Biosorption experiments were performed in a rotary shaker at 150 rpm using 250 mL Erlenmeyer flasks containing the desired biosorbent in 100 mL of different metal concentrations. After the equilibrium was reached, the mixture was centrifuged at 3000 rpm for 10 min. The metal content in the supernatant was determined using an atomic absorption spectrophotometer (AAS 6VARIO; Analytik Jena, Germany). The amount of metal biosorbed was calculated from the differences between the metal quantity added initially and the metal content of the supernatant, using the following Equation 7.1:

$$Q = V(C_0 - C_f) / M, \tag{7.1}$$

where Q is the metal uptake (mg/g); C_0 and C_f are the initial and equilibrium metal concentrations in the solution (mg/L), respectively; V is the solution volume (L); and M is the mass of biosorbent (g).

7.3.2 Effect of Pretreatment

In all of the brown algae used in the study, the light metals (Na^+, K^+, Ca^{2+}, and Mg^{2+}) were released during the sorption experiments. These light metal ions were acquired from seawater and bound to the acidic functional groups of the algae. When this native biomass was exposed to the metal solutions, the light metal ions were released and the pH tended to increase (Davis et al. 2003). In the protonated biomass, these light metal ions were replaced by protons acquired from the added acidic solution. In some cases, the protonation of the algal biomass may enhance biosorption. This is because some metal cations may prefer exchanging the sites with H^+ ions rather than light metal ions. However, acidification of the biomass may not be suitable for all marine algae. It sometimes deteriorates the algal structure and damages the binding sites, directly affecting the biosorption potential of the algae. Therefore, preliminary experiments were performed to examine the suitability of acidification for the selected algal species. Exposure of 0.1 *M* HCl to the *S. wightii* biomass resulted in a 30% weight loss. Also, the protonated *S. wightii* did not perform well in the cobalt and nickel biosorption compared with its native form. The difference in the capacities was more than 11%. By contrast, *T. ornata* and *T. conoides* actually performed well in their protonated forms compared with their native forms. The biomass weight loss was also insignificant. This may be due to the rigidity of *T. ornate* and *T. conoides*. The morphological characteristics of these algae enable them to survive in extreme environmental conditions. The algal tough thallus is able to withstand the high-energy hydrodynamics of the intertidal environment. The increase in the copper and lead sorption capacity after protonation could be due to the acid treatment, resulting in the production of new binding sites and removal of the ions blocking the binding sites. Considering the results obtained, protonated *T. ornata* (copper), native *S. wightii* (cobalt and nickel), and *T. conoides* (lead) were used in all further studies. *U. reticulate* (zinc) was not protonated, since there was a significant loss of weight in the biomass.

7.3.3 Influence of Particle Size

In general, smaller particles performed well with high uptake and removal efficiencies. In this study, different particle sizes of algae were used. It was found that the performance of the 0.767 mm algal particles performed was very close to that of the 0.456 and 0.598 mm particles. The removal efficiency and the uptake decreased only by 7%–8% in the case of the 0.767 mm algal particles compared with the 0.456 mm particles. Therefore, considering the rigidity and the strength, the 0.767 mm particles were subsequently selected for all further biosorption studies.

7.3.4 Influence of Biosorbent Dosage

The algal dosage strongly influenced both the metal uptake and the removal efficiency of the algae examined. The algae dosages varied between 1 and 10 g/L. In all cases, the uptake capacity decreased and the removal efficiency increased with an increase in the dosage. In the case of the algal biomass, at a high biosorbent dosage, the uptake capacity was severely affected. The effect was not well pronounced in the case of the removal efficiency and only a marginal increase was observed. Therefore, a 2 g/L biosorbent dosage was selected for further studies because it exhibited more than 3.2–3.6 times higher uptake values and only a 35%–38% reduction in the removal efficiencies as compared with 10 g/L.

7.3.5 Influence of Initial Concentration

The prediction of the batch biosorption kinetics is necessary for the design of industrial columns. The copper uptake by the *T. ornata* biomass was found to increase with time and attain maximum values at about 180 min; thereafter, it remained almost constant. Whereas for cobalt and nickel, *S. wightii* biosorbed almost all of the metal ions within 120 min. Similarly, the uptake of lead by the *T. conoides* biomass and the uptake of zinc by the *U. reticulata* biomass were found to increase with time and attain maximum values at about 120 min and, thereafter, proceeded at a slower rate and finally attained saturation. A higher sorption rate at the initial period (2 h) may be due to the increased number of vacant sites available at the initial stage, which results in an increased concentration gradient between the sorbate in the solution and the sorbate in the biosorbent surface. As time increased, this concentration gradient was reduced due to the sorption of the metal ions onto the vacant sites, leading to a decrease in the sorption rate at later stages. For copper, on changing the initial metal concentration from 250 to 1000 mg/L, the biosorbed metal increased from 51.0 to 116.7 mg/g. However, the removal efficiency decreased from 40.8% to 23.3% when the concentration increased from 250 to 1000 mg/L. The same trend was observed for cobalt, nickel, lead, and zinc.

7.3.5.1 Sorption Kinetics

The prediction of the batch sorption kinetics is necessary for the design of the industrial sorption columns. The nature of the sorption process depends on the physical or chemical characteristics of the biosorbent and also on the system conditions. In this study, the applicability of the pseudo-first-order and pseudo-second-order models has been tested for the sorption of all five metal ions onto the selected seaweeds. The best-fit model was selected based on the nonlinear regression coefficient, R^2 values.

The pseudo-first-order kinetic model has been widely used to predict the sorption kinetics (Ho et al. 2005). The expression of the pseudo-first-order kinetics is given by (Equation 7.2)

$$\ln(Q_e - Q_t) = \ln Q_e - k_1 t. \tag{7.2}$$

Thus, the rate constant k_1 (L/min) can be calculated from the slope of the plot of ln (Q_e − Q_t) versus time (t).

The pseudo-first-order model fitted the kinetic data reasonably well for all five metal ions, indicating that the sorption can be approximated to first-order kinetics; however, the model did not predict the equilibrium uptake values in any of the cases examined.

However, good fits were generally observed for the pseudo-second-order model, with correlation coefficients always greater than 0.99. The kinetic data were analyzed using pseudo-second-order kinetics (Ho and McKay 1998), which is represented by (Equation 7.3)

$$\frac{t}{Q_t} = \frac{1}{k_2 Q_e^2} + \frac{1}{Q_e} t, \tag{7.3}$$

where k_2 is the pseudo-second-order rate constant (g/mg min) and Q_e and Q_t represent the metal uptake at equilibrium and at any time t, respectively.

The model also predicted the equilibrium uptake values, which are in close agreement with the experimental values.

7.3.6 Sorption Isotherm

The resulting biosorption isotherms for the equilibrium metal concentration in the algal species at different pH conditions were examined. In all cases, the metal biosorption by the algal species was strongly dependent on the pH values. Among the temperature conditions examined, the room temperature (30°C) favored biosorption. Changing the temperature by ±5°C from room temperature resulted in a decreased metal biosorption capacity.

The solution pH affects the surface charge of the biosorbent, the degree of ionization, and the speciation of the surface functional groups (Reddad et al. 2002). Little uptake at low pH values may be an indication of proton competition with the chosen heavy metal ions. As the pH increases, the competition among the protons decreases and the surface functional groups are activated; thus, the biosorption increases. The lower uptake at higher pH values is probably due to the formation of anionic hydroxide complexes (Maquieira et al. 1994). Because of this effect at higher pH values, the ligands, such as carboxylate and sulfonate, could take up fewer metal ions (Kalyani et al. 2004). Also, during copper and lead biosorption, it was observed that the solution pH tends to decrease at the end of the sorption process. This is because the biomass is protonated, and since ion exchange is the major mechanism responsible for algae biosorption (Davis et al. 2003), the metal ion uptake through the cell wall is accompanied by the release of H^+ ions, which eventually decrease the solution pH.

It is also interesting to note that in the cases of the *S. wightii* biomass and zinc biosorption, the final pH tends to increase. This is because the biomass is used in its native form and the metal ion uptake is accompanied by the release of light metal ions (obtained from seawater), which eventually increase the solution pH. Therefore, the pH data obtained in this study support the fact that ion exchange may be the major mechanism responsible for metal biosorption by marine algae.

7.3.7 Batch Data Modeling

The biosorption isotherm data were fitted using the Langmuir (1918) and Freundlich (1907) models. The Langmuir sorption model was chosen to estimate the maximum metal biosorption by the biosorbent. The Langmuir isotherm can be expressed as (Equation 7.4)

$$Q = \frac{Q_{\max} b C_f}{1 + b C_f}, \tag{7.4}$$

where Q_{max} is the maximum metal uptake (mg/g) and b is the Langmuir equilibrium constant (L/mg).

The Freundlich model is represented by the Equation 7.5:

$$Q = K_F C_f^{1/n}, \tag{7.5}$$

where K_F and n are constants.

Both models fitted the data well, showing correlation coefficients greater than 0.9. For instance, the Langmuir model served to estimate the maximum metal uptake value, where this value could not be attained in the experiments. Both Q_{max} and b were observed to be maximum at optimum pH and room temperature (30°C) for all five metal ions. High values of b are reflected in the steep initial slope of a sorption isotherm, indicating desirable high affinity. Thus, for good biosorbents, a high Q_{max} and a steep initial isotherm slope (i.e., high b) are desirable.

The Freundlich model constant, K_F, which denoted the binding capacity, was observed to be maximum at optimum pH and room temperature (30°C) for all five metals. For the nickel biosorption data, the Langmuir model showed good agreement with the experimental data. The Freundlich model fitted well in the case of copper, lead, cobalt, and zinc. The overall results supported the fact that biosorption was a complex process involving more than one mechanism, where some mechanisms could be submechanisms of the overall mechanisms.

7.3.8 Desorption

The metal-loaded biomass after biosorption was contacted with different elutants in 250 mL Erlenmeyer flasks for 3 h on a rotary shaker (150 rpm) to study the elution of the biosorbed metal ions. The remaining procedure was the same as the biosorption equilibrium experiments. After desorption, the biomass was washed with distilled water, filtered, and dried overnight at 60°C. The loss in the biomass weight was calculated and the biomass was subsequently used for resorption studies.

The metal-loaded algal biomass was eluted using different mineral acids (HCl, H_2SO_4, and HNO_3) and a $CaCl_2$ solution. The seaweed biomasses loaded with approximately 116 mg/g (copper), 98 mg/g (cobalt), and 69 mg/g (nickel) were used separately in the elution experiments. All the mineral acids performed well in eluting the copper from the copper-loaded *T. ornata* biomass. The weight loss was insignificant for the biomass exposed to the mineral acids. This may be due to the high rigidity of the *T. ornata* biomass. By contrast, a mineral acid wash of the cobalt- and nickel-loaded *S. wightii* biomasses resulted in some weight loss even though they performed well in the cobalt and nickel elution. The biomass weight loss was over 30% in all cases at the end of the mineral acid elution process. This biomass weight loss was relatively high, especially when the regeneration process was considered because it affected the metal biosorption capacity of the subsequent cycle (Kuyucak and Volesky 1989b; Davis et al. 2000). The solution of 0.1 *M* $CaCl_2$ performed significantly well in metal elution, especially in the case of cobalt and nickel elution. The elutant, 0.1 *M* $CaCl_2$, exhibited 79.0% and 81.6% elution efficiencies during cobalt elution and nickel elution, respectively, with no significant biomass weight loss. By contrast, 0.1 *M* $CaCl_2$ exhibited only 58.3% copper elution efficiency. Davis et al. (2000) inferred that the elution efficiency of $CaCl_2$ was strictly pH-dependent. The elution performance of $CaCl_2$ was examined at different pH conditions adjusted using 0.1 *M* HCl. For all three metal ions, at lower pH values, $CaCl_2$ performed well. The results clearly suggested that the combined effect of Ca^{2+} and H^+ ions proved to be quite efficient in eluting the metal ions from the biomass. The results of the $CaCl_2$ elution at low pH values were comparable with that of the mineral acid elution.

In the case of the white metals (lead and zinc), the metal-loaded algal biomass was eluted using different mineral acids (HCl, H_2SO_4, and HNO_3) and a $CaCl_2$ solution. The seaweed biomasses, loaded with approximately 420.1 mg/g (lead) and 125.5 mg/g (zinc), were separately used in the elution experiments. Among the various mineral acids examined, HCl exhibited elution efficiencies greater than 99.63% and 99.6% up to a specific S/L ratio of the lead-loaded *T. conoides* biomass. A further increase in the S/L ratio resulted in a considerable reduction in the HCl elution efficiency. The other two acids, HNO_3 and H_2SO_4, were not able to elute all the lead ions from the biomass and also their elution efficiencies were severely affected by the S/L ratios. The weight loss was insignificant for the biomass exposed to the mineral acids. This may be due to the high rigidity of the biomass. By contrast, the mineral acid wash of a zinc-loaded *U. reticulata* biomass resulted in some weight loss. The biomass weight loss was over 15% in all cases of the mineral acid elution process. This biomass weight loss is relatively high, especially when the regeneration process is considered because it affects the metal biosorption capacity of the subsequent cycle (Kuyucak and Volesky 1989b; Davis et al. 2000). The solution of 0.1 *M* $CaCl_2$ at different pH conditions performed significantly well in zinc metal elution. The maximum elution efficiencies for zinc were observed in the pH range of 3–3.5. However, the biomass weight loss was 5.5% and 4% at a pH range of 3 and 3.5, respectively. Also, it is worth mentioning that the S/L ratio severely affects the elution efficiency of $CaCl_2$. For instance, at a pH of 3.5, the elution efficiency of 98% at 1 g/L S/L ratio was dropped to 62.5% at an S/L ratio of 10 g/L. However, it is desirable to use the smallest possible eluting volume in order to contain the highest concentration of the metal. At the same time, the volume of the solution is enough to provide maximum solubility for the desorbed metal. Considering the high elution efficiency and less biomass damage, the solution of 0.1 *M* $CaCl_2$ (in HCl, pH 3.5) at 4 g/L S/L ratio was identified as a practical elutant for zinc desorption.

7.3.9 Regeneration

The biosorption capacity of the regenerated seaweed biomass was examined in five sorption–desorption cycles. The *T. ornata* biomass maintained its copper biosorption capacity in all five cycles examined. Similarly, the *T. conoides* and *U. reticulata* biomasses maintained their lead and zinc biosorption capacity in all five cycles examined. In the case of cobalt and nickel, the regenerated *S. wightii* biomass slightly lost its original uptake capacity because of repeated use. This may be due to the slightly acidic nature of the elutant, which affects the metal binding sites of the biomass. However, the biomass weight loss was insignificant in both cases at the end of the fifth cycle. Thus, *T. ornata* (for copper), *S. wightii* (for cobalt and nickel), *T. conoides* (for lead), and *U. reticulata* (for zinc) were subsequently selected to examine their biosorption potential in column mode.

7.4 Column Studies

The batch experimental results served to study the fundamental information regarding the biosorbent's behavior and its performance in metal biosorption. However, they do not give accurate scale-up data for industrial treatment systems where a continuous flow system is normally employed (Wong et al. 2003). Hence, there is a need to perform biosorption experiments using columns. Column experiments were aimed at optimizing the bed height and the flow rate. Column regeneration experiments were performed to understand the sorption behavior of the regenerated biosorbent in repeated cycles. Column experiments on real industrial effluents were also studied.

In process applications, a packed bed column is effective for the cyclic sorption–desorption process (Yan and Viraraghavan 2001; Aksu and Gönen 2003). This operating mode ensures the highest possible concentration difference in the driving forces for metal biosorption. Starting at the inlet, the saturated solid sorbent zone gradually extends throughout the column, with the sorbent eventually breaking through the column. A record of the breakthrough usually gives a typical S-shaped breakthrough curve, whose shape and slope are mainly due to the equilibrium sorption isotherm and the mass transfer to and throughout the sorbent in the column (Silva et al. 2002).

A relatively low initial concentration of 100 mg/L (compared with the batch experiments) was used in all the column experiments, as a very high metal concentration was not likely to be present in the industrial effluents (Aksu and Kutsal 1998). A low initial concentration would also help obtain gentle breakthrough curves.

7.4.1 Experimental Setup

The continuous-flow sorption experiments were conducted in a glass column. The column was designed with an internal diameter of 2 cm and a length of 35 cm. Since the ratio of the column diameter to the particle diameter is high, the effects of channeling have a negligible effect. At the top of the column, an adjustable plunger was attached with a 0.5 mm stainless sieve. At the bottom of the column, a 0.5 mm stainless sieve was attached followed by glass wool. A 2 cm high layer of glass beads (1.5 mm in diameter) was placed at the column base in order to provide a uniform inlet flow of the solution into the column. Figure 7.1 shows the experimental setup of the column used in this research.

7.4.2 Experimental Procedure

A known quantity of biosorbent was placed in the column to obtain the desired bed height. A metal ion solution of known concentration and pH was pumped upward through the column at a desired flow rate by a peristaltic pump (pp40, Miclins). Samples were collected from the exit of the column at different time intervals and were analyzed for metal concentration using an AAS.

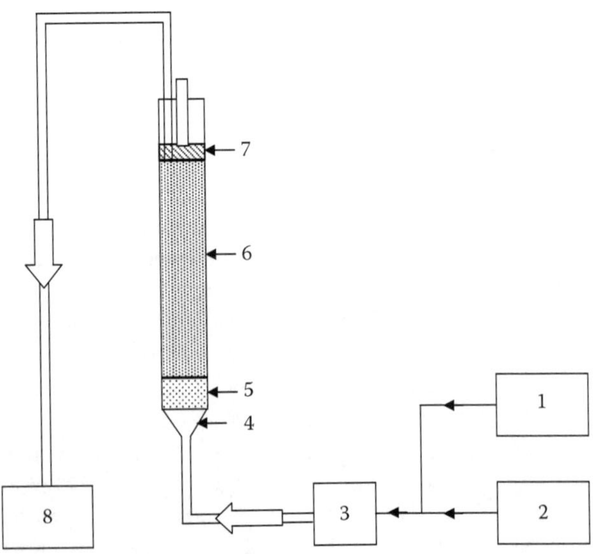

FIGURE 7.1
Experimental setup of the upflow packed bed column: (1) metal solution, (2) elutant, (3) peristaltic pump, (4) glass wool, (5) glass beads, (6) biosorbent, (7) adjustable plunger, and (8) effluent storage.

7.4.3 Analysis of Column Data

The quantity of metal retained in the column (m_{ad}) was calculated from the area above the breakthrough curve (C vs t) multiplied by the flow rate (Aksu et al. 2002). Dividing the metal mass (m_{ad}) by the sorbent mass (M) leads to the uptake capacity (Q) of the biosorbent (Volesky et al. 2003).

The breakthrough time (t_b; the time at which the metal concentration in the effluent reaches 1 mg/L) and the bed exhaustion time (t_e; the time at which the metal concentration in the effluent equals the inlet concentration) were used to evaluate the overall sorption zone (Δt) as follows (Volesky et al. 2003) (Equation 7.6):

$$\Delta t = t_e - t_b. \tag{7.6}$$

The effluent volume (V_{eff}) can be calculated as follows (Aksu and Gönen 2003) (Equation 7.7):

$$V_{eff} = F\,t_e, \tag{7.7}$$

where F is the volumetric flow rate (mL/min).

The total amount of metal ions sent into the column (m_{total}) can be calculated as follows (Aksu and Gönen 2003) (Equation 7.8):

$$m_{total} = \frac{C_0\,F\,t_e}{1000}. \tag{7.8}$$

The total percentage of metal removal with respect to the flow volume can be calculated as follows (Aksu and Gönen 2003) (Equation 7.9):

$$\text{Total metal removal (\%)} = \frac{m_{ad}}{m_{total}} \times 100. \tag{7.9}$$

The metal mass desorbed (m_d) can be calculated from the elution curve (C vs t). The elution efficiency (E) can be calculated as follows (Volesky et al. 2003) (Equation 7.10):

$$E(\%) = \frac{m_d}{m_{ad}} \times 100. \tag{7.10}$$

7.4.4 Column Data Modeling

7.4.4.1 Bed Depth Service Time Model

The analysis of the breakthrough curve was done using the bed depth service time (BDST) model. The BDST model is a simple model for predicting the relationship between the bed height (Z) and the service time (t) in terms of process concentrations and adsorption parameters (Zulfadhly et al. 2001) (Equation 7.11), which is given by

$$\ln\left(\frac{C_0}{C_b}-1\right)=\ln(e^{K_a N_0 Z/\upsilon}-1)-K_a C_0 t. \tag{7.11}$$

Hutchins (1973) proposed a linear relationship between the bed height and the service time, which is given by

$$t=\frac{N_0 Z}{C_0 \upsilon}-\frac{1}{K_a C_0}\ln\left(\frac{C_0}{C_b}-1\right), \tag{7.12}$$

where C_b is the breakthrough metal concentration (mg/L), N_0 is the sorption capacity of the bed (mg/L), υ is the linear velocity (cm/h), and K_a is the rate constant (L/mg h)

The model constants K_a and N_0 can be determined from the plot of Z against t in Equation 7.12.

7.4.4.2 Thomas Model

The successful design of the column sorption process requires the prediction of the concentration–time profile or the breakthrough curve for the effluent (Yan and Viraraghavan 2001). Various mathematical models can be used to describe the fixed bed adsorption. Among these, the Thomas model is simple and has been widely used by several researchers (Yan and Viraraghavan 2001; Aksu and Gönen 2003). The linearized form of the Thomas model is expressed as follows (Equation 7.13):

$$\ln\left(\frac{C_0}{C}-1\right)=\frac{k_{Th} Q_0 M}{F}-\frac{k_{Th} C_0}{F}V, \tag{7.13}$$

where k_{Th} is the Thomas model constant (L/mg h), Q_0 is the maximum solid-phase concentration of the solute (mg/g), and V is the throughput volume (L).

The model constants k_{Th} and Q_0 can be determined from a plot of $\ln[(C_0/C) - 1]$ against t at a given flow rate.

7.4.4.3 Effect of Bed Height

The accumulation of metals in a packed bed column is largely dependent on the quantity of the sorbent inside the column. In all cases, the influence of the bed height was well pronounced in terms of the breakthrough time (t_b) and the exhaustion time (t_e), as both increased with an increase in the bed height.

The slope of the S curve from t_b to t_e (dC/dt) decreased as the bed height increased from 15 to 25 cm, indicating that the breakthrough curve becomes steeper and the mass transfer zone (Δt) becomes shorter as the bed height decreases. Also, the metal uptake capacity and the metal removal percentage of seaweeds increased with an increase in the bed height due to the availability of more binding sites for sorption (Zulfadhly et al. 2001).

Thus, the BDST model plots for the copper (*T. ornata*), cobalt (*S. wightii*), nickel (*S. wightii*), lead (*T. conoides*), and zinc (*U. reticulata*) systems were found to be linear.

7.4.4.4 Effect of Flow Rate

Breakthrough and exhaustion occurred faster at higher flow rates. Also, as the flow rate increased, the metal concentration in the effluent increased rapidly, resulting in much sharper breakthrough curves. The flow rate severely influenced the copper uptake capacity of *T. ornate* and the lead uptake capacity of *T. conoides* at higher values, whereas the cobalt and nickel uptake capacity of *S. wightii* and the zinc uptake capacity of *U. reticulate* were least affected by the flow rate. This may be due to the diffusion limitations of the particular solute into the pores of the algae. Also, the Thomas model gave good fits to the copper, cobalt, nickel, lead, and zinc biosorption data at all flow rates examined.

7.4.5 Desorption

After the column reached exhaustion, the loaded biosorbent with metal ions was regenerated using a selected elutant. After elution, the distilled water was used to wash the bed until the pH of the wash effluent stabilized at around 7.0. Then, the column was fed again with the metal solution and the sorption studies were carried out. After bed exhaustion, the elutant was fed into the column and the regeneration studies were conducted. These cycles of sorption followed by desorption were repeated several times to evaluate the biosorbent resorption capacity.

7.4.6 Regeneration

Column regeneration studies were carried out for five cycles for copper, cobalt, and nickel and for three cycles for lead and zinc. The copper column was packed with 16.21 g of the *T. ornata* biomass, yielding an initial bed height of 25 cm and a bed volume of 78.5 mL with a packing density of 206.5 g/L. Both the cobalt and the nickel columns were packed with 11.73 g of the *S. wightii* biomass to yield a bed height of 25 cm and a bed volume of 78.5 mL with a packing density of 149.4 g/L. Similarly, the zinc column was packed with 6 g of the *U. reticulate* biomass, yielding an initial bed height of 25 cm and a bed volume of 78.5 mL with a packing density of 76.43 g/L.

The lead column was packed with 35.1 g of the *T. conoides* biomass, yielding an initial bed height of 25 cm and a bed volume of 78.5 mL with a packing density of 447.13 g/L. At the end of the fifth cycle, 15.4%, 20.1%, and 24.2% of biomass weight losses were observed in the copper, cobalt, and nickel columns, respectively. Similarly, at the end of the third cycle, 15.4%, 20.1%, and 24.2% of biomass weight losses were observed in the lead and zinc columns.

In general, a decreased breakthrough time and an increased exhaustion time were observed as the cycles progressed. In the case of *S. wightii*, comparing both metal ions, a breakthrough occurred earlier for nickel than for cobalt. Even though *S. wightii* slightly lost its cobalt and nickel biosorption capacity in subsequent cycles, its biosorption performance was always good, indicated by the removal percentage and the volume treated. In the case of the *T. ornata* biomass, it maintained a high copper biosorption capacity in all five cycles. The copper biosorption capacity was even slightly enhanced in the fifth cycle, probably due to the opening of new binding sites.

The elution curves for copper, cobalt, and nickel in all five cycles were examined. The solution of 0.1 *M* HCl always maintained copper elution efficiencies close to 100%. On other hand, 0.1 *M* $CaCl_2$ in HCl exhibited elution efficiencies above 98% for both cobalt and nickel. For nickel, the time for elution was shorter compared with cobalt. This may be due to the comparatively higher acidic nature of the elutant used in the nickel column. The copper, cobalt, and nickel elution processes were carried out at an average of 7.1, 5.3, and 3.7 h cycles. The elution curves observed in all the cycles for all three metal ions exhibited a similar trend, a sharp increase in the beginning followed by a gradual decrease. The solution of 0.1 *M* HCl exhibited elution efficiencies above 99.1% for chromium and lead. The solution of $CaCl_2$ (pH 3.5 HCl) was used for zinc. The elution curves observed in all cycles for all three metal ions exhibited a similar trend, a sharp increase in the beginning followed by a gradual decrease.

7.5 Application to Real Effluents

Any treatment process is only successful when it performs well under real conditions. Therefore, *U. reticulata* was examined in the present study and was checked for its compatibility with real industrial effluents for the removal of zinc. Two zinc-bearing, zinc-phosphating effluents from two different sources were collected. Effluent-1 was characterized by a considerable amount of light metals along with trace amounts of zinc. Effluent-2 was characterized by a relatively low conductivity, total dissolved solids, and total hardness compared with effluent-1 (Table 7.1).

FIGURE 1.1
Common Effluent Treatment Plant in Vatva Industrial Estate in Gujarat (India).

FIGURE 1.2
Krofta spiral scooper.

(a)

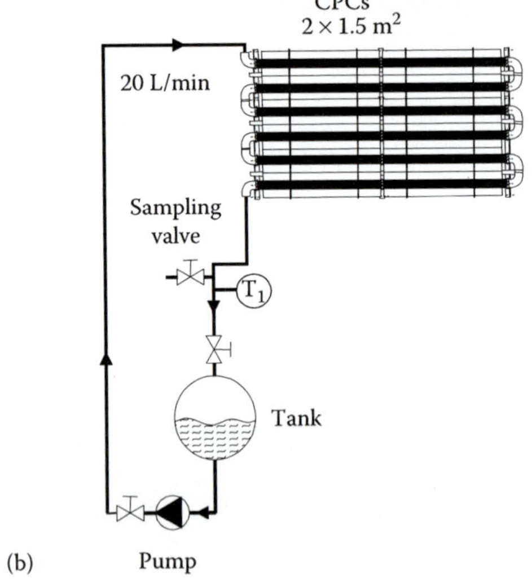

(b)

FIGURE 2.2
(a,b) Solar pilot plant: scheme and view of CPCs.

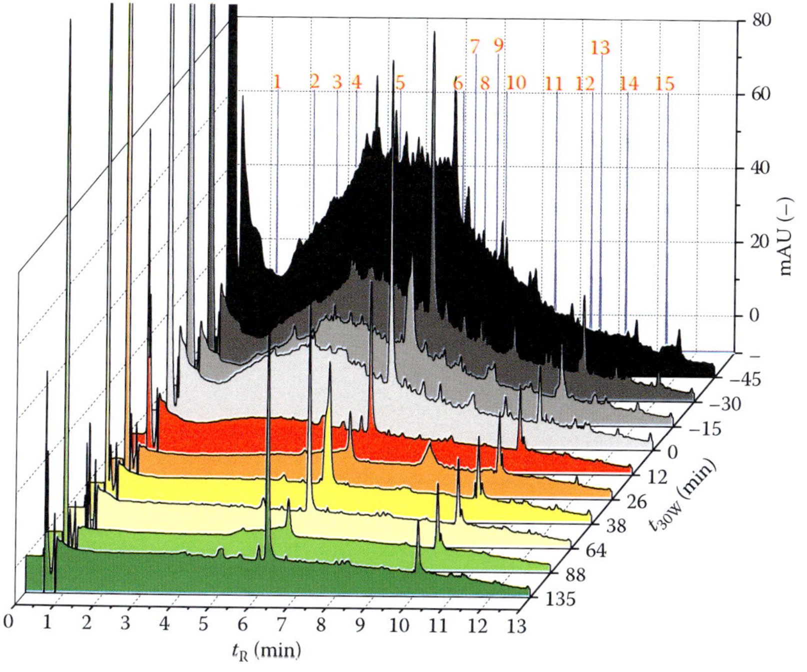

FIGURE 2.4
Degradation profile (HPLC-UV 245 nm chromatograms) of the 15 ECs (5 µg/L each) by photo-Fenton (5 mg/L Fe). 1 Acetaminophen, 2 caffeine, 3 ofloxacin, 4 antipyrine, 5 sulfamethoxazole, 6 carbamazepine, 7 flumequine, 8 ketorolac, 9 atrazine, 10 isoproturon, 11 hydroxybiphenyl, 12 diclofenac, 13 ibuprofen, 14 progesterone, and 15 triclosan.

FIGURE 3.7
Photograph of a CPC reactor at PSA, Spain.

FIGURE 3.8
Photograph showing the CPC with borosilicate glass tubes (PSA, Spain).

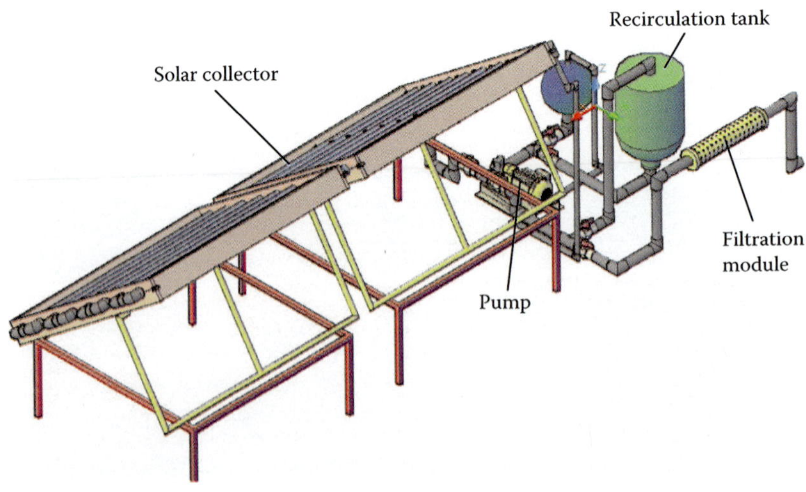

FIGURE 3.9
Schematic of a CPC photocatalytic water treatment system.

FIGURE 10.10
An activated sludge treatment plant. (Courtesy ARTES Ingegneria, Cannon, Italy.)

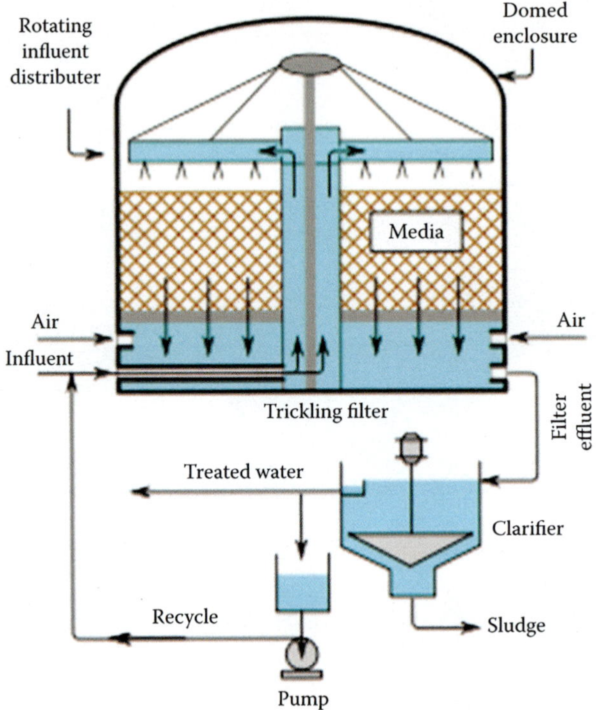

FIGURE 10.11
A trickling filter bed.

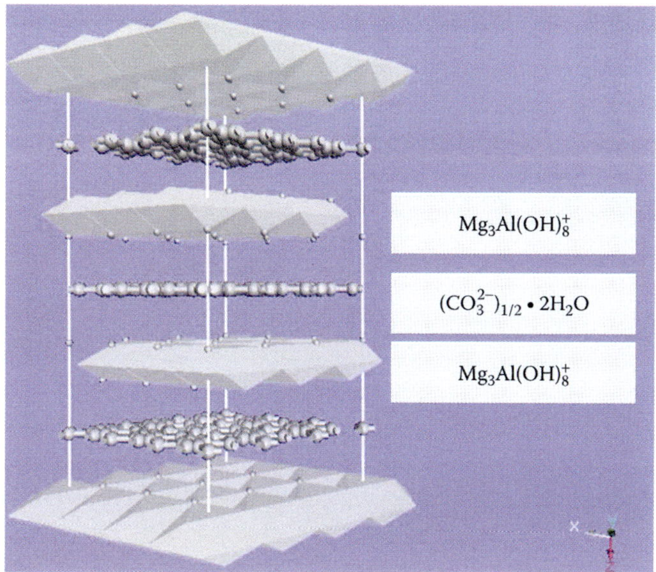

FIGURE 11.7
Schematized interlayer structure of an LDH.

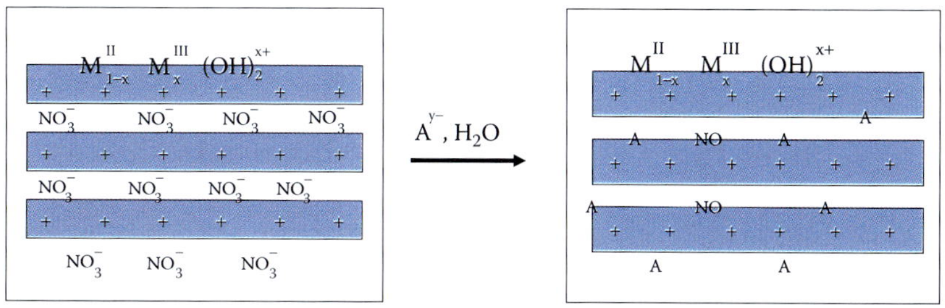

FIGURE 11.8
Anion exchange of LDH where the anion to be intercalated is denoted as A^{y-}.

FIGURE 13.2
Different types of inert carriers.

FIGURE 13.11
Photographs of biological solids attached to the anaerobic fixed bed media. (a) Noncolonized support, (b) colonized support, and (c) support after heat drying.

TABLE 7.1

Characteristics of Zinc-Phosphating Industrial Effluents

Parameter	Effluent-1	Effluent-2
pH	4.75	4.65
Conductivity (mS/cm)	13.86	10.23
Total dissolved solids (mg/L)	2736	2156
Total hardness (as $CaCO_3$; mg/L)	1154	745
Zinc (mg/L)	105	51
Sodium (mg/L)	480	285
Dissolved phosphate (mg/L)	0.25	<0.1
Chloride (mg/L)	605	310
Sulfate (mg/L)	258	198
Total chromium (mg/L)	<0.05	<0.05
Nickel (mg/L)	4.5	<0.01

7.5.1 *U. Reticulata*

Since the batch experimental results are useful for the study of the fundamental information regarding the biosorbent behavior and its performance in metal biosorption, further experiments were carried out in column mode. Zinc-bearing synthetic solutions, prepared from $ZnSO_4{\cdot}7H_2O$, were used to compare the sorption behavior of *U. reticulate* on real and synthetic zinc effluents. Relatively smooth breakthrough curves were observed for 51 and 105 mg/L synthetic zinc solutions. However, *U. reticulate* showed a slightly lower sorption capacity with real effluents. In the case of effluent-1, *U. reticulate* recorded a zinc uptake of 67.5 mg/g, whereas for the synthetic solution (105 mg Zn^{2+}/L of distilled water), *U. reticulate* recorded 77.1 mg/g. In terms of the percentage of zinc removed, *U. reticulate* exhibited 60.98% and 66.21% for effluent-1 and the synthetic solution, respectively. Zinc uptakes were 65.48 and 76.43 mg/g for effluent-2 and the synthetic solution (51 mg Zn^{2+}/L of distilled water), respectively. Also, *U. reticulate* maintained good zinc removal percentages of 83.26% and 85.59% for effluent-2 and the synthetic solution, respectively.

On comparing the two effluents on the basis of the percentage of zinc removed, it was observed that *U. reticulate* performed relatively better on effluent-2. It is a well-known fact that metal sorption largely depends on the solution chemistry of the metals and the competing ions (Volesky and Schiewer 1999). The presence of considerable amounts of nickel in effluent-1 may have had a negative effect on the zinc uptake, as they compete to occupy the binding sites. For the duration of the sorption experiments up to column exhaustion (in terms of zinc concentration), the exit concentrations of zinc were always below 100 µg/L. The excess amount of light metal ions (Na^+) and the total hardness (in terms of $CaCO_3$) in effluent-1 may have influenced the zinc binding. However, Volesky and Schiewer (1999) inferred that light metals generally bind less strongly than heavy metal ions and, therefore, do not strongly interfere with the heavy metal binding. The presence of anions can lead to the following: (1) formation of complexes that have a higher affinity to the sorbent than the free metal ions (i.e., an enhancement of sorption) and (2) formation of complexes that have a lower affinity to the sorbent than the free metal ions (i.e., a reduction of sorption) (Volesky and Schiewer 1999). However, in most cases of biosorption, metal binding tends to be reduced in the presence of anions (Kuyucak and Volesky 1989a; Ishikawa et al. 2002). Other parameters, such as conductivity and total dissolved solids, can also be cited as reasons for the significant deviation in the zinc uptake from the two effluents.

In regeneration studies, *U. reticulate* was reused for three sorption–desorption cycles. In practical applications, the loading of the biosorption column has to be stopped as soon as the metal ion concentration in the effluent exceeds the regulatory limit (Kratochvil and Volesky 1998). Therefore, in regeneration cycles, the column operation was stopped when the zinc concentration in the effluent exceeded 1 mg/L. Table 7.2 summarizes the breakthrough time and the zinc uptake obtained for the two effluents during the three cycles.

As the cycles progressed, decreased breakthrough time and exhaustion time were observed. *U. reticulate* maintained a relatively good zinc biosorption capacity for effluent-2 in the three cycles examined. A decline of 3.023% in the zinc uptake for effluent-2, compared with a 4% decline for effluent-1, was observed at the end of the third cycle. Also, the total volume of effluent-2 treated during the three cycles was nearly 1.4 times that of effluent-1. However, no major decrease in the bed height was observed at the end of the third cycle. This supports the fact that the loss of sorption performance was not primarily due to sorbent damage, but was due to the sorbing sites whose accessibility becomes difficult as the cycles progress (Volesky et al. 2003).

The elutant used, 0.1 *M* $CaCl_2$ (pH 3.5, HCl), provided elution efficiencies greater than 99%. The elution process resulted in very high concentrated metal solutions in the early part, followed by a gradual decrease in metal concentration. Similar trends were observed in all cycles for both the

TABLE 7.2

Column Parameters for Two Zinc Effluents during Three Regeneration Cycles Using *U. Reticulata*

Effluent	Cycle No.	t_b (h)	Uptake (mg/g)	Z (cm)	V_{eff} (L)	Elution Time (h)	Elution Efficiency (%)
Effluent-1	1	11.0	67.50	25.0	6.3	4.50	99.48
	2	10.0	66.20	25.0	6.0	4.25	99.8
	3	9.5	64.80	24.9	5.7	4.00	99.5
Effluent-2	1	20.0	65.48	25.0	9.0	4.00	99.9
	2	19.5	64.00	25.0	8.4	3.75	99.4
	3	18.0	63.50	24.9	8.1	3.50	99.7

effluents. The time for elution decreased as the cycles proceeded, indicating that less metal ions were available for elution and also that they were loosely bound to the sorbent in successive cycles. The overall achievement of the biosorption process is to concentrate the metal solution. This is assessed by expressing a simple overall process parameter, the concentration factor (Volesky and Schiewer 1999). The concentration factor is defined as the ratio of the total volume of effluent treated (in the sorption process) to the total volume of desorbent used (in the elution process). The overall concentration factors for the entire three cycles were 4.7 and 7.5 in the case of effluent-1 and effluent-2, respectively. This result was as expected, because the concentration factor strongly depends on the initial metal concentration. The higher the initial metal concentration, the lower is the concentration factor, and vice versa. Thus, this study proved that *U. reticulate* can efficiently remove zinc from zinc-phosphating industrial effluents.

7.6 Conclusions

The evidence from this research suggests that biosorption is a viable process for the removal of copper, cobalt, nickel, lead, and zinc from aqueous solutions. The biosorbents examined in the study exhibited a superior biosorption capacity in the batch and column modes of operation. The findings of this research correspond well and, in some cases, are even superior to the results found in the literature. In the case of seaweeds, interesting results were obtained during screening experiments. Brown seaweeds used in the study always excelled over green and red seaweeds for copper, cobalt, nickel, and lead metal ions. Green seaweeds excelled for zinc compared with brown and red seaweeds. Due to the low M/G ratio, *T. conoides* performed very well in lead biosorption. Due to its good biosorption capacity, *U. reticulata* performed very well in zinc biosorption.

Cost-effectiveness is the main attraction of metal biosorption. A sorbent can be assumed to be of "low cost" if it requires little processing, is abundant in nature, or is a by-product or waste material from another industry. Seaweeds are plentiful, fast growing, and exist in many parts of the world's oceans. However, some seaweeds possess commercial importance, and care must be taken in selecting seaweeds for metal biosorption. In this study, the selected seaweeds are either commercially unimportant or are plentiful. In some cases, the biomass was acid-washed. Therefore, the biosorbents used in this study are inexpensive, effective, and readily available.

Thus, in this study, different biosorbents were identified based on their biosorption capacity. The application of the selected biosorbents for the removal of zinc from two different zinc-phosphating industrial effluents was successfully demonstrated. Hence, biosorption can act as a potential weapon against metal-bearing industrial effluents.

References

Aksu, Z. and F. Gönen. 2003. Biosorption of phenol by immobilized activated sludge in a continuous packed bed: Prediction of breakthrough curves. *Process Biochem.* 39: 599–613.

Aksu, Z., F. Gönen, and Z. Demircan. 2002. Biosorption of chromium(VI) ions by Mowital B30H resin immobilized activated sludge in a packed bed: Comparison with granular activated carbon. *Process Biochem.* 38: 175–186.

Aksu, Z. and T. Kutsal. 1998. Determination of kinetic parameters in the biosorption of copper(II) on *Cladophora* sp. in a packed bed column reactor. *Process Biochem.* 33(1): 7–13.

Davis, T.A., B. Volesky, and A. Mucci. 2003. A review of the biochemistry of heavy metal biosorption by brown algae. *Water Res.* 37: 4311–4330.

Davis, T.A., B. Volesky, and R.H.S.F. Vieira. 2000. *Sargassum* seaweed as biosorbent for heavy metals. *Water Res.* 34: 4270–4278.

Dodge, J.D. 1973. *The Fine Structure of Algal Cells*. London: Academic Press.

Figueira M.M., B. Volesky, V.S.T. Ciminelli, and F.A. Roddick. 2000. Biosorption of metals in brown seaweed biomass. *Water Res.* 34(1): 196–204.

Fourest, E. and B. Volesky. 1997. Alginate properties and heavy metal biosorption by marine algae. *Appl. Biochem. Biotechnol.* 67: 33–44.

Freundlich, H. 1907. Ueber die adsorption in *loesungen. Z. Phys. Chem.* 57: 385–470.

Hashim, M.A. and K.H. Chu. 2004. Biosorption of cadmium by brown, green, and red seaweeds. *Chem. Eng. J.* 97: 249–255.

Ho, Y.S., T.H. Chiang, and Y.M. Hsueh. 2005. Removal of basic dye from aqueous solution using tree fern as a biosorbent. *Proc. Biochem.* 40: 119–124.

Ho, Y.S. and G. McKay. 1998. Sorption of dye from aqueous solution by peat. *Chem. Eng. J.* 70: 115–124.

Hutchins, R.A. 1973. New method simplifies design of activated carbon systems. *Chem. Eng.* 80: 133–138.

Ishikawa, S., K. Suyama, K. Arihara, and M. Itoh. 2002. Uptake and recovery of gold ions from electroplating wastes using eggshell membrane. *Biores. Technol.* 81: 201–206.

Kalyani, S., P. Srinivasa Rao, and A. Krishnaiah. 2004. Removal of nickel(II) from aqueous solutions using marine macroalgae as the sorbing biomass. *Chemosphere* 57: 1225–1229.

Kratochvil, D. and B. Volesky. 1998. Advances in the biosorption of heavy metals. *TIBTECH* 16: 291–300.

Kuyucak, N. and B. Volesky. 1989a. Accumulation of cobalt by marine alga. *Biotechnol. Bioeng.* 33: 809–814.

Kuyucak, N. and B. Volesky. 1989b. Desorption of cobalt-laden algal biosorbent. *Biotechnol. Bioeng.* 33: 815–822.

Langmuir, I. 1918. The adsorption of gases on plane surfaces of glass, mica and platinum. *J. Am. Chem. Soc.* 40: 1361–1403.

Maquieira, A., H.A.M. Elmahadi, and R. Puchades. 1994. Immobilized cyanobacteria for online trace metal environment by flow injection. Atomic absorption spectrometry. *Anal. Chem.* 66: 3632–3638.

Reddad, Z., C. Gerente, Y. Andres, and P.L. Cloirec. 2002. Adsorption of several metal ions onto a low-cost biosorbent: Kinetic and equilibrium studies. *Environ. Sci. Technol.* 36: 2067–2073.

Silva, B., H. Figueiredo, I.C. Neves, and T. Tavares. 2009. The role of pH on Cr(VI) reduction and removal by *Arthrobacter Viscosus*. *Int. J. Chem. Biol. Eng.* 2(2): 100–103.

Silva, E.A., E.S. Cossich, C.R.G. Tavares, L.C. Filho, and R. Guirardello. 2002. Modeling of copper(II) biosorption by marine alga *Sargassum* sp. in fixed column. *Process Biochem.* 38: 791–799.

Vieira, R.H.S.F. and B. Volesky. 2000. Biosorption: A solution to pollution. *Int. Microbiol.* 3: 17–24.

Volesky, B. and S. Schiewer. 1999. Biosorption of metals. In *Encyclopedia of Bioprocess Technology*, eds. Flickinger, M. and S.W. Drew. New York: John Wiley, pp. 433–453.

Volesky B. 2001. Detoxification of metal-bearing effluents: Biosorption for the next century. *Hydrometallurgy* 59: 203–216.

Volesky, B., J. Weber, and J.M. Park. 2003. Continuous-flow metal biosorption in a regenerable *Sargassum* column. *Water Res.* 37: 297–306.

Wong, K.K., C.K. Lee, K.S. Low, and M.J. Haron. 2003. Removal of Cu and Pb from electroplating wastewater using tartaric acid modified rice husk. *Process Biochem.* 39(4): 437–445.

Yan, G., T. Viraraghavan, and M. Chen. 2001. A new model for heavy metal removal in a biosorption column. *Adsorp. Sci. Technol.* 19(1): 25–43.

Yang, J. and B. Volesky. 1999. Biosorption of uranium on Sargassum biomass. *Water Res.* 33(5): 3357–3363.

Zulfadhly, Z., M.D. Mashitah, and S. Bhatia. 2001. Heavy metals removal in fixed-bed column by the macro fungus *Pycnoporussanguineus*. *Environ. Pollut.* 112: 463–470.

8

Microbial Treatment of Heavy Metals, Oil, and Radioactive Contamination in Wastewaters

Sourish Karmakar, Arka Pravo Kundu, Kanika Kundu, and Subir Kundu*

CONTENTS

8.1 Introduction

Since the early twentieth century, there has been a rapid growth in industrialization. Simultaneously, the world population has increased many folds. Each of these has contributed to large amounts of waste that are harmful to the environment as well as to the human population. The mining

* Professor (Dr.) Subir Kundu, School of Biochemical Engineering, Institute of Technology, Banaras Hindu University, Varanasi 221005, India. E-mail: subirbhu@gmail.com; skundu.bce@itbhu.ac.in

industry produces a large amount of toxic dumps that need immediate attention. In the mid-twentieth century, the world experienced an increase in the use of nuclear energy and polymers. These technologies lead to unknown toxic wastes.

These toxic wastes require longer periods of time to decompose. As no technology has been developed to date to efficiently treat these wastes, underground/deep-sea disposal is carried out. Today, there is an urgent need to develop technology to treat these wastes before they become a serious environmental hazard. Nuclear technology has tremendous potential; however, there is no technology available to prevent widespread radioactive leakage accidents. The nuclear accidents at Hiroshima, Nagasaki, Chernobyl, Three Miles Island, and the recent Fukushima nuclear plant have left humankind worried about nuclear safety. The world has also faced major disasters through oil spills. The bombardment of the oil wells in Iraq and the oil leakage of the British Petroleum facility in the Mexican Gulf are among the examples that have endangered the nearby ecosystems. Such incidents pose a major danger to the existence of the human race.

The current chapter focuses on the treatment of these wastes through biological processes. A large number of studies on the treatment of these wastes are ongoing. However, most of the time, the process is either too costly or it is time-consuming. As the handling of these wastes is also very tricky, often the research remains in the hands of governments. Very few of these researches are released into the public domain.

Chemolithotrophic and chemoorganotrophic are two domains that are mainly involved in wastewater treatment. They use chemical energy instead of sunlight to function. There are facultative anaerobes and obligate anaerobes in extremophiles. Most of the time, the facultative anaerobes are used in wastewater treatment because they can use chemical sources instead of oxygen as an electron acceptor, which work well in wastewater conditions because of their high chemical oxygen demand (COD). Another advantage of using the facultative anaerobes is their ability to adapt to an aerobic environment within a few hours, which makes them highly useful even if the aerated wastewater treatment processes are used along with the biological treatment processes. The primary focus of this chapter is the treatment of three kinds of contamination in wastewater, mining waste-contaminated wastewater, oil sludge-contaminated wastewater, and radioactive mineral-contaminated wastewater, using these extremophiles. All these wastes have a high COD value, making conventional chemical treatment very costly. Moreover, the excessive amount of these wastes makes the treatment more costly. The chemolithotrophic and chemoorganotrophic microorganisms work in tandem to remove these contaminations. The use of extremophiles makes it easier and cheaper to treat these wastes as they require minimal conditions to survive and can grow in conditions of extreme pH and high mineral toxicity.

8.2 Mining Wastes

Mining techniques have greatly developed, and even the lowest-grade ore is now used for extraction. The wastes generated thus vary considerably both in volume and in their properties because of the different mining conditions employed, the composition of the mined ore or rock, and the techniques used to extract the metals. Some of the environmental threats from the mine wastes are the releases from the deposits. These are acid or alkaline drainage, toxic substances (metals, radionuclide), dust, and suspended solids. These cause difficulty in revegetating the area because of the deficiency in nutrients or an excess of toxic substances, apart from the terrestrial difficulties. The contaminants from the ore and rock and the process chemicals (flotation reagents, surfactants, leaching agents, oxidants, extractants, etc.) are also detrimental to the natural flora and fauna of the environment.

Microorganisms can affect the mobility of the metals in the mine waste via metal mobilization or metal retardation. The effect of microorganisms on metal mobility is dependent on various environmental conditions. These types of effects also need special attention as they create major metal displacement in mine waste environments.

Mining waste falls into two categories: mine tailing, which is generated during ore processing, and the volume of rocks produced because of the mining operations. The process of extraction involves grinding, which generates volumes of wastes. The accumulated wastes are released as slurry in a nearby retention pond. In the case of low-grade mining ore, almost the complete volume extracted becomes tailings. Tailings are detrimental to the environment and need to be treated or, even better, prevented (Ritcey 1989). The other form of waste is rocks, which are extensively used in construction works. However, ore containing sulfide and other toxic residues needs treatment. Older waste deposits often have higher amounts of this valuable element because of the poor extraction techniques used in earlier days. These ores can be reutilized to extract the residual amount of minerals using modern techniques.

Carbon is the basic element of life. It has the ability in making covalent bonds as well as sharing electron pairs with other elements. The only natural way that new fixed carbon is formed is via autotrophy, which fixes atmospheric carbon as carbon dioxide to a fixed organic carbon. When consumed by other organisms (heterotrophy), these autotrophs or their products replenish the carbon, allowing further growth and the formation of newer cells. Some bacteria and algae also have this property. These are the primary fixers, which convert inorganic carbon to organic carbon, aiding the evolution of more complex carbon-dependent life forms. Unlike plants, they mostly do not require sunlight for fixing and are known as chemoautotrophs. They are basic producers or fundamental life forms as they fix energy from the sunlight to organic carbon. Table 8.1 lists the types of microorganisms found in

TABLE 8.1

Types of Microorganisms Involved in Typical Waste Treatment

Carbon Source	Types of Organisms Based on Energy Source	Electron Source	Examples
Inorganic carbon (CO_2)	Photolithotrophic autotrophs (light-based energy source)	Water	Algae and cyanobacteria
		H_2S, sulfur, oxygen, hydrogen	Purple and green bacteria (sulfur based)
	Chemolithotrophic autotrophs (chemical-based energy source)	Metal ions, H_2S, sulfur, hydrogen, ammonium ion, nitrite ions	Nitrogen ion oxidizing bacteria, metal-oxidizing bacteria, sulfur-oxidizing bacteria, hydrogen-oxidizing bacteria, and methanogens
Organic compounds	Photoorganotrophic heterotrophs (light-based energy source)	Organic compound	Purple and green bacteria (nonsulfur)
	Chemoorganotrophic heterotrophs (chemical-based energy source)		General bacteria and fungi

Source: Pirt, S.J., *Principles of Microbe and Cell Cultivation*, Blackwell Scientific, Oxford, 1975.

typical mine sites. Table 8.1 clearly indicates that the utilization of inorganic carbons by microorganisms may facilitate the uptake of metals from wastes as their electron source. The above property can affect the metal mobilization.

As there is a scarcity of light sources in almost all parts of the mine waste environments, the production of new organic compounds mostly occurs via chemical-dependent pathways. The heterotrophic microorganism carries the reverse process from organic to inorganic carbon form or the carbon dioxide production, also sometime called mineralization. These microorganisms include mostly bacteria as they can survive in a diverse range of environments. These bacteria can convert almost all the carbon sources taken as the energy source to atmospheric carbon dioxide.

8.2.1 Selection of Microorganisms for Wastewater Treatment

Microorganisms have a wonderful ability to adapt themselves to the environment. For billions of years, the Earth's atmosphere was made up of hydrogen sulfide, methane, cyanides, ammonia, CO, CO_2, and other gases. It took billions of years to reach today's oxygen-rich environment. Growing under extreme conditions can be an excellent method for selection. As the genetics of microorganisms holds the same capability to survive in the extreme, they can be selected to survive by utilizing and transforming the compounds present in mining wastewater. The selection also ensures their survivability among toxicants. The concentration of toxic materials is very high in the mining environment while it is scarce in organic matter. Another extremity is due to the acid mine drainage (AMD; Ledin and Pedersen 1996). Both aerobic and anaerobic conditions can prevail, depending on the mining type. Underground mines are abundant in methane but lack oxygen. However, open mines do not have oxygen limitations. Therefore, if the microorganisms are properly selected, they can be an efficient tool for the treatment of wastewater. It is always better to use groups of microorganisms as this may promote mutualism for common survival. These microorganisms can survive in the presence of each other, forming complex pathways among themselves.

8.2.2 Types of Mining Wastes: Inorganic

A major group of microorganisms depends on oxygen as the electron acceptor. However, very few can adapt to survive on ferric and nitrate ions. They are called chemolithotrophs and are abundant in the mining environment (Kuenen and Bos 1989). The mining environment has a scarcity of oxygen and an abundance of methane and other gases, which preferably select these microorganisms based on their survivability. They are nature's most potent way of controlling inorganic wastes by oxidizing them for their main energy source. Most of these can fix carbon dioxide from the environment to organic carbon.

8.2.3 Types of Mining Wastes: Organic

The heterotrophic organism consumes organic carbon as its main energy source. It is degraded to inorganic compounds and CO_2. Under an abundance of oxygen, these microorganisms can completely convert organic carbon to CO_2. Under mining conditions, the matter differs completely. The scarcity of oxygen shifts them from aerobic to anaerobic mode. Biomineralization is carried out through complex pathways with many interactions among the microorganisms. The rate of each step varies with the growth rate of the previous class of microorganisms involved in the complex pathway. This is similar to human beings living in a society where each group performs specific roles by virtue of which the whole society benefits. The organic carbon can be converted to inorganic CO_2 via oxidative pathways. Under an oxygen-deficient condition, organic wastes are converted to methane or acetate via redox pathways. The product formation completely depends on the microbial flora of the environment.

8.2.4 Types of Mining Wastes: Metal Excess

The metals in mines stay either in the mobile phase or in the immobile phase. Mining wastewater contains excess metals, which makes it an important part of the removal process. Microorganisms highly affect this phenomenon and thus can be utilized in the waste treatment in metal dumps. Microorganisms can reduce the flow of the mobile metals washed out of the metal dumps to prevent contamination in nearby water resources. Metals can be in a mobile phase or a stationary phase, and the types of microorganisms are to be suitably selected. Microorganisms living as biofilms are slow in adsorbing metals, but free-living microorganism can have a higher adsorbing power. Some bacteria also produce compounds such as sulfides that can precipitate the metal ions and remove them from the mining tail (Widdel 1988). Complex processes may occur while metal removal takes place. The microorganism may be directly involved with the metal uptake and metal transformations. Microorganisms have many functional groups on their cell surface that can passively adsorb metal ions. In the neutral pH range, cells are generally negative in charge that facilitates cation binding. The intracellular uptake of metals via adenosine triphosphate (ATP)-depending pathways has also been reported (Beveridge 1986). Some chemolithotrophs produce bulk organic compounds that can bind with metal ions that are present in wastewater and can make them immobile, which can be separated out. These types of metallo-organic compounds are tough to degrade. Therefore, either a conventional treatment for metal removal can be carried out or these compounds are treated with multiple microorganism systems with high metal-tolerating microorganisms that can take up the metals in one chamber, with other microorganisms oxidizing the organic compounds in a highly aerated chamber. Acidophilic algae, *Euglena* sp., are abundantly available in the drainage waters from mine waste in Ontario, Canada (Mann et al. 1989a). These microorganisms are capable of taking up the aqueous metal forms. Fe, Ni, and Cu were found in high concentrations in these microorganisms. The sediments also had a higher concentration of these metal ions. Microorganisms were responsible for the nucleation of iron oxide and oxyhydroxide minerals in iron-rich sediments (Mann et al. 1989b). These bioprecipitated Ti oxides, Fe oxides, and oxyhydroxides acted in the removal of heavy metals, such as copper, lead, nickel, and thorium (Mann and Fyfe 1989). Alkylation or reduction/oxidation can also be used for metal transformation. The transformation of iron and sulfur also affects the mobility of metals. Microbial alkylation of mercury is a major mechanism that prevents the escape of volatile elements. Acidophilic *Thiobacillus ferrooxidans* carry out the oxidation of iron to use as their sole energy source. They are found in an acid-rich environment and can survive in the highly acidic conditions of AMD from mine tails. They can also survive by utilizing the reduced forms of sulfur, such as H_2S and metal sulfides. They can use CO_2 as a carbon source and ammonia or nitrate as a nitrogen source. They can be both mesophilic and thermophilic. *Leptospirillum ferrooxidans* belong to a class of mesophiles that can survive on ferrous ion. Thermophilic *Sulfobacillus thermosulfidooxidans* can utilize Fe^{2+}, S^0, or metal sulfides for their energy (Harrison 1984; Golovacheva and Karavaiko 1978).

In sulfate-rich areas, a sulfate-reducing bacteria in an anoxic environment interacts with metal ions to make insoluble compounds. HPO_4^{2-} producing *Citrobacter* sp. can precipitate out the metals, thus helping in their complete removal (Brierley and Lanza 1985). The acid-producing microorganism can do the reverse, which can be used to treat the wasteland. Microorganisms produce two main types of metal-complexing agents. Either they are the byproducts of the metabolic systems or they are the exudates that are released due to the scarcity of the metal needed in microorganism's metabolism. Recently, siderophores have gained attention due to the metal complexing property. They have a high affinity to chelate Fe^{3+} and a low affinity to chelate Fe^{2+}. Despite gaining high popularity as an iron-chelating compound, siderophores can also chelate out many metals and facilitate their uptake in cells (Bossier et al. 1988).

8.2.5 Isolation of Microorganisms for Wastewater Treatment

Microorganism-utilizing metals can be easily found in metal dumps and mine drainage reservoirs, which are rich in all types of flora, ranging from archaic to high-metal-tolerating acidophiles and halophiles, as shown in Table 8.2. These types of mine sites are rich in microorganisms that can be utilized for metal wastewater treatment. Some of these can survive completely in the presence of inorganic wastes, while some can utilize the organic wastes. The main factor of diversity is the metal ion availability; thus, each type of dump has different flora and water availability. The amounts are drastically reduced in summer due to the lack of water.

8.2.6 Wastewater Treatment of Mining-Generated Wastes

Mining wastewater treatment is a collection of processes. It has separate processing mills and stations. Many of the mining companies recycle the same water to treat and process the metals, thereby reducing the drainage to normal inland water resources. There is no hard and fast protocol for the treatment. It completely depends on the mineral composition of the wastewater. There are certain conditions when distributing the mine tails in fractions that can also help in effective wastewater treatment. The acidity can be reduced by removing the sulfide compounds in the mill itself before releasing it into the tails. These sulfides can be stored in underground dumps or they can be used to produce sulfuric acid with the help of acidophiles. The microbial oxidation of the iron pyrite–rich coal dumps can produce both sulfuric acid and iron. Metal treatment can be done in both naturally and artificially induced conditions.

Nature has an effective way of converting iron pyrites to Fe^{2+}, SO_4^{2-}, and H^+ in the presence of a microorganism, such as *Thiobacillus* sp., and diffused oxygen (Figure 8.1). This is the main cause for the formation of AMD (Berthelin 1983).

There are several ways to stop the cycle to reduce the acidity of the AMD and the iron loss. The foremost effective way is by chemical treatment (Pulford 1991) (Figures 8.2 and 8.3) in which an iron-binding chemical is poured into the

TABLE 8.2
Distribution of Metal-Tolerant Microorganisms

Microorganism(s)	Environmental Status	Availability	Reference
Acidiphilium sp.	Aerated as well as inorganic in commensalism with autotrophs	Acid mine drainage, mine dumps of low pH in the range of 2–3	Wichlacz and Unz (1981)
Acidianus sp.	Acidophilic environment surviving on hydrogen oxidation	Acid mine drainage	Kelly (1988)
Cladosporium, *Penicillium*, *Trichosporon*, and *Rhodotorula*	Heterotrophic organisms	Solid waste dumps, waste rocks, coal mines, mine tails	Groudev and Groudeva (1993)
Leptospirillum ferrooxidans and *T. ferrooxidans*	Aerated environment	Waste rock piles, mining tails rich in iron	Guay et al. (1992)
Metallogenium sp.	Acidophilic environment	Acid mine drainage, mine dumps of pH 3–4	Hughes and Poole (1989)
Psuedomonas, *Bacillus*, *Aerobacter*, and *Caulobacter*	Heterotrophic organisms	Mine tails, acid mine drainage	Groudev and Groudeva (1993)
Radium-leaching sulfate-reducing bacteria	Acid-rich environment	Topsoil of uranium mine tails	Goodman et al. (1981)
Sulfobacillus sp.	Thermophilic environment rich in reduced sulfur and ferrous ion	Self-heating ores and coal mines	Ehrlich (1990)
Sulfolobus sp.	Highly acidophilic environment	Hot springs and coal dumps, acid mine drainage, and dumps in the range of pH 1–2	Hughes and Poole (1989)
Sulfobacillus thermosulfidooxidans	Thermophilic and chemolithotrophic environment	Pyrite rich dump	Ehrlich (1990)
Thiobacillus sp.	Aerated environment	Waste rock piles, mining tails rich in iron	Goodman et al. (1981) and Johnson et al. (1979)
Thermoplasma	Anaerobic thermophilic environment	Coal mine, iron pyrite mines, copper pyrite mines	Prescott et al. (1993)
T. thiooxidans	Inorganic wastes environment	Sulfur-rich mine tails, waste rock piles, iron pyrite ores	Yates and Holmes (1986) and Johnson et al. (1979)

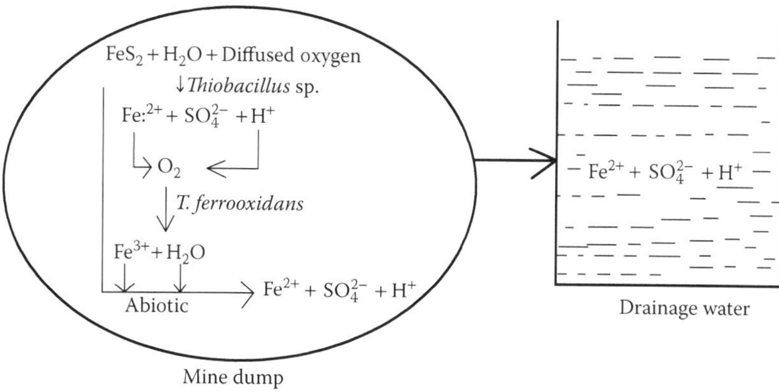

FIGURE 8.1
A natural process occurring in a microbe-rich environment in proper aeration causing acidic drainage.

ore dumps that does not allow free Fe^{2+} and Fe^{3+} ions. Sometimes, alkalis are poured into the dump to reduce the acidity. The main advantage of the process is its simplicity and it is less time-consuming. The process also has disadvantages, for example, there is no certainty that the iron-binding compounds will reach all the areas of the dump. The alkali treatment of ore reduces the alkalinity of water drainage, which is itself another major problem.

Another process is the treatment of AMD with an alkali such as limestone (Figure 8.3), which is carried out in most mines (EPS 1987). Even though it is effective, the maintenance requirements are high due to the voluminous

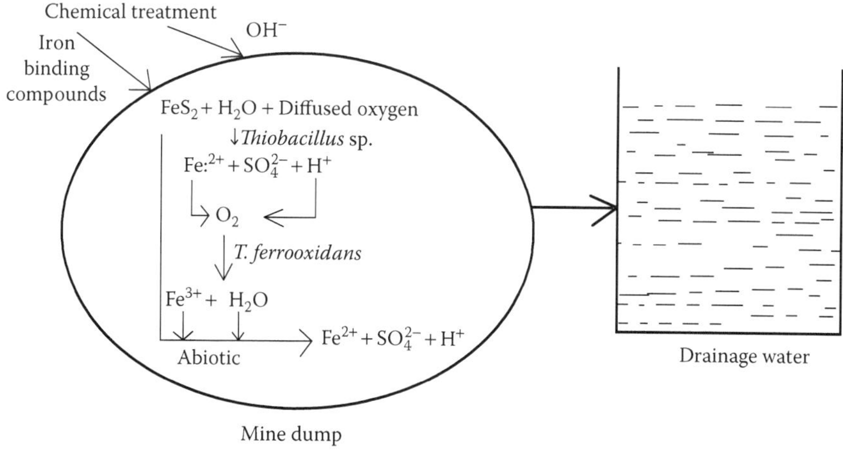

FIGURE 8.2
The addition of iron-binding chemicals and alkali to make the dump alkaline and remove the iron ions preventing drainage to the water tail.

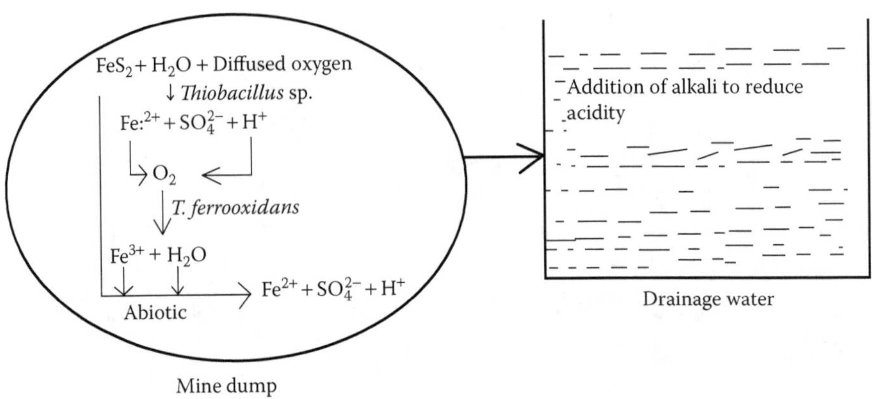

FIGURE 8.3
The chemical process of adding alkali to acid mine drainage to reduce the acidity.

sludge, which is composed mainly of calcium sulfate and hydroxides of metals. In addition, the limestone is covered with a film in an oxidizing environment, which quickly makes it ineffective. The use of bactericides can be combined with a chemical process that can cause direct inhibition of the microorganisms responsible for sulfur and iron oxidation (Sobek et al. 1990).

Other methods include physical methods (Figure 8.4) in which ore dumps are covered with vegetation and inoculated with other microorganisms that

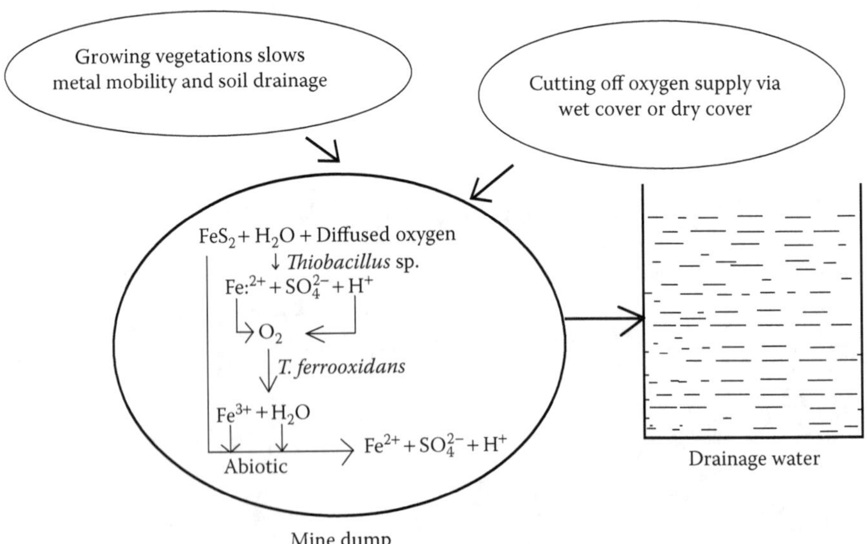

FIGURE 8.4
Physically blocking off the oxygen supply by the land mass (wet or dry cover) or slowing the metal mobility with the growth of vegetation or a combination of both.

prevents oxygen access and the flow of water through the deposit (EPS 1987). Metal-tolerant plants can be used for vegetation, a special cover layer where plants are sown, by the addition of fertilizers, mycorrhizal inoculation, and growing plants in symbiosis with a nitrogen-fixing bacteria. It also prevents soil erosion and thus supports the ecosystem.

The other process includes covering the dump with dry soil or wet soil to reduce the oxygen diffusion and water inflow. However, dry soil is prone to rain-washing while wet soil can dry up under sunlight creating cracks that may facilitate diffusion of oxygen (EPS 1987). These processes effectively reduce the AMD, but they are not completely effective in all conditions and thus need constant monitoring. Some artificial materials, such as plastic membranes, have a higher impenetrability than natural materials that can be used effectively. The process can be accomplished by covering the spoil material with a polyethylene membrane prior to covering the capping material. However, the high cost of the process makes it difficult to implement.

Microorganisms can play an effective role in the reduction of AMD and the removal of metal ions from the water drains. Metal-binding microorganisms (Figure 8.5) can be an effective way to remove the Fe^{2+} and Fe^{3+} ions from the dump. The process does not ensure complete binding, but it is cheap and can be effectively used to reduce the drainage.

Microorganisms capable of sulfur transformation can also be used to prevent the drainage (Figure 8.6). No bacteria alone can reduce the sulfides to elemental sulfur. However, the task can be accomplished by using mixed cultures. The mixed culture consists of a sulfate-reducing bacterium and a sulfide-oxidizing bacterium that transform sulfate to sulfide

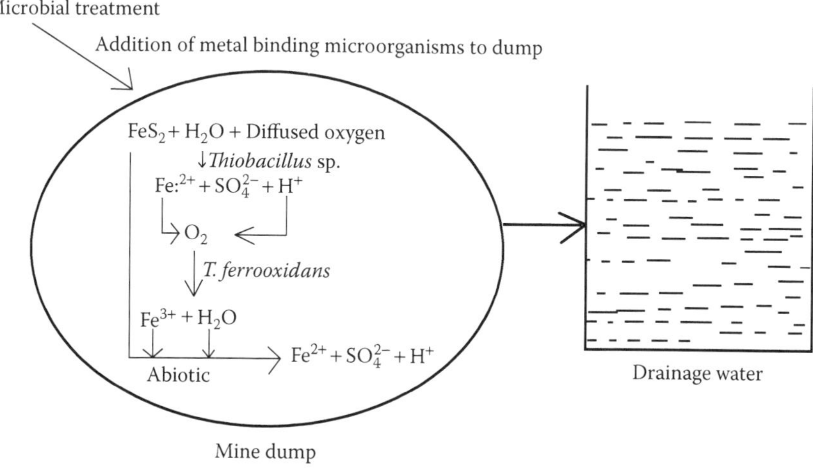

FIGURE 8.5
Microbial treatment by inoculating the mine dump with metal-binding microorganism, thus reducing the metal mobility.

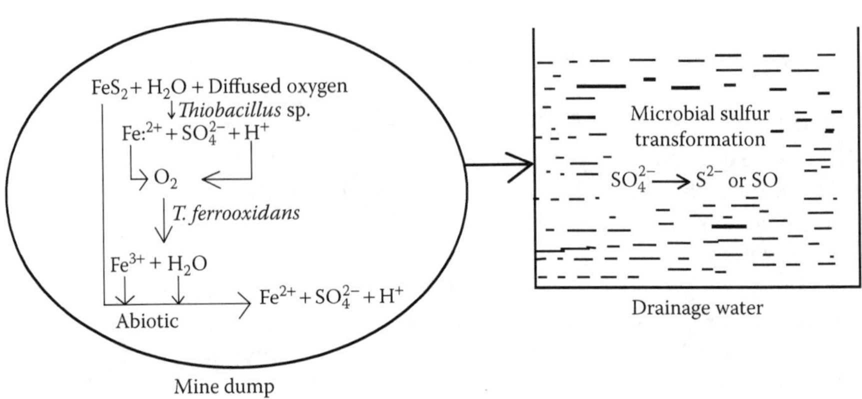

FIGURE 8.6
Microbial sulfur transformation of the drainage reducing the acidity.

and then to elemental sulfur. There have been many studies on this matter. Two-stage anaerobic processes were developed in which first *Desulfobacter postgateii* converted sufate to sulfide and degraded the acetate and then the anaerobic photoautotroph *Chlorobium 1 imicola* converted the sulfide to sulfur. The process is slow at start; however, it can be effectively repeated to remove the acidity from the mine tails (Cork et al. 1986). An upflow anaerobic packed-bed reactor was also developed to enhance the bioremoval of sulfates from the effluent from platinum mines, using molasses as the organic carbon substrate (Maree and Strydom 1987). *Thiobacillus* sp. are also reported to be effective in the removal of sulfur-containing salts from mining waste effluent.

The other process includes metal transformation (Figure 8.7). The solubility of metals can be changed by microbial transformation to precipitate the metal ions from the mine tails, thereby hindering the formation of acids. The combination of the above two processes can be an effective way of tackling mine tail issues for a large-scale treatment. The support of the alkali-generating microbes can also be combined to reduce the acidity of these mine tails. Many microorganisms are reported to extract metals from heavily salted waters too. In one study, dispersed and immobilized forms of *Chlorella vulgaris* and *Spirulina platensis* reversibly adsorbed gold (Damall et al. 1989). They were able to extract gold even when the concentration was in parts per billion (ppb) using the pH changing techniques. One such system was designed to use an artificial meander system of algae to remove metals and sulfate-reducing bacteria to change the pH (Sterritt and Lester 1979). The system requires the external addition of nutrients such as nitrogen compounds for algae and a carbon source for the bacterial growth.

Mining wastes are not only naturally formed, but are also formed due to the release of mine processing chemicals. Many microorganisms are capable

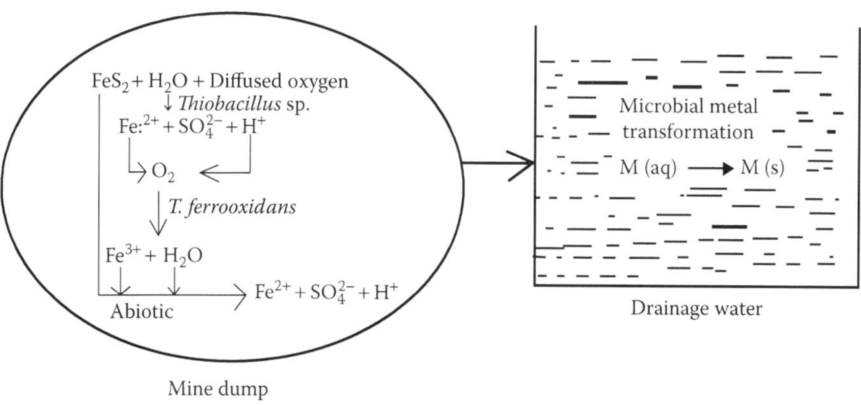

FIGURE 8.7
The microbial metal transformation of the drainage, reducing the metal content by precipitating them by reducing the solubility.

of enzymatic oxidation or detoxification of cyanide, which is used in processing gold ores. Flotation chemicals are also found to be reduced when treated with the less harmful and easily available bacteria species (*Bacillus subtilis, Saccharomyces cerevisiae, Bifidobacterium, Lactobacillus,* and *Acinetobacter calcoaceticus*) (Ghiani et al. 1993). Microorganisms can be used to oxidize both thiocyanate and cyanide to nitrogen. A multistage reactor system can use a mutant strain of *Pseudomonas paucimobilis* to convert thiocynate and cyanide to ammonia, which can be converted to nitrogen after treatment with nitrification–denitrification by *Nitrosomonas* sp. and *Nitrobacter* sp. (Mudder and Whitlock 1984).

8.3 Wastewater Treatment of Oil Wastes

Oil is an important part of life. Crude oil from oilfields, cooking oil from domestic households, motor oils from garages and auto repairing, drilling mud in drilling operations, and accidental oil spills are common sources of wastewater contamination. The application of various oil and chemical treatments is too costly; moreover, oil sludge from domestic households is usually less treated before releasing it into the sewage system and subsequently into rivers. Recently, there has been a surge in the research of oil-utilizing microorganisms, to biologically treat the oil spills and sludge from refineries.

Wastewater from oilfields, oil mills, petroleum industry, and domestic households is a major threat to the environment. There are several ways of treating wastewater to extract the oil. Crude oil from water can be removed by filtration. Oil in water emulsions can be treated with physical and chemical processes such as centrifugation. However, several other chemicals that are dissolved in oil droplets or in water are causing serious issues of toxicity. In such water, the entire COD is due to these wastes. These types of wastes can be effectively treated biologically. There are several reported studies where microorganisms have been found to survive in crude oil, utilizing it as a carbon source. There are studies of using an immobilized cell for the continuous removal of waste oil from water. Bioremediation of these toxic sites is very effective in completely degrading the hydrocarbon to CO_2, H_2O, and biomass. There are basically two ways to treat these wastes: (a) directly inoculating the microorganisms into the pollution sites and allowing them to grow, which is known as bioaugmentation; or (b) aerating the sites and adding high nitrogen and phosphorous contents such as fertilizer, so that indigenous microorganisms grow and effectively degrade the oil wastes. The rate of metabolizing hydrocarbon depends on many factors, including the pH, the oxygen content, the water content, the temperature, the concentration of oil, and the bioavailability. Oil bioremediation is greatly affected by the availability of nitrogen and phosphorous, lacking in hydrocarbons, which are needed by the bacteria. The supply of these two essential elements as water-soluble salts does not serve the purpose as they are rapidly diluted and easily consumed by non–oil-degrading bacteria. The use of hydrophobic sources of nitrogen and phosphorous can solve these problems.

8.3.1 Drilling Fluid–Contaminated Wastewater Treatment

The mining and oil exploration industries use drilling on a large scale. The drilling process utilizes drilling mud or drilling fluid for the smooth operation of the drills. These fluids facilitate the shaft movement and help remove the drill cuttings of the drilling holes. If left untreated, these can cause contamination of the oil, the detergents, and the heavy metals in the swamps and dumps and in the nearby water sources. The chemical treatment of these drilling muds is tedious as it requires a high amount of surfactant and water for dilution before a proper slurry can be made, which can be fed to a chemical reactor. These oil wastes are very detrimental to the aquatic flora and fauna.

Microorganisms have been reported to utilize the oils in these swamps. Several species of oil-consuming microorganisms have been identified. The bacterial species are mostly facultative anaerobic with the genera *Staphylococcus*, *Nocardia*, *Clostridium*, *Enterobacter*, *Klebsiella*, and *Pseudomonas* predominating the oil decomposition. A few fungal species (*Penicillium*, *Cladosporium*, and *Fusarium*) were also isolated from oil-rich swamps and may have a role in the microbial decomposition of oil (Benka-Coker and Olumagin 1995). These species can be utilized for waste sludge treatment as they have a comparatively higher capacity to utilize hydrocarbons. Using a

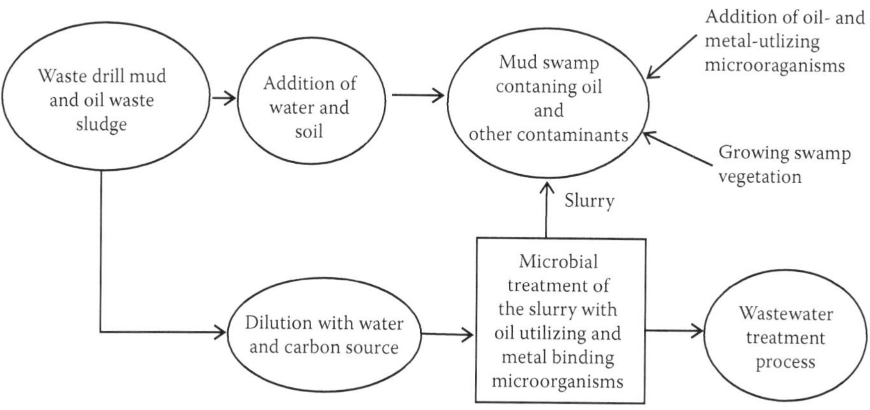

FIGURE 8.8
Drilling mud or drilling fluid wastewater treatment.

mixed culture of these strains can effectively remove the heavy metal contamination as well as the oil contents from the sludge (Benka-Coker and Olumagin 1995; Amund and Igiri 1990).

The flowchart (Figure 8.8) describes the process of the proposed treatment protocols. The direct wastewater treatment of the drilling fluid involves the dilution with water and the subsequent addition of a suitable carbon source. Metal-treating microorganisms and oil-utilizing microorganisms can be directly inoculated. The slurry is then filtered out and dumped back into the swamp. The water can be reused or passed for further treatment before being dumped into the water sources. The other process includes the dumping of the drilling fluids into the swamps and inoculating the swamp with a mixed culture of oil-utilizing, nitrogen-fixing, and metal-treating microorganisms. The process can be assisted with surviving vegetation such as mangrove. These vegetation can prevent soil mobility. They can also provide growth support to the microorganisms. These can survive in an extreme environment along with the microflora (Benka-Coker and Olumagin 1995).

8.3.2 Edible Oil–Contaminated Wastewater Treatment

Another form of oil in water contamination is from domestic households. Oil is frequently used for cooking purposes. One household may not cause a major threat, but if we combine the effects of even a small locality, the threat is raised several fold and becomes a major source of water contamination. The food industry, such as cakes, pastries, and French fries, generates oily waste that must be discarded. Even the oil mills release a lot of oil-contaminated wastewater. The sources of vegetable oil manufacturing are seeds and flowers, some nuts, soybeans, and sunflowers. The waste generated is mostly solid oil-rich wastes and wastewater. The cleaning operation also generates oil-rich wastewater. It is mostly acidic and rich in oil and colloidal particles. The basic protocol is to combine the physicochemical (flocculation, coagulation, flotation) and biological processes for the treatment of such wastewater. Oil mill wastewaters (OMWs) constitute a major environmental problem. They are produced in large quantities and there is a high content of toxic phenolic compounds present. OOMW (Olive oil mill waste) has a very high COD value and a comparatively low pH with suspended solids with some biorecalcitrant and inhibiting compounds, such as polyphenols, which make the traditional bioremediation less effective (Martirani et al. 1996). Most of these aromatic compounds can be broken down to the components of lignin. Microorganisms, mainly "white-rot" basidiomycetes or *Pleurotus ostreatus*, are able to degrade lignin by means of oxidative reactions catalyzed by phenol-oxidizing enzymes (Martirani et al. 1996). There are many structural similarities between the substrates of these enzymes and the aromatic compounds of oil wastes. The ligninolytic microorganism that produces such enzymes can be used in the effective treatment of these kinds of substrates (Borja et al. 2006). *Trichoderma viride* is also reported to reduce the carbon compounds by 60% and the phenolic compounds by 50% in olive oil wastes. The sequencing bioreactor can be effective in the treatment of OOMWs. These wastes are centrifuged, followed by filtering with microfilters, thereby removing most of the solid materials. These wastes are reported to be useful for energy recovery and hence can be used for combustion purposes.

8.3.3 Petroleum-Contaminated Wastewater Treatment

Petroleum hydrocarbons contain a complex mixture of various hydrocarbons, which can be categorized as saturates, aromatics, resins, and asphaltenes. The saturates are composed of alkanes, branched alkanes, and cyclic alkanes. Benzene, toluene, xylene, and other polyaromatic hydrocarbons with aromatic sulfur compounds (thiophenes and dibenzothiophenes) dominate the aromatic category. These polyaromatic hydrocarbons are major threats as they are carcinogenic in nature. The aromatic hydrocarbons that are soluble in water can attack the lipid-containing membranes surrounding the cells, making them detrimental to marine life. The resins and asphaltenes fractions consist of polar molecules with a high content of nitrogen, sulfur, and oxygen. The former are small molecules dissolved in oil, while the latter mostly stay as a colloidal suspension in oil.

Oil spills in the oceans are a major source of trouble both for aquatic life and for the humans dependent on them. The oil spills can be from damaged oil tankers or uncapped oil wells, as was the case in the Mexican gulf oil rig accident of a BP oil well in 2010. There are some man-made disasters too. A famous case was when 240 million gallons of oil spilled from oil terminals and tankers during the Gulf War. Another spill occurred over a 10-month period (June 1979 to February 1980) when 140 million gallons spilled out of the damaged Ixtoc I well in the Gulf of Mexico.

Crude oil normally contains large amount of asphaltenes and causes major havoc in the marine ecosystem. The marine ecosystem, from the fish to the oceanic birds, suffer the greatest blow. Until now, there is no suitable process for the treatment of these kinds of oil spills.

There are two principles of non–bioremediation-based oil waste treatment. First is removal and the second is dispersive dilution (Figure 8.9). The major process in the treatment of any typical oil spill is the siphoning of the oil from the slicks to the oil tankers. The biggest problem is that oil is lighter than water, and it spreads over a vast area due to oceanic activity. Therefore, a lot of water and oil are pumped in the reservoir. The tankers that float on the principle of buoyancy take very little oil in one run. Thus, the process becomes cumbersome and the recovery of oil becomes too costly. The other process is the dispersion of the slick with detergents. The detergents have the ability to bind with the oil molecules, segregating them from each other by forming micelles. This allows the oil slick to spread over vast areas of the ocean. However, these detergents with oil destroy the local marine ecosystem, before it spreads into the ocean. The size of the destruction varies with the conditions of the oceanic weather, as sometimes it takes weeks before the oil detergents dilute the local area.

Bioremediation can play an effective role in the removal of these slicks (Figure 8.9). Several microorganisms have been reported to be effective in oil removal. These are sometimes termed oil-eating microbes (OEMs). These are oleophilic in nature and can take up these forms of organic carbons (Atlas 1981). The OEMs have the potential to break almost all forms of

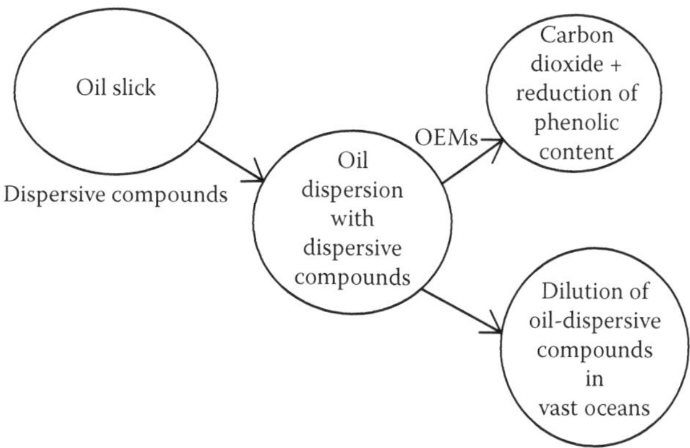

FIGURE 8.9
Treating oil slicks with and without bioremediation with OEMS.

organic carbon predominant in any oil slick (Spormann and Widdel 2000). They break it down into fatty acids or carboxylic acids that are further broken down for energy via Krebs cycle to harmless carbon dioxide and water. The OEMs can work extensively in a wide range of temperature and oceanic conditions (Diaz et al. 2002). If proper nutrients are incorporated to ensure proper oxygenation, these microorganisms can remove the oil completely.

Anaerobic bacteria, such as *Clostridia*, sulfate reducers, methanogens, *Desulfovibrio* species, and Fe(III)-reducing bacteria, can reduce nitroaromatic compounds. These can also be used to reduce the explosive grade nitrocarbon compounds (Lazar et al. 1995). The α, ε, and γ subclasses of *Proteobacteria* (α-*Proteobacteria*, *Pseudomonas*, and *Cycloclasticus*) are also reported to be playing a role in the bioremediation of oil slicks. Even the *Alcanivorax* genus bacteria showed accelerated growth in such conditions. *Methanosaeta* spp., *Methanospirillum*, *Desulfotomaculum*, and *Syntrophus* spp. are also effective in anaerobic conditions. *Geobacter* spp. is reported to be effective against benzene. *Acinetobacter calcoaceticus* is also effective against mineral oil, whereas *P. putida* and *Pseudomonas* sp. can be used to treat surfactants (Turkovskaya et al. 1999). *Acinetobacter* sp. has been proved effective against long-chain n-paraffin from engine oils. Car engine oil is composed of a base oil and additives, and the base oil mainly consists of hydrocarbons with a C_{16}–C_{36} chain length. The strains are reported to reduce the waste oil content (1% w/v) by 20% even without the biosurfactants (Daisuke et al. 2000). *Candida catenulata* is an excellent example of bioaugmentation. It is reported to decompose 84% of petroleum hydrocarbons in a sludge dump containing a mixture of diesel-contaminated soil and food wastes (4:1), in 2 weeks, compared with a nearly 50% decomposition with the existing microbes present in the sludge (Joo et al. 2008).

A mixture of aerobic and anaerobic OEMs, some nutrients and oil-dispersing compounds (e.g., surfactants or biosurfactants) can be sprayed on an oil slick. These compounds disperse the oil slick, increasing the surface area of the oil–microbe suspension, thereby increasing the oxygen contact for the aerobic OEMs. In the presence of other nutrients, microbes grow on this oil, consuming/decomposing the oil chunks with less or no oxygen content. In this way, a large amount of oil slick can be treated. These microorganisms remain adhered to the oil droplets; therefore, until they consume the oil contents completely, they continue to work the bioremediation. The process of dispersion and complete consumption to harmless compounds is complete within a week or two. The process is true for all forms of petroleum-based contamination, be it domestic, accidental, or natural. The sludge from automobile repair shops and oil processing industries can be treated with biosurfactants and water, and the subsequent addition of OEMs can be helpful to reduce the bulk oil content of the sludge. The processed sludge can then be further treated with other decontamination processes before releasing it to the environment.

8.4 Radioactive-Contaminated Wastewater Treatment

Uranium (U), technetium (Tc), thorium (Th), neptunium (Np), and plutonium (Pu) are the most common wastes generated during coal burning in thermal plants, oil exploration, mining, nuclear power generation, medical wastes from radiotherapy of cancer patients, etc. The mining industries are the major source of environmental radioactive contamination. Most of the ores contain limited yet nonextractable radioactive minerals. Mining companies do not have any use for them, so they extract the required mineral and dump the rest. These operations lead to the accumulation of these wastes that were previously in low quantities, contaminating the local soil and water resources, causing several diseases. In their reduced state, they are geochemically inactive and insoluble. The higher oxidation states are normally avoided because of their high toxicity. They can cause severe damage to human health and aquatic life. They can also affect the groundwater composition, rendering the entire place uninhabitable for centuries because of their accumulative radioactivity and enhanced solubility (Nelson and Lovett 1978). This is the reason for the mobility of radioactive materials from dumps to water sources. There have been some studies in this field to tackle the issue with biotechnology. There are some scientific reports on the anaerobic reduction of these wastes to convert these toxic elements into their nonsoluble state. The microbial treatment of radionuclides is achieved using three methods: bioaccumulation, biosorption, and biotransformation. Biotransformation has shown major promise in the treatment of radioactive wastes, utilizing them as their energy resource. The other two processes are not as helpful.

8.4.1 Microorganisms Involved in Radioactive Waste Treatment

Few anaerobic microorganisms can enzymatically reduce the higher state of radionuclides to their lower state, such as U, Np, Tc, and Pu. The mesophilic sulfate-reducing microorganisms, thermophilic microorganisms, mesophilic Fe(III)-reducing microorganisms, fermentative microorganisms, alkalophilic microorganisms, and acid-tolerant microorganisms are among a few of the physiological groups that are responsible for the dissimilatory reduction of radionuclide elements (Lovley and Coates 2000; Lovley et al. 1991, 1993a,b; Lovley 2003; Lovley and Phillips 1992; Tebo and Obraztsova 1998; Suzuki et al. 2003, 2005; Coates et al. 2001). Actinides can be solubilized or precipitated by either enzymatic-based or chemical-based reactions. The process can be based on the following (Francis 1998):

1. Oxidation–reduction reactions via enzymatic treatment (oxidoreductase)
2. Changing solubility by altering the pH
3. Chelating with binder molecules

4. Biosorption by a biomass and biopolymers formed by microorganisms
5. Formation of stable minerals by chemical means or biomineralization
6. Biodegradation of complex actinides with the organic waste constituents

The typical reactions in the removal of radioactive decontaminants are as follows:

$$UO_2^{2+}{}_{(soluble)} + H_2 \rightarrow UO_{2(insoluble)} + 2H^+$$
$$UO_2^{2+}{}_{(soluble)} + CH_2O + H_2O \rightarrow UO_{2(insoluble)} + 2H^+$$
$$TcCO^-_{4(soluble)} + 3e^- + 4H^+ \rightarrow TcO_2{\cdot}nH_2O_{(insoluble)} + (2-n)H_2O$$
$$NpO^{2+}{}_{(soluble)} + 1e^- + 4H^+ \rightarrow Np^{4+}{}_{(insoluble)} + (2-n)H_2O$$
$$Pu^{5+\text{ or }6+}{}_{(soluble)} \rightarrow Pu^{4+}{}_{(insoluble)}$$

8.4.2 Radioactive Wastewater Treatment

Uranium reductase reduces U(VI) (uranyl ion: UO_2^{2+}) to U(IV) (uraninite: UO_2) in the presence of hydrogen or organic compounds such as electron donors. The *Geobacter, Desulfovibrio,* and *S. putrefaciens* species are reported to have uranium reductase. The *Geobacter* species have reductase in the periplasmic space while the *Desulfovibrio vulgaris* have this enzyme in soluble fraction requiring cytochrome C3 and hydrogenase. *S. putrefaciens* and *S. oneidensis* are reported to have U(VI) reduction in the decaheme outer membrane cytochrome c and the tetraheme c-type cytochromes with the help of other nitrite-reducing enzymes (Khijniak et al. 2005). Technetium (Tc) can be reduced under all conditions through microorganism changes. The *Geobacter* and *Desulfovibrio* species, *S. putrefaciens,* and *E. coli* can reduce Tc in neutral pH. *Acidithiobacillus ferrooxidans* and *A. thiooxidans* can reduce Tc in acidic conditions, while halophilic microorganisms reduce Tc in basic conditions. Since the precipitation of TcO_2 was at the periphery of the cell, it is presumed that the enzyme for reduction is in the cell surface. The c-type cytochrome and hydrogenase act as the electron donor (Lloyd et al. 1997). *S. putrefaciens, S. oneidensis,* and *G. metallireducens* reduce the Pu(V) or Pu(VI) to Pu(IV) under normal environmental conditions. Special care needs to be taken to prevent the formation of soluble Pu(III). *Bacillus polymyxa, B. circulans, Clostridium* sp., and some strains of *Geobacter* and *S. oneidensis* can reduce the Pu(IV) to Pu(III) (Macaskie and Basnakova 1998). Neptunium (Np) is found in spent fuel rods produced mainly through the disintegration of ^{241}Am and ^{241}Pu. NpO_2^+ does not form complexes with any ligands, which makes it difficult to remove from a solution by physiochemical methods. It emits the α-particle, making it biologically toxic. The reduction can be carried out by ascorbic acid and/or by enzymes produced by *S. putrefaciens.* Though the process of sedimentation occurs only in the presence of *Citrobacter* sp., the sulfate-reducing microbes, such as *D. desulfuricans,* and *D. gigas,* are also reported to reduce

Np(V) by an unknown mechanism, as it seems that more than one enzyme system participates in the process. Recently. *G. metallireducens* and *S. oneidensis* have been reported to reduce Np in the presence of lactate and acetate (Macaskie et al. 1996, 1998; Lloyd et al. 2000).

These microorganisms can be used to treat mine dumps and spent fuel waste dumps. They can well ensure that no amount of radioactive nuclide can escape into the water sources from these dumps. Sometimes, the reduction also ensures the stopping of radioactive disintegration. However, this is difficult to understand because radioactive disintegrations are nuclear phenomena while reduction is a nonnuclear event. Changing the solubility helps in cleaning the contaminated groundwater or wastewater from these dumps. The wastewater can be treated in a tank with these microorganisms before it goes for further clarification/purification. The radionuclide is settled while the rest of the wastewater is decontaminated. The settled radionuclide(s) need to be handled with care and can be dumped in suitable containers to prevent further leakage.

8.5 Conclusion

Several million gallons of potable water are contaminated every year. Either they are dumped in the oceans via rivers and channels, or they form large water bodies with higher level of toxicity. Every year, tons of ores are dumped near water sources and often this leads to water contamination due to the lack of knowledge that metals have higher mobility. Several million gallons of oil are leaked, spilled, or discarded as domestic waste in water bodies. Radioactive wastes that are dumped even deep into the soil can migrate via metal mobility. All these conditions create a major threat to the environment. Microorganisms are indeed wonderful machines. With proper handling, not only can they survive the toughest conditions, but they can also remove radioactive elements, mining wastes, and oil in all forms from water. They can be an effective tool for wastewater treatment. The treatments using microorganisms are cheaper, faster, and without toxic residues, unlike most chemical treatments. They can be effective even when all the known chemical processes fail to remove the contaminations.

Acknowledgment

The authors acknowledge Banaras Hindu University, Varanasi, and the University Grant Commission, New Delhi, India, for their extensive support in pursuing the work in the field of advanced chemistry and biochemical engineering.

References

Amund, O. and C.O. Igiri. 1990. Biodegradation of petroleum hydrocarbons under tropical estuarine conditions. *World J. Microb. Biotechnol.* 6: 255–262.

Atlas, R.M. 1981. Microbial degradation of petroleum hydrocarbons: An environmental perspective. *Microbial. Rev.* 45: 180–209.

Benka-Coker, M.O. and A. Olumagin. 1995. Waste drilling-fluid-utilising microorganisms in a tropical mangrove swamp oilfield location. *Biores. Technol.* 53: 221–215.

Berthelin, J. 1983. Microbial weathering processes. In *Microbial Geochemistry*, ed. W.E. Krumbein. Oxford: Blackwell Scientific, pp. 223–262.

Beveridge, T.J. 1986. The immobilization of soluble metals by bacterial walls. *Biotechnol. Bioeng. Symp.* 16: 127–139.

Borja, R., B. Rincon, and F. Raposo. 2006. Anaerobic biodegradation of two-phase olive mill solid wastes and liquid effluents: Kinetic studies and process performance. *J. Chem. Technol. Biotechnol.* 8: 1450–1462.

Bossier, P., M. Hofte, and W. Verstraete. 1988. Ecological significance of siderophores in soil. In *Advances in Microbial Ecology*, ed. K.C. Marshall. New York: Plenum Press, pp. 385–414.

Brierley, C.L. and G.R. Lanza. 1985. Microbial technology for aggregation and dewatering of phosphate clay slimes: Implications on resource recovery. In *Soil Reclamation Processes*, eds. Tate III, R.L. and D.A. Klein. New York: Marcel Dekker, pp. 243–277.

Coates, J.D., V.K. Bhupathiraju, L.A. Achenbach, M.J. McInerney, and D.R. Lovley. 2001. *Geobacter hydrogenophilus* and *Geobacter chapellei* and *Geobacter grbiciae*, three new, strictly anaerobic, dissimilatory Fe(III)-reducers. *Int. J. Syst. Evol. Microbiol.* 51: 581–588.

Cork, D.J., D.E. Jerger, and A. Maka. 1986. Biocatalytic production of sulfur from process waste streams. *Biotechnol. Bioeng.* 16: 149–162.

Daisuke, K., F. Hasumi, E. Yamamoto, T. Ohta, S.Y. Chung, and M. Kubo. 2000. Biodegradation of long-chain n-paraffins from waste oil of car engine by *Acinetobacter* sp. *J. Biosci. Bioeng.* 91(1): 94–96.

Damall, D.W., R.M. McPherson, and J. Gardea-Torresdey. 1989. Metal recovery from geothermal waters and ground waters using immobilized algae. *Biohydrometallurgy* 341–348.

Diaz, M.P., K.G. Boyd, S.J. Grigson, and J.G. Burgess. 2002. Biodegradation of crude oil across a wide range of salinities by an extremely halotolerant bacterial consortium MPD-M, immobilized onto polypropylene fibers. *Biotechnol. Bioeng.* 79(2): 145–153.

Ehrlich, H.L. 1990. *Geomicrobiology*, 2nd edn. New York: Marcel Dekker.

EPS. 1987. Mine and mill wastewater treatment. Environmental Canada Report EPS 2 MM 3.

Francis, A.J. 1998. Biotransformation of uranium and other actinides in radioactive wastes. *J. Alloys Compd.* 271–273: 78–84.

Ghiani, M., G. Loi, N. Passarini, P. Trois, and G. Rossi. 1993. Microbial purification technique of mineral dressing plants reject waters. *FEMS Microbial. Rev.* 11: 153–157.

Golovacheva, R.S. and G.I. Karavaiko. 1978. A new genus of thermophilic sporeforming bacteria, *Sulfobacillus*. *Microbiology* 47: 815–822.

Goodman, A.E., A.M. Khalid, and B.J. Ralph. 1981. Microbial ecology of rum jungle. Part 1. Environmental study of sulphidic overburden dumps, environmental heap-leach piles and tailings dam area. Aust. AEC AAEC/E53 1.

Groudev, S.N. and V.I. Groudeva. 1993. Microbial communities in four industrial copper dump leaching operations in Bulgaria. *FEMS Microbial. Rev.* 11: 261–268.

Guay, R., S. Dufresne, E.R. Desjardins, and P. Gklinas. 1992. Microbiological modeling of acid mine drainage production in pyritic waste rock pile. *Abstr. Gen. Meet. Am. Sot. Microbial.* 344.

Harrison, A.P.J. 1984. The acidophilic *thiobacilli* and other acidophilic bacteria that share their habitat. *Annu. Rev. Microbiol.* 131: 68–76.

Hughes, M.N. and R.K. Poole. 1989. *Metals and Micro-Organisms*. London: Chapman & Hall.

Johnson, D.B., W.I. Kelso, and D.A. Jenkins. 1979. Bacterial streamer growth in a disused pyrite mine. *Environ. Pollut.* 1X: 107–118.

Joo, H.S., P.M. Ndegwa, M. Shoda, and C.G. Phae. 2008. Bioremediation of oil-contaminated soil using *Candida catenulata* and food waste. *Environ. Pollut.* 156: 891–896.

Kelly, D.P. 1988. Evolution of the understanding of the microbiology and biochemistry of the mineral leaching habitat. In *Biohydrometallurgy*, eds. Norris, P.R. and D.P. Kelly. Kew, UK: Science and Technology Letters, pp. 3–14.

Khijniak, T.V., A.I. Slobodkin, V. Coker, J.C. Renshaw, F.R. Livens, E.A. Bonch-Osmolovskaya, N.K. Birkeland, N.N. Medvedeva-Lyalikova, and J.R. Lloyd. 2005. Reduction of uranium (VI) phosphate during growth of the thermophilic bacterium Thermoterrabacterium ferrireducens. *Appl. Environ. Microbiol.* 71: 6423–6426.

Kuenen, J.G. and P. Bos. 1989. Habitats and ecological niches of chemolitho(auto)trophic bacteria. In *Autotrophic Bacteria*, eds. Schlegel, H.G. and B. Bowien. Berlin: Springer, pp. 53–80.

Lazar, I., A. Voicu, S. Dobrota, M. Stefanescu, L. Sandulescu, G. Archir, I.G. Lazar, D. Mucenica, A. Balalia, and C. Nicolescu. 1995. Investigations on potential bacteria for the bioremediation treatment of environments contaminated with hydrocarbons. In *The Fifth International Conference on Microbial Enhanced Oil Recovery and Related Biotechnology for Solving Environmental Problems*, eds. Bryant, R. and R.L. Sublette. National Technical Information Service, Dallas, TX, pp. 535–547.

Ledin, M. and K. Pedersen. 1996. The environmental impact of mine wastes – Roles of microorganisms and their significance in treatment of mine wastes. *Earth Sci. Rev.* 41: 67–108.

Lloyd, J.R., J.A. Cole, and L.E. Macaskie. 1997. Reduction and removal of heptavalent technetium from solution by *Escherichia coli*. *J. Bacteriol.* 179: 2014–2021.

Lloyd, J.R., P. Young, and L.E. Macaskie. 2000. Biological reduction and removal of Np(V) by two microorganisms. *Environ. Sci. Technol.* 34: 1297–1301.

Lovley, D.R. 2003. Cleaning up with genomics: Applying molecular microbiology to bioremediation. *Nat. Rev. Microbiol.* 1: 35–44.

Lovley, D.R. and J.D. Coates. 2000. Novel forms of anaerobic respiration of environmental relevance. *Curr. Opin. Microbiol.* 3: 252–256.

Lovley, D.R. and E.J.P. Phillips. 1992. Reduction of uranium by *Desulfovibrio desulfuricans*. *Appl. Environ. Microbiol.* 58: 850–856.

Lovley, D.R., E.J.P. Phillips, Y.A. Gorby, and E. Landa. 1991. Microbial reduction of uranium. *Nature* 350: 413–416.

Lovley, D.R., E.E. Roden, E.J.P. Philips, and J.C. Woodward. 1993a. Enzymatic iron and uranium reduction by sulfate-reducing bacteria. *Mar. Geol.* 113: 41–53.

Lovley, D.R., P.K. Widman, J.C. Woodward, and E.J.P. Phillips. 1993b. Reduction of uranium by cytochrome c3 of *Desulfovibrio vulgaris*. *Appl. Environ. Microbiol.* 59: 3572–3576.

Macaskie, L.E. and G. Basnakova. 1998. Microbially-enhanced chemisorption of heavy metals: A method for the bioremediation of solutions containing long-lived isotopes of neptunium and plutonium. *Environ. Sci. Technol.* 32: 184–187.

Macaskie, L.E., J.R. Lloyd, R.A.P. Thomas, and M.R. Tolley. 1996. The use of micro-organisms for the remediation of solutions contaminated with actinide elements, other nuclides, and organic contaminants generated by nuclear fuel cycle activities. *Nucl. Energ.* 35: 257–271.

Mann, H. and W.S. Fyfe. 1989. Metal uptake and Fe-, Ti-oxide biomineralization by acidophilic microorganisms in mine-waste environments, Elliot Lake, Canada. *Can. J. Earth Sci.* 26: 2731–2735.

Mann, H., W.S. Fyfe, R. Kerrich, and M. Wiseman. 1989a. Retardation of toxic heavy metal dispersion from nickel copper mine tailings, Sudbury District, Ontario: Role of acidophilic microorganisms: I. Biological pathway of metal retardation. *Biorecovery* 1: 155–172.

Mann, H., K. Tazaki, W.S. Fyfe, and M. Wiseman. 1989b. Retardation of toxic heavy metal dispersion from nickel copper mine tailings, Sudbury District, Ontario: Role of acidophilic microorganisms: II. Structure and microanalysis of bioprecipitants. *Biorecovery* 1: 173–188.

Maree, J.P. and W.F. Strydom. 1987. Biological sulphate removal from industrial effluent in an upflow packed bed reactor. *Water Res.* 21: 141–146.

Martirani, L., P. Giardina, L. Marzullo, and G. Sannia. 1996. Reduction of phenol content and toxicity in olive oil mill waste waters with the ligninolytic fungus *Pleurotus ostreatus*. *Water Res.* 30(8): 1914–1918.

Mudder, T.I. and J.L. Whitlock. 1984. Biological treatment of cyanidation wastewaters. *Miner. Metal Process.* I: 161–165.

Nelson, D.M. and M.B. Lovett. 1978. Oxidation state of plutonium in the Irish Sea. *Nature* 276: 599–601.

Prescott, L.M., J.P. Harley, and D.A. Klein. 1993. *Microbiology*. Dubuque, IA: WCB.

Pirt, S.J. 1975. *Principles of Microbe and Cell Cultivation*. Oxford: Blackwell Scientific.

Pulford, I.D. 1991. A review of methods to control acid generation in pyritic coal mine waste. In *Land Reclamation*, ed. M.C.R. Davies. London: Elsevier, pp. 269–278.

Ritcey, G.M. 1989. Effluent treatment for environmental control. Tailings management: Problems and solutions in the mining industry. *Proc. Metal. Rep.* 6: 411–574.

Sobek, A.A., D.A. Benedetti, and V. Rastogi. 1990. Successful reclamation using controlled release bactericides: Two case studies. In *Proceedings of the 1990 Mining and Reclamation Conference and Exhibition*, eds. Skousen, J., J. Sencindiver, and D. Samuels. West Virginia University Publication Services, Charleston, WV, April 23–26, pp. 33–41.

Spormann, A.M. and F. Widdel. 2000. Metabolism of alkylbenzenes, alkanes, and other hydrocarbons in anaerobic bacteria. *Biodegradation* 11: 85–105.

Sterritt, R.M. and L.N. Lester. 1979. The microbiological control of mine waste pollution. *Miner. Environ.* 1: 45–47.

Suzuki, Y., S.D. Kelly, K.M. Kemner, and J.F. Banfield. 2003. Microbial populations stimulated from hexavalent uranium reduction in uranium mine sediment. *Appl. Environ. Microbiol.* 69: 1337–1346.

Suzuki, Y., S.D. Kelly, K.M. Kemner, and J.F. Banfield. 2005. Direct microbial reduction and subsequent preservation of uranium in natural near-surface sediment. *Appl. Environ. Microbiol.* 71: 1790–1797.

Tebo, B.M. and A.Y. Obraztsova. 1998. Sulfate-reducing bacterium grows with Cr(VI), U(VI), Mn(IV), and Fe(III) as electron acceptors. *FEMS Microbiol. Lett.* 162: 193–198.

Turkovskaya, O.V., A.Yu. Muratova, and L.V. Panchenko. 1999. Selection and study of surfactant and mineral oil-degrading microbial consortia. *Resour. Conserv. Recycl.* 27: 169–178.

Wichlacz, P.L. and R.F. Unz. 1981. Acidophilic, heterotrophic bacteria of acidic mine waters. *Appl. Environ. Microbial.* 41: 1254–1261.

Widdel, F. 1988. Microbiology and ecology of sulfate- and sulfur-reducing bacteria. In *Biology of Anaerobic Microorganisms*, ed. J.B. Zehnder. Wiley Series in Ecological and Applied Microbiology. New York: Wiley, pp. 469–585.

Yates, J.R. and D.S. Holmes. 1986. Molecular probes for the identification and quantitation of microorganisms found in mines and mine tailings. *Biotechnol. Bioeng.* 16: 301–310.

9

Anaerobic Wastewater Treatment in Tapered Fluidized Bed Reactor

R. Parthiban

CONTENTS

9.1 Anaerobic Microbiology

Anaerobic treatment converts the organic pollutants (chemical oxygen demand [COD] and biochemical oxygen demand [BOD]) in wastewater into a small amount of sludge and a large amount of biogas (methane and carbon dioxide). In the absence of oxygen, many different groups of anaerobic

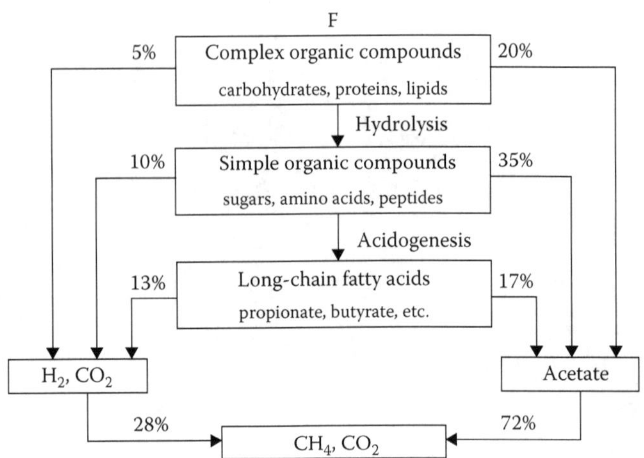

FIGURE 9.1
A schematic diagram of the patterns of carbon cycle in anaerobic processes.

bacteria "work" together to degrade the complex organic pollutants into methane and carbon dioxide. The microbiology of anaerobic processes is more complex and delicate than that of aerobic processes, where most bacteria "work" individually. As a result, the anaerobic systems require more control and monitoring systems to operate successfully. The biological degradation of complex organic compounds takes place in several consecutive biochemical steps (chain reaction), each performed by different groups of specialized bacteria, which have been studied in detail (Dunn et al. 1992; Quasim 1999; Speece 1983).

The biochemical steps can occur simultaneously in one bioreactor ("one-phase systems"), or they can occur in two consecutive tanks ("two-phase systems"). Several intermediate products are continuously generated and are immediately processed further again. To simplify the degradation process, four major steps can be distinguished, as shown in Figure 9.1. In practice, it is important to realize that all the steps have to occur at matching rates, in order to avoid a buildup of the intermediate products. Without the good "teamwork" of all the microbial communities involved, no complete degradation is possible.

The consortium of anaerobic organisms that work together to bring about the conversion of organic sludge and wastes (Holland et al. 1987; Wiesmann 1988) can be grouped as follows:

- Organisms responsible for hydrolyzing the organic polymers and lipids to basic structural building blocks, such as monosaccharides, amino acids, and related compounds. This step is carried out by the extracellular enzymes of facultative or obligate anaerobic bacteria, for example, *Clostridium* (degrading the compounds containing cellulose and starch) and *Bacillus* (degrading proteins and fats).

- Anaerobic bacteria, which ferment and break down products to simple organic acids, the most common of which, in an anaerobic digester, is acetic acid (acidogens or acid formers). These bacteria are described as nonmethanogenic and can be *Clostridium* spp., *Peptococcus anaerobius*, *Bifidobacterium* spp., *Desulfovibrio* spp., *Corynebacterium* spp., *Lactobacillus*, *Actinomyces*, *Staphylococcus*, and *Escherichia coli*.
- Organisms that convert the hydrogen and acetic acid formed by the acid formers to methane gas and carbon dioxide, designed as methanogens or methane formers. The most important microorganisms that have been identified include the rods (*Methanobacterium* and *Methanobacillus*) and the spheres (*Methanococcus* and *Methanosarcina*).

Engineered anaerobic consortia are needed to expand the catabolic diversity of the sludge and to shorten the adaptation to recalcitrant and toxic substrates. The specialized microbial consortia can be "biochemically rerouted" by the induction of a desirable biochemical pathway and the repression of an undesirable pathway. Therefore, a narrow thermodynamic window must be maintained with respect to the hydrogen concentration in order to permit methane formation but not to inhibit propionic acid degradation. This bottleneck can best be prevented by using a microorganism population that is particularly active with respect to hydrogen utilization. The measurement of the hydrogen concentration can provide important information on the interaction of the various types of microorganisms and may be used for process control (Archer et al. 1986).

A number of contributions report the start-up studies and also the transition behavior of the anaerobic reactors with the aid of mathematical models for the essential degradation steps of acid hydrolysis and methane formation (Kleinstreuer and Poweigha 1982; Bryers 1985).

Although the anaerobic digestion process involves several intermediate and simultaneous reactions, the process can be simplified and, in general terms, described by the following overall biochemical reaction. Some of the other end products are the gases such as H_2S and H_2 (Equation 9.1).

$$\begin{pmatrix}\text{Organic}\\ \text{matter}\end{pmatrix} \xrightarrow{\text{Anaerobic microbes}} \begin{pmatrix}\text{New}\\ \text{cells}\end{pmatrix} + \begin{pmatrix}\text{Energy}\\ \text{for cells}\end{pmatrix} + CH_4 + CO_2 + \begin{pmatrix}\text{Other end}\\ \text{products}\end{pmatrix}. \quad (9.1)$$

Most models developed for this reaction type considered the four overall steps of the process (Yang and Okos 1987; Denac and Dunn 1988; Figure 9.1):

- The enzyme-mediated hydrolysis of carbohydrates, proteins, and/or lipids, resulting in monomer compounds such as amino acids, sugars, and fatty acids
- The fermentation of organic acids (bacterial assisted), through which organic acids (acetic, propionic, and butyric) can be obtained (acidogenic step)

- The conversion (bacterial assisted) of organic acids containing more than three carbon atoms to acetic acid and hydrogen (acetogenic step)
- The conversion of acetic acid to methane and the reduction of carbon dioxide with hydrogen to methane (methanogenic step)

The necessity of an interlinkage of sequential degradation steps in the anaerobic process by various microorganisms means that the various steps must proceed at the same speed in order to avoid disturbances. This explains the higher sensitivity of the anaerobic processes.

9.2 Factors Affecting Anaerobic Treatment

Under proper environmental conditions, anaerobic bacteria will continually produce biogas. The most important factors affecting the rate of digestion and biogas production are temperature, pH, organic loading rate (OLR), hydraulic retention time (HRT), nutrient concentrations, and reactor design and operation. All factors must be considered in the design and operation of an anaerobic process for the successful treatment of wastewater.

9.2.1 Temperature

Temperature has a major influence on the effectiveness of biological systems, affecting the metabolic rate, ionization equilibria, solubility of the substrates and fats, and bioavailability of iron (Speece 1996). Research has shown that anaerobic microorganisms will function effectively over two temperature ranges, the mesophilic range (29°C–38°C) and the thermophilic range (49°C–57°C) (Eckenfelder 1999; Speece 1996). Anaerobic digestion is a function of temperature, where the rate of decomposition increases as the temperature increases until the optimum growth temperature is reached. At temperatures above and below the optimum growth temperature, the metabolic activity decreases, resulting in a decrease in the reactor kinetics. Full-scale anaerobic treatment systems typically operate in the mesophilic range because maintaining higher operating temperatures is usually not economically justifiable. An exception may be for domestic sludge digestion because of the more rapid pathogen inactivation in thermophilic systems (Speece 1996).

9.2.2 pH

The pH within the reactor of a system greatly affects the rate of methane production and the overall success of the anaerobic digestion process. Research has shown that the methane-producing microbes function effectively in a pH range of approximately 6.5 to 8.2, with an optimum level of function near a pH of 7.0 (Eckenfelder 1999; Speece 1996). In an anaerobic system, the acidogenic bacteria convert organic matter to organic acids, possibly decreasing the pH and reducing the rate of methane production and thus the overall anaerobic process if the acids are not quickly consumed by methanogens. At pH levels above 8.2 and below 6.5, unacclimated microorganisms begin to die as the microbial growth is inhibited and the conditions become toxic to the existing population. Therefore, attention to the pH is essential for the successful operation of anaerobic systems.

9.2.3 Organic Loading Rate

The organic loading that may be treated efficiently and effectively in an anaerobic system will depend primarily on the biomass concentration in the reactor and the characteristics of the waste in addition to the system design parameters (i.e., reactor volume and HRT) (Evans 2001). Typically, the volatile solids (VS) concentration is used as a measure of the content of organic matter in waste, and thus, the OLR for biological systems is stated in terms of VS per reactor volume per unit time (i.e., kg_{VS}/m^3 day). Anaerobic systems with a high OLR depend on large bacterial populations to achieve rapid treatment and can generate enough methane to be self-sufficient (producing enough energy to operate the system). Generally, the anaerobic reactors can sustain much higher OLRs than the aerobic systems because they are not limited by the lack of oxygen in the system or the oxygen transfer rate.

9.2.4 Hydraulic Retention Time

As with all biological systems, the microorganisms require a certain amount of time to digest the organic matter and to achieve the desired level of treatment. The HRT is defined as the amount of time that the waste will be retained in the reactor to be digested and is equal to the volume of the reactor divided by the daily feed rate. The required HRT will depend primarily on the rate of digestion, which is dependent on the waste characteristics, the operating temperature, the availability of microorganisms, the species in the bacterial population, the reactor design, and the level of treatment required. The control of the HRT is important in anaerobic systems to prevent cell washout of the slow-growing methanogenic microorganisms (Shieh and Nguyen, 2000). In addition to the HRT, the solids retention time (SRT) is used for controlling the growth rate of the microbes in the reactor and is the average time a solid particle, such as a microbe, is present in the reactor. It is calculated by dividing the mass of the solids in the reactor by the mass of the solids removed from the system each day. "SRT determines which organisms can replicate and predominate within the system as well as what biomass inventory (biological safety factor) can be maintained" (Speece 1996).

9.2.5 Nutrient Concentrations

To ensure optimal digestion of the organic matter, traces of inorganic elements, such as iron, nickel, cobalt, and zinc, are required in low concentrations to stimulate fermentation and aid in the metabolism of the organic matter. Some wastes are deficient in these nutrients, and they must be added to the system for it to function properly. However, determining the specific nutrient requirements for a biological system can be difficult, and it depends on the waste characteristics, the availability of the nutrients within the system, the design of the system, and other parameters (Malaspina et al. 1995). In general, anaerobic systems require higher concentrations of iron and other inorganic elements, such as nickel, cobalt, and zinc. The essential nutrients for the anaerobic digestion of a specific wastewater can be determined by conducting a biochemical methane potential (BMP) assay to evaluate the effect of the addition of different nutrient solutions on the methane production rate from the waste.

9.3 Anaerobic Fluidized Bed Reactors

The anaerobic fluidized bed reactors (AFBRs) were originally a chemical engineering tool used for performing phase transformations, reactions, and diffusions of various chemicals existing in the solid, liquid, and vapor phases. With the concepts of maximum diffusion and maximum chemical reaction within a minimum volume in mind, the AFBRs have been adopted to treat biological wastewaters and are utilized in several process configurations.

When there is an increasing flow of liquid through a bed of particles, the bed initially expands and the particles are then suspended and are free to move with respect to the others. The bed is then fluidized. After passing the flow threshold, which causes fluidization, two different types of behavior can occur:

- The bed expands, increasing the distances between the particles.
- There is excess flow through the bed, forming bubbles.

These fluidization types are known as particulate and aggregative, respectively. Solid–liquid fluidized beds of low-density solids, such as glass, sand, or granular activated carbon (GAC), can be considered ideal homogenous particulate beds. The bed itself behaves as a fluid; it obtains a new set of physical properties (density and viscosity) and follows the hydrostatic and hydrodynamic fluid laws. In order to get a fluidized bed, an increasing flow of liquid is applied through a settled bed of particles, which form a fixed bed. During the progressive increase of the flow, the bed starts to expand. At this moment, all equations that apply to the fixed bed describe the situation. If the flow rate is increased, a transition occurs and the particles start to move suspended on the upflow liquid separated from the other particles. At this very moment, the fixed bed laws are still followed. If the flow rate is increased, the particles are more separated from each other; their hydrodynamic behavior resembles the behavior of the particles settling. The limit of this phenomenon is called fluid transport, where the particles are carried out of the bed by the liquid flow.

9.3.1 Fluidized Bed Characteristics

Pressure losses (Δp) in a fixed bed can be described by the Kozeny–Carman equation:

$$\frac{\Delta P}{L} = \frac{150\,\bar{V}_0\mu}{\Phi_s^2 D_p^2} \frac{(1-\varepsilon)^2}{\varepsilon^3}, \tag{9.2}$$

where Δp is the pressure drop, L is the total height of the bed, $\bar{V}_0$ is the superficial or "empty-tower" velocity, μ is the viscosity of the fluid, ε is the porosity of the bed, Φ_s is the sphericity of the particles in the packed bed, and D_p is the diameter of the spherical particle.

Equation 9.2 holds for flow through packed beds with particle Reynolds numbers up to approximately 1.0, after which point the frequent shifting of the flow channels in the bed causes considerable kinetic energy losses.

There is a linear relationship between the pressure losses and the upflow liquid velocity. At the onset of fluidization, the weight of the particles is just equal to the fluid drag force; the particles flow freely and the bed becomes fluidized.

The oldest application of fluidization is in the area of nitrification and denitrification of wastewater. Nitrification is the biological oxidation of ammonia to nitrate. The primary concern of any anaerobic treatment is the removal of the COD from the wastewater. The COD sources in wastewater can act as both a carbon source and an electron donor to a microbial consortium in many AFBR treatment processes. In this treatment, the presence of high biomass concentrations on a carrier material allows for faster utilization of the COD per unit volume than many other types of biological treatment (Heijnen et al. 1989).

The performance of biological fluidized bed reactors was first recognized for the aerobic oxidation of dissolved organic carbon and the denitrification and biological degradation of refractory organics such as phenol and chlorinated hydrocarbons in wastewater. In the AFBR, the fine carrier matrices are used for the microbial film development. These particles, with an entrapped biomass, are fluidized by high upflow fluid velocities that are created by the combination of the influent and recirculated treated wastewaters.

9.3.2 Advantages of Fluidized Bed Technology

The advantages of fluidized bed technology for anaerobic digestion include the following:

- A high treatment efficiency.
- A high rate of mass transfer under isothermal operating conditions is attainable due to good mixing. Low concentration gradients are established around the fluidizing particles, and thus it applies to the treatment of low strength waste.
- There are no moving parts because the reactor is not mechanically agitated, resulting in low maintenance costs.
- The reactor is mounted vertically and therefore has a small footprint.
- A skilled reactor operator is unnecessary.
- The system offers ease of control even for large-scale operations.
- The multiple stages of this kind of reactor are easily implemented, resulting in easy and effective control of the solids and the fluid residence times.
- A high concentration of biomass, attached to a dense carrier, which cannot be easily washed out from the reactor.
- A very large surface area for biomass attachment.
- An initial dilution of the influent with the effluent, which provides alkalinity; thus, neutralization, to a certain degree, reduces the substrate concentration (important for high COD wastes) and contributes to the reduction of the shock effect of toxicant spikes.
- No plugging, channeling, or gas holdup.
- The ability to control and optimize the biofilm thickness.

There are, however, a number of disadvantages, including

- A high-energy consumption due to a very high liquid recirculation ratio
- An equal cross-sectional velocity (i.e., equal drag force for all bioparticles), which frequently causes a washout of the bioparticles
- A reduction in the specific gravity of the bioparticles, causing a washout of the bioparticles owing to biogas entrapment

9.4 Anaerobic Tapered Fluidized Bed Reactors

Scott and Hancher (1976) modified the geometrical circular fluidized bed configuration (identical cross-sectional area) along its length to a tapered form. The wastewater flow is relatively stable throughout the reactor as the entry at the tapered cross section is sufficiently small and the expansion is gradual (an angle of few degrees). Hence, both the superficial velocity and the drag force exerted by the fluid on the bioparticles in the upper parts of the anaerobic tapered fluidized bed reactors (ATFBRs) are small. This prevents the washout of bioparticles from the reactor (Parthiban et al. 2007a, 2008a,b).

The distribution of the bioparticles in the ATFBR is more uniform than that of the AFBR. The arithmetic means of the biofilm thickness, the specific surface area, and the specific biomass in the ATFBR are larger than those in the AFBR and are increased especially in the ATFBR with a taper angle of 5°.

The attachment media commonly used in AFBRs are GAC, anthracite, and sand. The important properties of these media include minimum variation in size and shape, low cost, appropriate specific gravity for good fluidization, good abrasive resistance, and a surface conducive to cell immobilization. Some of the advantages of GAC as a carrier matrix in the AFBRs over other media are its availability over a wide range of particle sizes and its relative hardness and resistance to abrasion in an anaerobic environment. The exterior surface roughness of GAC renders it superior to most other media in microbial sheltering and attachment. The adsorptive properties of GAC increase the concentration of the soluble organic matter at the interface, thereby stimulating biological growth and assimilation. In addition, the pits and crevices of GAC allow the biofilm to be firmly attached to the particles, offering high attrition resistance.

The biofilm thickness (δ) is found to be inversely proportional to the specific energy dissipation rate (ω) (Huang et al. 2000). A small ω value assists in the growth of a thick biofilm. When the flow rate is less than the critical flow rate (Q_{cr}), ω increases with the increasing flow rate, resulting in poor attachment. If the flow rate is greater than Q_{cr}, ω decreases with the increase in the flow rate, resulting in better biofilm attachment. In the AFBR, the bioparticles are in a dynamic equilibrium state in which the upward drag force (exerted by the upwardly flowing water surrounding the bioparticles) is balanced by the negative buoyant force from gravity. With an increase in the biofilm thickness or accumulation, the bioparticles are fluidized to a greater extent, owing to the decrease in specific gravity. A smaller erosive effect on the biofilm surface exists due to a smaller ω value in the surroundings. As the biofilm grows thicker, the bioparticles also move to the upper part of the reactor. The bioparticles with a thinner biofilm thickness do not move upward and stay in the lower part of the reactor (Wu and Huang 1996). The distribution of the bioparticles in the ATFBR is more uniform than that of the AFBR. The arithmetic means of the biofilm thickness, the specific surface area, and the specific biomass in the ATFBR are larger than those in the AFBR and are increased especially in the ATFBR with a taper angle of 5° (Wu and Huang 1996; Huang et al. 2000).

The ATFBR has the following shortcomings:

- A major problem associated with this reactor is the mechanical design, since the ATFBR is more difficult to build with the desired taper angle than the conventional AFBR.
- The scale-up of the systems is very difficult because it requires many compromises between the technical and the economic aspects.

9.5 Anaerobic Treatment of Synthetic Sago Wastewater: A Case Study

The manufacturing industry of sago, the common edible starch, is predominantly clustered in the Salem district, Tamil Nadu, India, which uses the tubers of tapioca as the raw material. The growth of the starch processing industry has resulted in large-scale water pollution, as it generates a huge volume of wastewater with an extremely high concentration of organic pollutants. The wastewater discharged is acidic, foul smelling, and highly organic in nature (Sastry and Mohan 1963). The washing and extracting processes constitute the main sources of wastewater discharged from the industry. Studies on the tapioca processing industry reported that about 12 m^3 of wastewater is discharged per ton of starch produced (Hien et al. 1999; Mai et al. 2004). The wastewater is characterized by a pH of 4.5–5.0, a BOD_5 of 6,200–23,077 mg/L, a COD of 7,000–41,406 mg/L, and a suspended solids concentration of 4,200–7,600 mg/L. These values indicate that the wastewater is highly biodegradable with a high methane potential. Many anaerobic technologies have been implemented for the treatment of wastewater, each having its own inherent problems. The untreated or partially treated wastewater discharged from the sago wastewater results in excessive environmental damage, especially in the pollution of rivers, wells, and groundwater (Parthiban et al. 2007a, 2008b).

9.5.1 Characteristics of Synthetic Sago Wastewater

The synthetic sago wastewater was prepared by mixing a finely ground sago powder (–500 + 400 μm DIN standard sieve size) with ordinary tap water to represent the typical composition of industrial sago wastewater. The trace elements, including sulfide, were added to the synthetic

TABLE 9.1

Composition of Stock Solution of Nutrients and Trace Elements

Nutrients and Trace Elements	Concentration
Macronutrients	
NH_4Cl	170 g/L
KH_2PO_4	37 g/L
$CaCl_2 \cdot 2H_2O$	8 g/L
$MgSO_4 \cdot 4H_2O$	9 g/L
Trace Elements	
$FeCl_3 \cdot 4H_2O$	2000 mg/L
$CaCl_2 \cdot 6H_2O$	2000 mg/L
$MnCl_2 \cdot 4H_2O$	500 mg/L
$CuCl_2 \cdot 2H_2O$	30 mg/L
$ZnCl_2$	50 mg/L
H_3BO_3	50 mg/L
$(NH_4)_6Mo_7O_{24} \cdot 4H_2O$	90 mg/L
$Na_2SeO_3 \cdot 5H_2O$	100 mg/L
$NiCl_2 \cdot 6H_2O$	50 mg/L
EDTA	1 mL/L
HCl 35.5%	500 mg/L
Resazurin	500 mg/L
Sulfide	
$Na_2S \cdot 9H_2O$	100 g/L (prepared and added freshly)
Yeast extract	0.2 g yeast extract/L of synthetic wastewater

wastewater at a concentration of 1 mL/L and the macronutrient concentration was 2 mL/L, as given in Table 9.1. They were added for the growth and activity of the anaerobic bacteria, which ensured that there was no nutrient limitation. The pH of the synthetic wastewater was maintained by adding ammonium carbonate or ammonium dihydrogen phosphate.

The characteristics of the synthetic sago wastewater are presented in Table 9.2.

The parameters of the ATFBR, such as the temperature, pH, COD, BOD, biogas production rate, total volatile fatty acids (VFA), alkalinity, and oxidation reduction potential (ORP), were monitored daily for the influent and treated wastewater from the ATFBR. The VFA composition was measured using gas chromatography (Netel, India). The components of the VFA were acetic acid, propionic acid, and butyric acid. All analytical determinations of the parameters COD, BOD, total VFA, and bicarbonate alkalinity were performed according to *Standard Methods for the Examination of Water and Wastewater* (APHA, AWWA, WEF 2005).

TABLE 9.2

Characteristics of Synthetic Sago Wastewater

	Influent Value[a]	
Parameter	**Minimum**	**Maximum**
pH	6.8 ± 0.2	7.0 ± 0.2
COD	1100	7000
BOD_5	870	5500
Alkalinity (as $CaCO_3$)	350	870
Total solids	1400	1910
Total dissolved solids	1000	1200
Total suspended solids	400	710
Volatile suspended solids	300	570
Total phosphorus (as P)	5	21
Kjeldahl nitrogen (as N)	8	40

[a] All values except the pH are expressed in milligrams per liter.

9.5.2 Experimental Setup

The experimental setup, the materials used, and the methodology followed are illustrated in the following sections. Having fixed the objectives of the research work on the basis of a literature review, a detailed experimental program and the methods of analysis are presented.

9.5.2.1 Equipment Specification

A schematic diagram of the experimental setup is shown in Figure 9.2. The ATFBR consisted of a conical-shaped acrylic column having a 5° taper angle with a total volume of 7.8 L and the volume of the tapered section was 1.5 L. The reactor column was 290 mm in height with a progressive increase in the diameter from 46.6 mm at the base to 91.5 mm at the top. An upper settling section was attached to it, which was 1073 mm in height and 91.5 mm in diameter. A bed of mesoporous activated carbon (MAC) was used as the fluidized biomass carrier matrix.

The effluent was recycled using a magnetic-driven polypropylene centrifugal pump. Complete fluidization of the MAC was achieved by operating at a constant rate. The settlement zone of the reactor contained a conical gas–liquid separator to allow venting of the biogas produced.

Ports were provided along the column length to measure the pressure drop during the operation. Synthetic sago wastewater was applied continuously at the bottom of the reactor, using a peristaltic pump for low flow rates and a magnetic-driven polypropylene centrifugal pump for higher flow rates. The treated wastewater was collected from an outlet located in the cylindrical section at a distance of 55 mm below the top of the column. The biogas

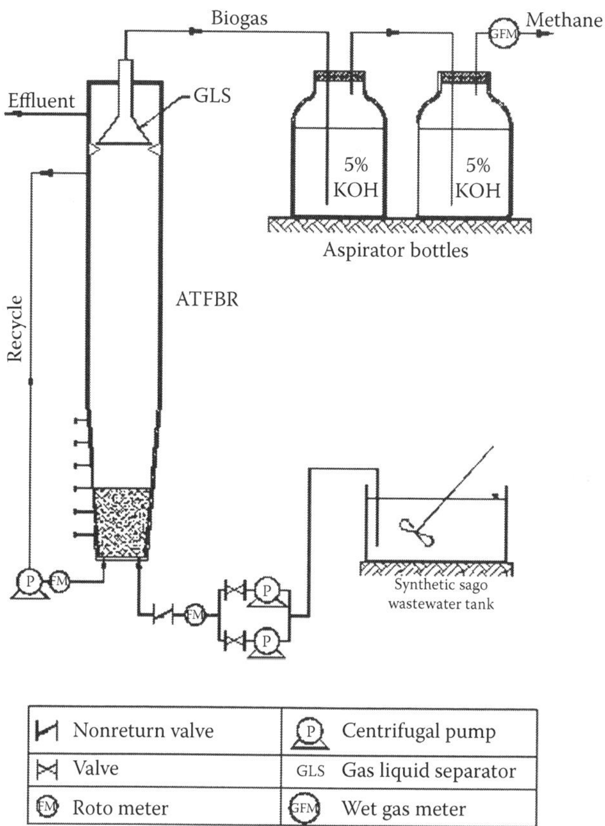

FIGURE 9.2
A schematic diagram of the anaerobic tapered fluidized bed reactor experimental setup.

produced from the ATFBR was scrubbed in 5% KOH solution contained in two aspirator glass bottles of capacity 20 L (borosilicate glass) arranged in series. The scrubbed gas, mainly CH_4, was measured using a wet gas meter (Racine, India).

9.5.3 Results and Discussion

9.5.3.1 Determination of Minimum Fluidization Velocity

The curve shown in Figure 9.3 can be classified into two sections. In the first section (velocity 0–67 m/h), the pressure drop increased linearly from 0 to 1800 Pa with an increase in the superficial liquid velocity. This may be attributed to the high resistance offered by the packed bed.

In the second section, the pressure drop remained constant at 1600 Pa, despite the increase in the superficial liquid velocity from 67.0 to 91.0 m/h.

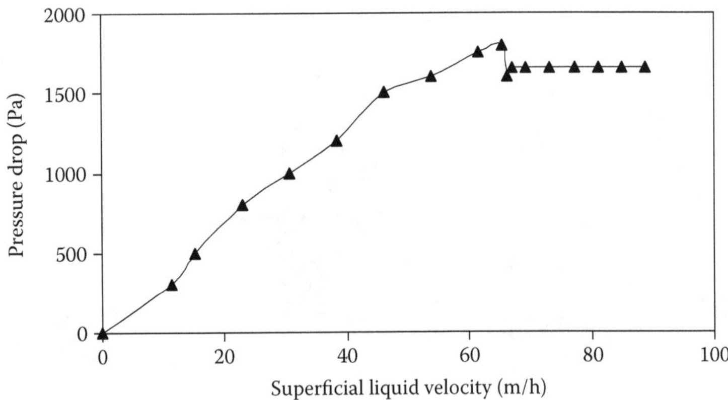

FIGURE 9.3
The pressure drop as a function of the superficial liquid velocity in ATFBR.

At a velocity of 67.0 m/h, the MAC particles in the bed were about to be separated from each other; however, the particles were still in contact with one another. As the velocity was further increased beyond 67.0 m/h, the particles were separated and true fluidization began.

This characteristic fluidization behavior of MAC that was observed in the current research resembled the observation reported by Peng and Fan (1995) and Senthilkumar et al. (1997).

9.5.3.2 Performance of ATFBR during Start-Up Period

The ATFBR was operated at ambient conditions. The room temperature was recorded daily, and it fluctuated between 28°C and 35°C during the entire period of the study. The experimental results obtained during the start-up of the reactor are presented in Figures 9.4 through 9.7.

Figure 9.4a shows the COD concentration in the influent and the treated wastewater. The COD (outlet) values decreased with an increase in the period of operation during the start-up of the reactor. This observation corroborates with that of Hsu and Sheih (1993), treating acetic acid as the substrate. It is known that during the start-up period, the dissolved organics contributed by starch were hydrolyzed by hydrolytic enzymes, acidified by acidogenic bacteria, and converted into CH_4/CO_2 by methanogenic bacteria, contributing to a decrease in COD. After reaching the steady state, there was no appreciable increase in the COD removal.

Figure 9.4b shows the profile of the influent BOD applied to the reactor and BOD of the treated wastewater as a function of time (days). Figure 9.4b illustrates the response of the reactor to the influent BOD from 860 to 140 mg/L over a period of 50 days, and it resembles the COD profile.

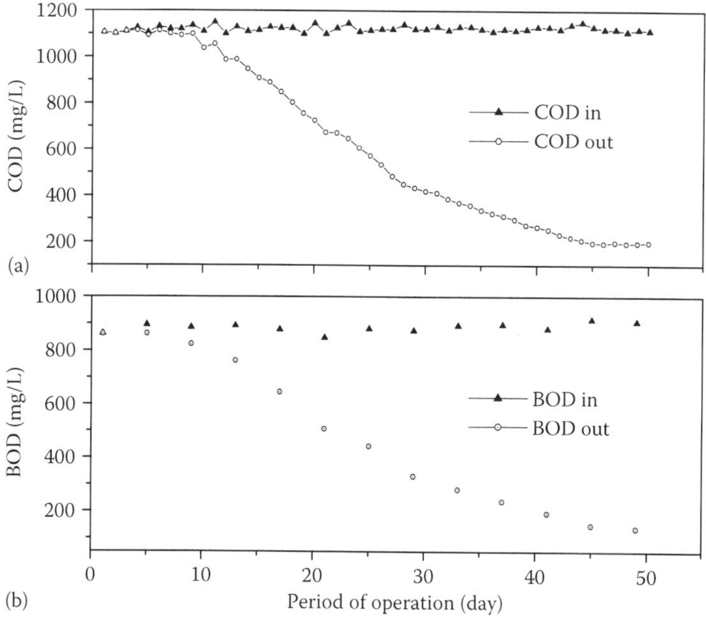

FIGURE 9.4
A profile of (a) the COD inlet and COD outlet and (b) the BOD inlet and BOD outlet as a function of the period of operation during the start-up of the reactor.

Figure 9.5a shows the profile of the influent OLR applied to the ATFBR during the start-up period. The reactor was started with a low OLR of 0.3 kg/m^3 day. A very low OLR was preferred to prevent the washout of the inoculated biomass (Hickey et al. 1991). OLR was increased in a stepped manner to prevent the shock loading of the reactor up to the maximum of 1.00 kg/m^3 day. The difference in OLR for successive stages was 0.5 kg/m^3 day. From Figure 9.5b, it can be observed that the percentage of COD removal was very low in the initial phase for about 10 days and it gradually increased with the period of operation and stabilized after 45 days. This is due to the acclimatization of the immobilized biomass for the destruction of the organic compounds in the wastewater. The removal of COD was 82% after 45 days of the start-up of the reactor. The start-up period was assumed to be completed at the end of the 45th day because there was no further increase in COD removal even after 50 days of operation of the reactor. COD removal was 85% during the start-up period of 45 days in the current research when compared to 83% COD removal in a circular AFBR for synthetic sago wastewater. This is in agreement with the findings of Saravanane et al. (2001).

The methane gas formed as a function of time, in the current research, is shown in Figure 9.5c. It was observed that the methane yield increased with an increase in time and stabilized at a value of 2.10×10^{-3} m^3/day

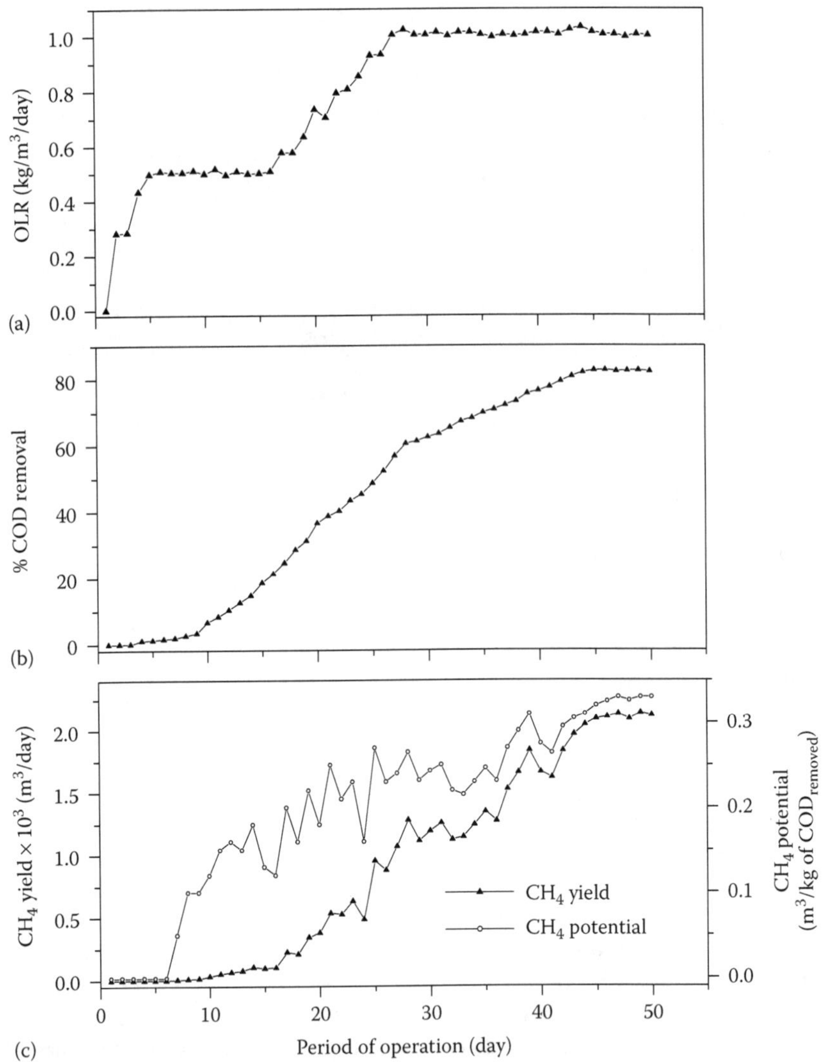

FIGURE 9.5
The variation of (a) OLR, (b) % COD removal, and (c) CH_4 yield and potential with the period of operation during the start-up of the reactor.

(0.330 m^3/kg of $COD_{removed}$). The formation of VFAs was very high in the initial phase of the start-up period, even though COD removal was less.

The variation in the VFA of the treated wastewater with the period of operation is shown in Figure 9.6a. The total VFA increased from 140 mg/L on 0th day to 680 mg/L on the 19th day. The higher levels of VFA in the treated wastewater during the initial phase of the operation indicate a faster rate of fermentation of the dissolved organics in the influent (Van Haandel and

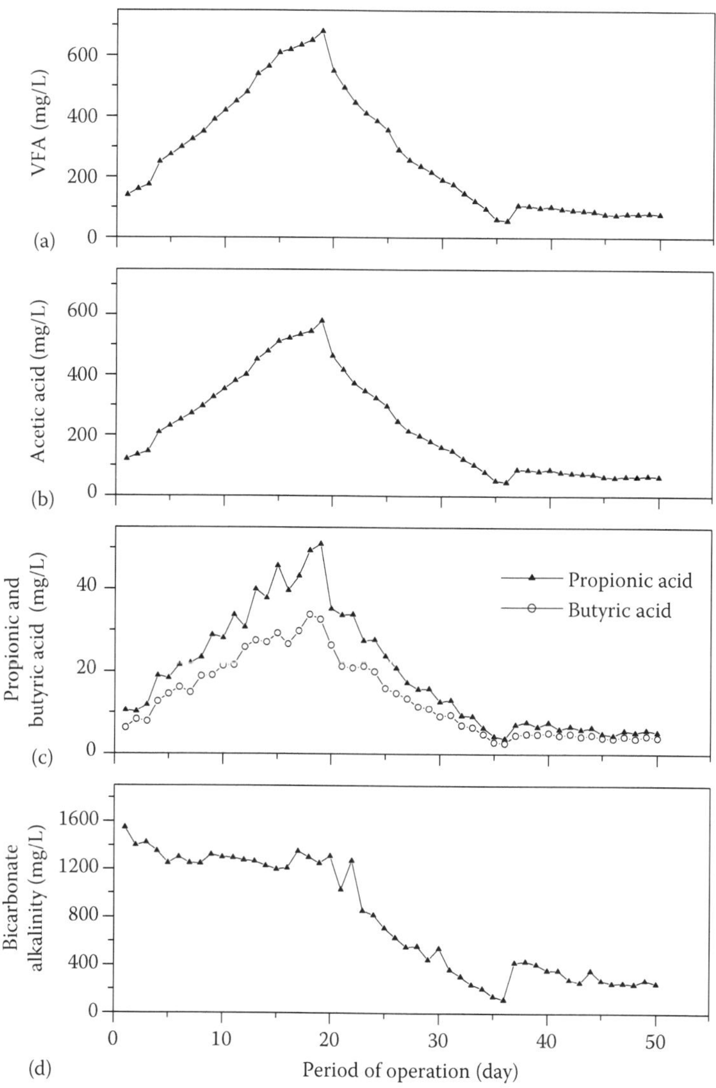

FIGURE 9.6
The variation of (a) VFA, (b) acetic acid, (c) propionic acid and butyric acid, and (d) bicarbonate alkalinity with the period of operation during the start-up of the reactor.

Lettinga 1994). There was a decrease in the VFA concentration to 105 mg/L on the 37th day, and it stabilized at this value, which could be attributed to the onset of methane gas formation, indicating a healthy anaerobic environment and an increased methanogenic activity in ATFBR. A similar profile of acetic acid, propionic acid, and butyric acid as that of VFA can be observed in Figures 9.6b and 9.6c, respectively. It is expected that the excessive production

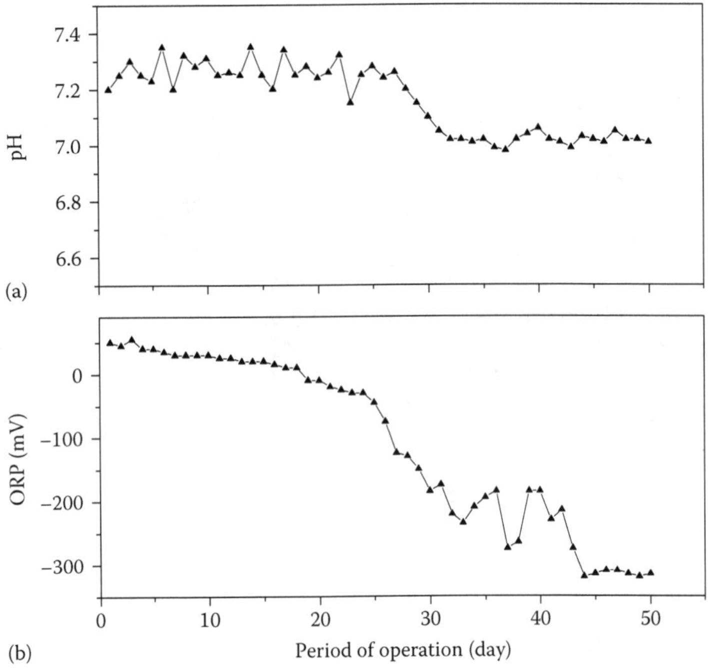

FIGURE 9.7
The profile of (a) pH and (b) ORP as a function of the period of operation during the start-up of the reactor.

of VFA in the initial phase of the start-up period (0–20 days) generally leads to a decrease in the pH such that the survival of the methanogenic bacterial species and, further, the acidogens becomes difficult. The drastic decrease in the pH is offset by the development of the buffering system in the reactor as a consequence of the establishment of the following equilibrium reaction (Equation 9.3):

$$H_2CO_3 \rightleftharpoons HCO_3 \rightleftharpoons CO_2. \tag{9.3}$$

The onset of equilibrium is indicated by the generation of bicarbonate alkalinity (Parthiban et al. 2007b,c).

The variation in the bicarbonate alkalinity of the treated water is shown in Figure 9.6d. The bicarbonate alkalinity in the treated wastewater varied between 1550 and 250 mg/L during the start-up of the reactor. These values are in agreement with the observations made by Rajesh Banu et al. (2006) on the treatment of synthetic sago wastewater in a hybrid upflow anaerobic sludge blanket reactor (HUASB). The bicarbonate alkalinity was stabilized at a range of 250–350 mg/L from the 40th day onward.

The decrease in VFA is due to its dissociation into methane and CO_2. The CO_2 undergoes a reduction to form methane, performed by the methanogenic bacteria and the hydrogenotrophic bacteria present in the system. The formation of methane was manifested in the form of COD removal (Figure 9.5c).

The variation in the pH of the treated wastewater as a function of time is shown in Figure 9.7a. The pH of the treated water was in the range between 7.3 and 7.4 until the 30th day, and it stabilized at a value of 6.9 ± 0.1 after 35 days.

Figure 9.7b illustrates the variation of ORP as a function of the period of operation of the reactor. ORP reached a value of –320 mV from +50 mV over the period of 44 days.

9.5.3.3 ***Optimization of Fluidization Velocity in ATFBR***

After the start-up operation, continuous experiments were carried out to optimize the MAC bed height and the fluidization velocity with regard to the percentage of COD removal and methane yield for synthetic sago wastewater.

After the start-up process, the continuous flow experiments in ATFBR were initiated at a fluidization velocity of 70.32 m/h. It was gradually increased to a fluidization velocity in the order of 82.08, 87.94, and 90.87 m/h. The COD concentration at the outlet for each of the fluidization velocities was measured to determine the percentage of COD removal in ATFBR. The variation in the methane yield (m^3/kg of $COD_{removed}$) was recorded for each fluidization velocity. The unsteady-state behavior was observed during the transition from one fluidization velocity to another. The reactor was operated continuously until the steady state was attained for each fluidization velocity at a given COD concentration.

Figure 9.8 illustrates the percentage of COD removal and the methane yield under the steady-state condition for a period of 10 days with a MAC bed height of 0.17 m (H_1) for different fluidization velocities. COD removal was 84% and the methane yield was 2.4×10^{-3} m^3/day at 87.94 m/h for an inlet COD concentration of 3000 mg/L. COD removal and methane yield increased to 89% and 2.4×10^{-3} m^3/day, respectively, at 87.94 m/h. A further increase in the fluidization velocity retarded the efficiency of the COD removal to 80% and the methane yield was reduced to 1.4×10^{-3} m^3/day. This could be attributed to the lack of contact time for the dissolved organics with the immobilized biomass. Hence, the fluidization velocity of 87.94 m/h was considered as the optimum velocity for the removal of COD and methane gas generation. The corresponding fluidization velocity at the circular cross section of the ATFBR was 22.81 m/h, which is in agreement with the range of values (15–30 m/h) reported in the literature for AFBR (Heijnen et al. 1989).

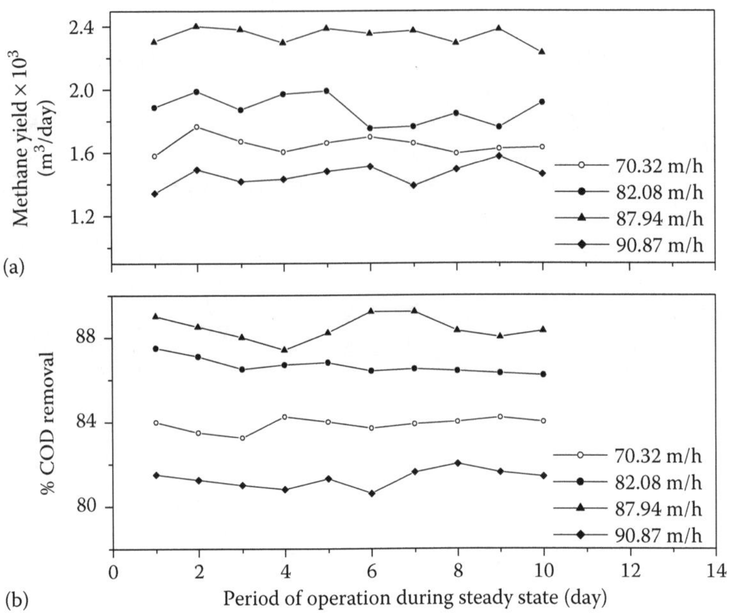

FIGURE 9.8
(a) The variation of the methane yield with the fluidization velocity and (b) the variation of the COD removal (%) with the fluidization velocity.

9.5.3.4 Determination of Optimum OLR for ATFBR

In order to determine the optimum OLR for ATFBR, the reactor was started afresh at a flow rate of 7×10^{-3} m³/day and an influent COD concentration of 1100 mg/L. The flow rate was gradually increased in steps of 10% until the flow rate was 16×10^{-3} m³/day. Then, keeping the flow rate at 16×10^{-3} m³/day, the influent COD concentration was increased gradually in steps of 10% to 7000 mg/L (Figure 9.9a). This took 120 days and from the 125th day onward, the influent COD concentration was maintained at 7000 mg/L over the period of 535 days.

Figure 9.9b shows the profile of the influent BOD applied to the reactor and the BOD of the treated wastewater as a function of time (days). Figure 9.9b illustrates the response of the reactor to the stepped increase in the influent BOD from 860 to 5580 mg/L over the period of 535 days.

From the 125th day onward, the influent COD concentration was maintained at 7000 mg/L and the flow rate was increased gradually by 10%, resulting in a reduction in HRT, and consequently there was an increase in the OLR. The incremental increase of 10% in OLR between the two successive applications was kept constant for 3–5 days to stabilize the reactor. OLR was increased in a stepped manner up to 85.44 kg/m³ day over the period of 535 days (Figure 9.10a).

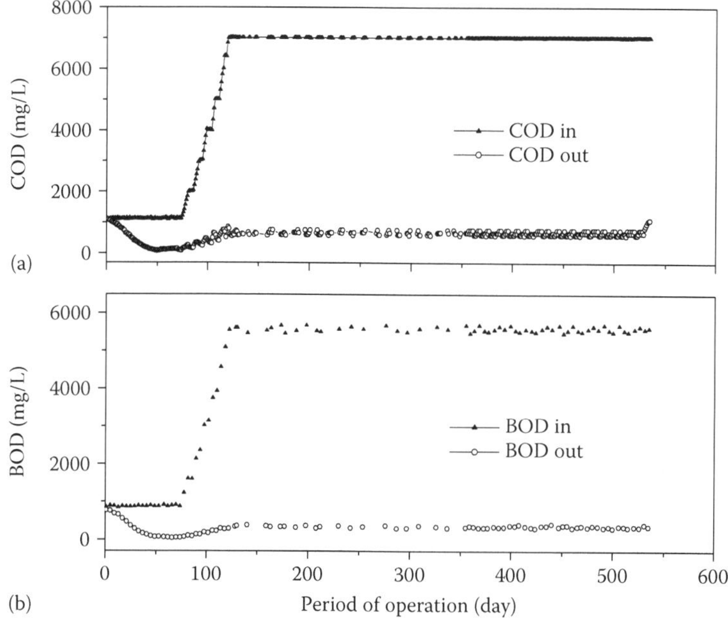

FIGURE 9.9
A profile of (a) the COD inlet and COD outlet and (b) the BOD inlet and BOD outlet as a function of the period of operation.

It is evident from Figure 9.10b that the maximum COD removal was 92% for an OLR of 83.7 kg/m^3 day. The methane production dropped significantly for the OLR beyond 83.7 kg/m^3 day (Figure 9.10c).

The methane production also increased gradually and attained the maximum value of 0.34 m^3/kg COD$_{removed}$, which corresponds to 25.36 m^3/m^3 of the reactor per day for an OLR of 83.7 kg COD/m^3 day (Figure 9.10c), for the maximum COD reduction of 92% (Figure 9.10b). The production of methane after the start-up period for an OLR of 1.0 kg/m^3 day was 0.345 m^3/kg of COD$_{removed}$. The system was disturbed to a lesser extent by an increase in OLR and resulted in a COD removal efficiency of 87%–89%, and consequently the methane generation was reduced. The methane gas generation resumed to the maximum value in the range 0.335–0.345 m^3/kg of COD$_{removed}$ when the reactor was operated at a constant OLR for 2–3 days continuously.

One of the factors that contributed to the poor efficiency of the conventional ATFBR is the soluble biogas in the wastewater. The soluble biogases are released into the atmosphere during subsequent aerobic treatment and also contribute to global warming. The suggested methodology to remove the soluble biogas in ATFBR was by installing agitators. This adds to the investment cost. In the current research, MAC has been incorporated into

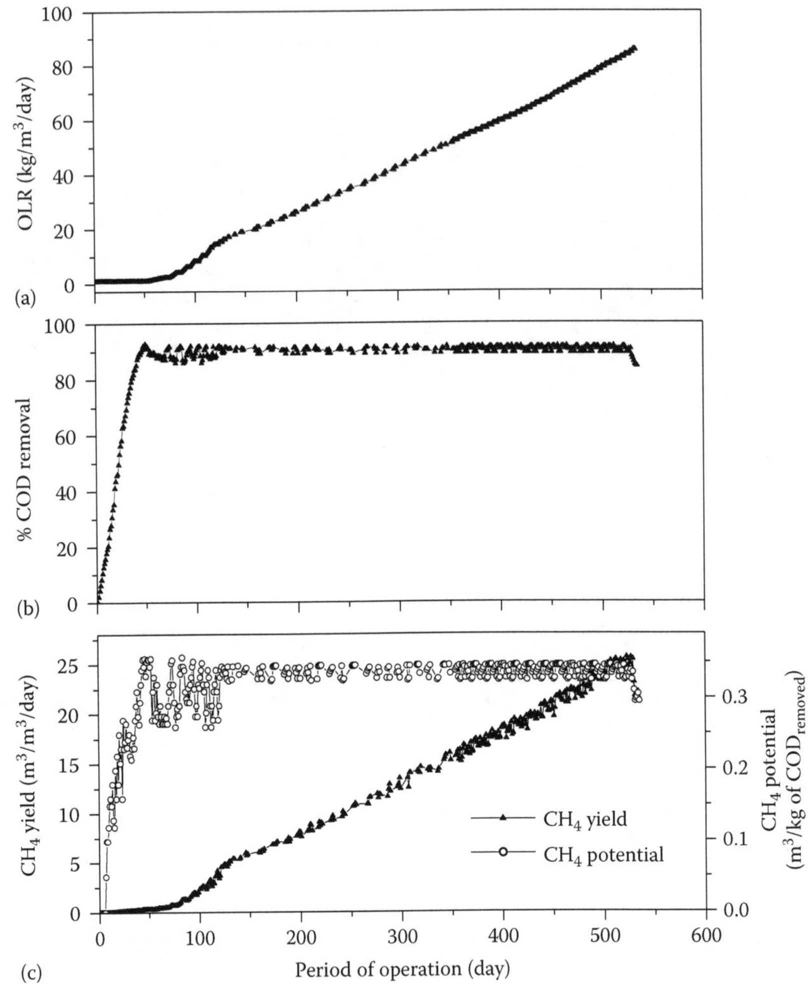

FIGURE 9.10
The variation of (a) OLR, (b) % COD removal, and (c) CH_4 yield and potential with the period of operation.

AFBR to facilitate the removal of the biogas from the wastewater. MAC serves two purposes in increasing the efficiency of ATFBR. MAC is used as a carrier matrix for anaerobic bacteria and as a reservoir for storing methane gas (Liu et al. 2006) and hydrogen gas (Erdogan and Kopac 2007). The methane gas formed during the anaerobic treatment of the wastewater is stored in the pores of the MAC, which increases the buoyancy (lowering of apparent density) of the activated carbon particles, so that it rises to the surface of the liquid medium where it is released. The MAC particles come down due to gravity settling. This transport of carbon particles creates a

TABLE 9.3

Comparison of OLR and Methane Yield with the Literature

Literature Cited	Reactor	Wastewater	Carrier	OLR (kg/m³ day)	% COD Removal	Methane Gas Yield (m³/kg) $COD_{removed}$
Saravanane et al. (2001)	AFBR	Synthetic sago	GAC	66.3	82.5	0.2–0.25
Rajesh Banu et al. (2006)	HUASB	Synthetic sago	PUF	23.5	87	0.06–0.11
Chen et al. (1997)	AFBR	Hog	GAC	10.4	80	0.35
Chen et al. (1998)	AFBR	Tannery	GAC	—	75	0.22
Present study	ATFBR	Synthetic sago	MAC	87.3	92	0.335–0.345

secondary swirling motion over and above the fluidized motion. The biofilm attached to the MAC also follows the secondary motion, thereby mixing homogeneously with the wastewater. The second major advantage of MAC is that it immobilizes the bacterial consortia present in the wastewater, and the immobilized bacteria degrade the organic components of the wastewater at a faster rate.

The results obtained in the current research were compared with the values reported in the literature on methane gas generation and COD removal efficiency for the treatment of synthetic sago wastewater and hog and tannery wastewater (Table 9.3). It was observed that the COD removal efficiencies and methane production were less when compared with the ATFBR.

VFA was observed to increase in the treated synthetic wastewater as OLR was increased and the maximum concentration of 345 mg/L was recorded at an OLR of 85.44 kg COD/m³ day. The impact of VFA accumulation was reflected in the marked decrease in COD removal efficiency from 92% to 85% for an increase in OLR from 83.70 kg COD/m³ day, as evident from Figure 9.11a. When there is an increase in OLR, the acetic acid concentration decreases (Figure 9.11b) and the propionic acid and butyric acid concentrations increase (Figure 9.11c); therefore, there was a considerable reduction in the methane gas generation. A similar observation was reported by Fang et al. (1994) that the VFA (accumulated) concentration was responsible for souring the UASB reactor while working on the synthetic dairy wastewater.

Bicarbonate alkalinity was decreased from a value of 1550 to over 750 mg/L over the period of 535 days, as shown in Figure 9.11d. The steady decrease in the bicarbonate alkalinity value was observed despite the increase in OLR during the operation of the reactor.

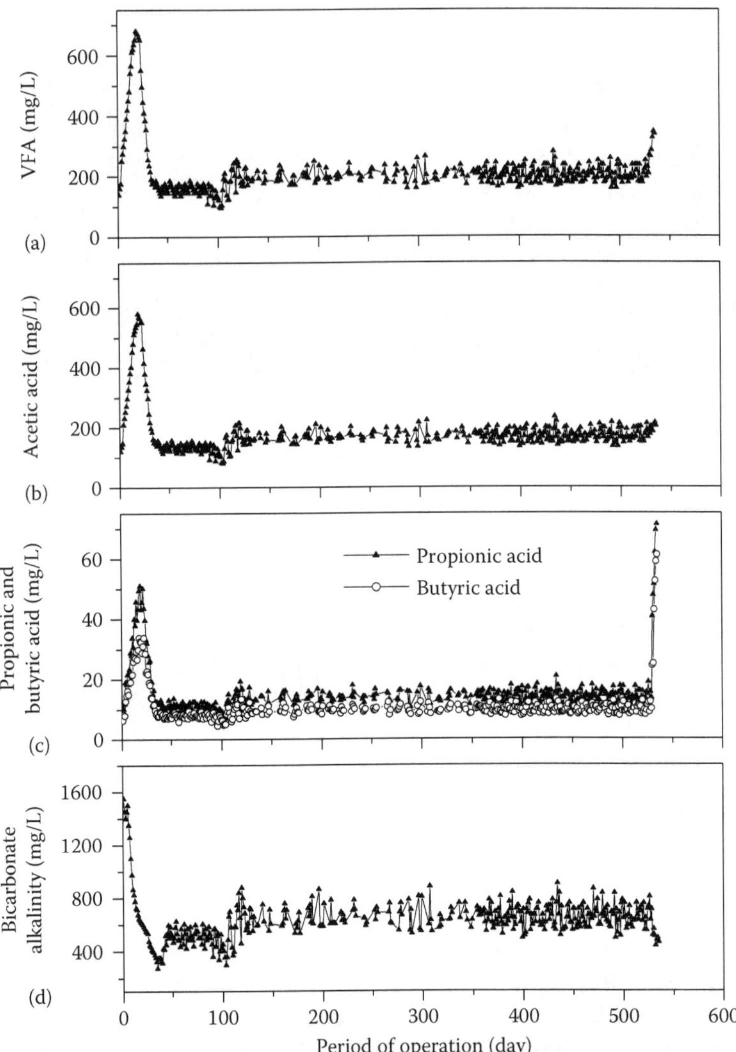

FIGURE 9.11
The variation of (a) VFA, (b) acetic acid, (c) propionic acid and butyric acid, and (d) bicarbonate alkalinity with the period of operation.

Figure 9.12a depicts the pH profile of ATFBR over the period of 85 days for the application of OLR from 1.0 to 85.44 kg/m^3 day. The pH was decreased from 7.4 to 7.0 during the period up to 40 days, and it was maintained at a value of 7.0 ± 0.1 for the rest of the period. The pH was constant at a value of 7.0 ± 0.1 after the start-up period of 45 days.

Figure 9.12b illustrates the variation of ORP as a function of the period of operation of the reactor. ORP reached a value of –550 mV from +45 mV on the

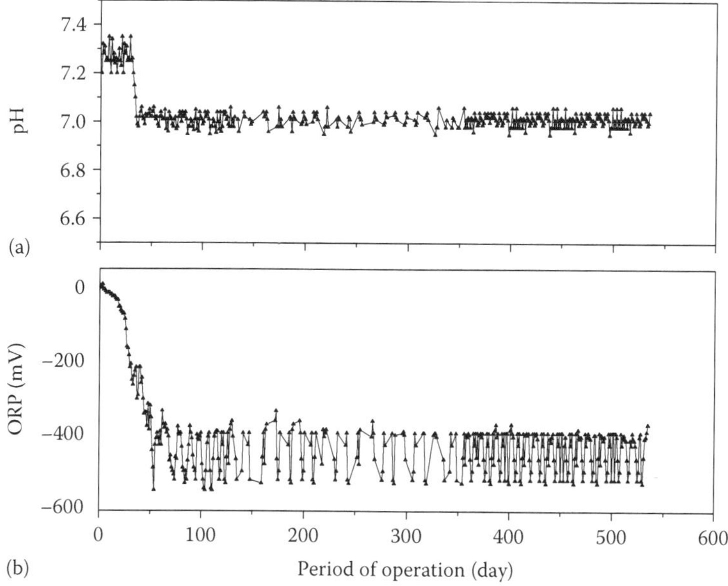

FIGURE 9.12
A profile of (a) pH and (b) ORP as a function of the period of operation.

54th day. It was observed that the ORP value fluctuated between −375 and −550 mV in response to an increase in OLR during the period of operation between 54 and 535 days.

9.6 Conclusions

The minimum fluidization velocity of ATFBR was found to be 67.0 m/h. The start-up period of the reactor was 45 days at a flow rate of 7×10^{-3} m^3/day for a COD concentration of 1100 mg/L. The optimum OLR of the reactor was 83.70 kg COD/m^3 day. The COD removal efficiency was 92% for all the flow rates with a methane yield of 0.340–0.345 m^3/kg of $COD_{removed}$. The COD removal efficiency and methane formation were substantially affected by an increase in the OLR beyond 83.70 kg COD/m^3 day.

9.7 Scope and Directions of Future Work

The present study can be extended to the anaerobic treatment of the wastewater discharged from the sago industries and also of the industrial wastewaters containing natural and synthetic chemicals. The possible effect of inhibition due to sulfate and sulfite on the generation of methane and thus on the efficiency of the reactor needs to be studied using the ATFBR. Studies can be extended to high-strength wastewaters discharged from the distillery, tannery, pulp and paper, and pharmaceutical industries. The treatment of low-strength wastewaters discharged from the domestic sector using the ATFBR can be another area for further study. The fundamental aspects, such as biofilm development, its morphology, the enzyme release kinetics, and the substrate utilization mechanism in ATFBR, need to be addressed. The mechanism for the in situ turbulence caused by MAC in the reactor needs to be addressed. Studies on the effect of inverse fluidization using particles of density less than that of the industrial wastewaters can form a logical extension to this work. The present study can be extended to study the online control of the anaerobic fluidized bed process using the ANFIS model developed, since the adaptive control is capable of predicting the effects of small changes in the control variables on the response of the microbial systems.

References

APHA, AWWA, and WEF. 2005. *Standard Methods for the Examination of Water and Wastewater*, 21st edn. Washington, DC: American Public Health Association.

Archer, D.B., M.G. Hilton, P. Adams, and H. Wiecko. 1986. Hydrogen as a process control index in a pilot scale anaerobic digester. *Biotechnol. Lett.* 8: 197–202.

Bryers, J.D. 1985. Structural modeling of the anaerobic digestion of biomass particulate. *Biotechnol. Bioeng*. 27: 638–649.

Chen, C.Y., C.T. Li, and W.K. Shieh. 1997. Anaerobic fluidized bed pretreatment of hog wastewater. *J. Environ. Eng. – ASCE.* 123: 389–394.

Chen, S.J., C.T. Li. and W.K. Shieh. 1998. Anaerobic Fluidized Bed Treatment of a Tannery Waste Water. *Chem. Eng. Res. Des.* 66a: 518–523.

Denac, M. and I.J. Dunn. 1998. Modeling dynamic experiments on the anaerobic degradation of molasses wastewater. *Biotechnol. Bioeng*. 31: 1–10.

Dunn, I.J., E. Heinzle, J. Ingham, and J.E. Prenosil. 1992. *Biological Reaction Engineering – Principles Applications and Modeling with PC Simulation*. Weinheim: VCH.

Eckenfelder, W.W. 1999. *Industrial Water Pollution Control*. Boston, MA: McGraw-Hill.

Erdogan, F.O. and T. Kopac. 2007. Dynamic analysis of sorption of hydrogen in activated carbon. *Int. J. Hydrogen Energy* 32: 3448–3456.

Evans, G. 2001. *Biowaste and Biological Waste Treatment*. London: Earth Scan.

Fang, H.H.P., G. Liu, Z. Jinfu, C. Bute, and G. Guowei. 1994. Treatment of brewery effluent by UASB process. *J. Environ. Eng. ASCE* 116: 454–460.

Heijnen, J.J., A. Mulder, W. Enger, and F. Hoeks. 1989. Review on the application of anaerobic fluidized bed reactors in wastewater treatment. *Chem. Eng. J.* 41: B37–B50.

Hickey, R.F., W.M. Wu, M.C. Veiga, and R. Jones. 1991. Start-up, operation, monitoring and control of high-rate anaerobic treatment systems. *Water Sci. Technol.* 24: 207–255.

Hien, P.G., L.T.K. Oanh, N.T. Viet, and G. Lettinga. 1999. Closed wastewater system in the tapioca industry in Vietnam. *Water Sci. Technol.* 39: 89–96.

Holland, K.T., J.S. Knapp, and J.G. Shoesmith. 1987. *Anaerobic Bacteria.* New York: Chapman and Hall.

Hsu, Y. and W.K. Sheih. 1993. Start up of anaerobic fluidized bed reactors with acetic acid as the substrate. *Biotechnol. Bioeng.* 41: 347–353.

Huang, J.S., J.L. Yan, and C.S. Wu. 2000. Comparative bioparticles and hydrodynamic characteristics of conventional and tapered anaerobic fluidized bed bioreactors. *J. Chem. Technol. Biotechnol.* 75: 269–278.

Kleinstreuer, C. and T. Poweigha. 1982. Dynamic simulator for anaerobic digestion processes. *Biotechnol. Bioeng.* 24: 1941–1951.

Liu, X., L. Zhou, J. Li, Y. Sun, and Y. Zhou. 2006. Methane sorption on ordered mesoporous carbon in the presence of water. *Carbon* 44: 1386–1392.

Mai, H.N.P., H.T. Duong, T.T.T. Trang, and N.T. Viet. 2004. UASB treatment of tapioca processing wastewater in South Vietnam. In *Proceedings of the 10th World Congress – Anaerobic Digestion 2004*, 4, pp. 2420–2424. National Research Council, Canada, QC.

Malaspina, L.S., C.M. Cellamare, and A. Tilche. 1995. Cheese whey and cheese factory wastewater treatment with a biological anaerobic-aerobic process. *Water Sci. Technol.* 32: 59–72.

Parthiban, R., P.V.R. Iyer, and G. Sekaran. 2007a. Anaerobic tapered fluidized bed reactor for starch wastewater treatment and modeling using multilayer perceptron neural network. *J. Environ. Sci.* 19: 1416–1423.

Parthiban, R., P.V.R. Iyer, and G. Sekaran. 2007b. Anaerobic digestion in a tapered fluidized bed reactor and modeling using radial basis function neural network. *Biotechnology* 6: 534–542.

Parthiban, R., P.V.R. Iyer, and G. Sekaran 2007c. Time lagged recurrent neural network modeling for an anaerobic tapered fluidized bed reactor – Treatment of sago industry effluent. *Indian J. Environ. Prot.* 27: 839–848.

Parthiban, R., P.V.R. Iyer, and G. Sekaran. 2008a. Tapered fluidized bed reactors for wastewater treatment – A critical review. *J Indian Public Health Eng.* 8: 30–34.

Parthiban, R., P.V.R. Iyer, and G. Sekaran 2008b. Anaerobic tapered fluidized bed reactor for the treatment of sago industry effluent. *Indian Chem. Eng.* 50: 323–333.

Peng, Y. and L.T. Fan. 1995. Hysteresis in liquid solid tapered fluidized beds. *Chem. Eng. Sci.* 50: 2668–2671.

Quasim, S.R. 1999. *Wastewater Treatment Plants.* Lancaster-Basel: Technomic.

Rajesh Banu, J., S. Kaliappan, and D. Beck. 2006. High rate anaerobic treatment of sago wastewater using HUASB with PUF as carrier. *Int. J. Environ. Sci. Technol.* 3: 69–77.

Saravanane, R., D.V.S. Murthy, and K. Krishnaiah. 2001. Anaerobic treatment and biogas recovery for Sago wastewater management using a fluidized bed reactor. *Water Sci. Technol.* 44: 141–147.

Sastry, C.A. and G.J. Mohan. 1963. Anaerobic digestion of industrial wastes. *Indian J. Environ. Health* 5: 20–25.

Scott, C.D. and C.W. Hancher. 1976. Use of a tapered fluidized bed as a continuous bioreactor. *Biotechnol. Bioeng*. XVIII: 1393–1403.

Senthilkumar, R., M. Hari, D. Prakash, M. Lima Rose, and M. Velan. 1997. Studies in tapered fluidized bed. *Hung. J. Ind. Chem*. 25: 281–286.

Shieh, W.K and V.T. Nguyen. 2000. Anaerobic Treatment. In *Wastewater Treatment*, eds. Liu, D.H.F. and B.G. Lipták, 208–214. Boca Raton, FL: CRC Press LLC.

Speece, R.E. 1983. Anaerobic biotechnology for industrial wastewater treatment. *Environ. Sci. Technol*. 17: 416a–427a.

Speece, R.E. 1996. *Anaerobic Biotechnology for Industrial Wastewaters*. Nashville, TN: Archae Press, pp. 394–400.

Van Haandel, A.C. and G. Lettinga. 1994. *Anaerobic Sewage Treatment: A Practical Guide for Regions with a Hot Climate*. Chichester, UK: Wiley.

Wiesmann, U. 1988. Kinetics and reaction engineering of anaerobic sewage treatment. *Chem. Ing. Technol*. 60: 464–474.

Wu, C.S. and J.S. Huang. 1996. Bioparticle characteristics of tapered anaerobic fluidized-bed bioreactors. *Water Res*. 30: 233–241.

Yang, S.T. and M.R. Okos. 1987. Kinetic study and mathematical modeling of methanogenesis of acetate using pure cultures of methanogens. *Biotechnol. Bioeng*. 30: 661–667.

10

Treatment of Effluent Waters in Food Processing Industries

D. G. Rao, N. Meyyappan, and S. Feroz

CONTENTS

10.1 Introduction

The food processing industry (FPI) is a typical processing industry in which water plays a vital role, such as the following:

1. As one of the ingredients in processing
2. For potable purpose by the plant personnel
3. For cleaning various items, such as raw materials, finished products, packaging materials, and containers and for washing and rinsing the process equipment
4. For associated operations, such as cooling water in air conditioning and refrigeration systems and as boiler feedwater for steam generation

Thus, the process industries consume a huge amount of water. Even a conservative estimate puts the water consumption rate in most of the food industries in the order of 1 million gallons of potable water per day. The typical rates of water consumption in various process industries are shown in Table 10.1. Much of it is discharged in the form of *effluent*, which needs to be processed before being discharged or reused. No effluents are discharged from previously mentioned applications (1), (2), and (4). However, those discharged from (3) require processing. The US Environmental Protection Agency (EPTS 1974) indicated the possible quantities of wastewater that are generated in 10 specialty FPIs (Table 10.2).

10.1.1 FPIs and Their Characterization

Water is typically used in the process industries for various specific and general purposes, such as washing and cleaning the process halls and the factory premises. Incidentally, such waters are not heavily contaminated and can be re-treated for further use. Similarly, the water used for cleaning raw materials such as fruits and vegetables (other than animal-based raw materials) can also be easily treated. At the time of harvest, based on the harvesting pattern, fruits and vegetables catch dust and dirt, which

TABLE 10.1

Water Use in Various Food Processing Industries

Industry	Water Use (gal/ton of Production)
Fruit and vegetable processing	
Green beans	12,000–17,000
Peaches and pears	3,600–4,800
Other fruits and vegetables	960–8,400
Food and beverage industries	
Beer	2,400–3,800
Bread	480–960
Meat packing	3,600–4,800
Milk products	2,400–4,800
Whisky	14,400–19,200

Source: http://www.p2pays.org/ref/09/08853.htm. Clean Technologies in US Industries: Focus on Food Processing. USAEP.

are removed by transporting the fruits and vegetables in a stream of water (known as *hydraulic transportation*). Such waters can be reused several times. If water becomes hard, it needs to softened, lest the hardness would affect the texture of the fruits and the vegetables at the time of blanching. Another very classical use of water in food processing is in process operations, for instance, water is used to make sugar syrups in the beverage industry. In all these operations, water needs to be free from salts because the salts present

TABLE 10.2

Wastewater Generation in Some Specialty Food Industries

Industry Category	Quantity of Wastewater Generated (tons/ton of Production)
Prepared dinners	12
Frozen bakery products	11
Dressings, sauces, and spreads	2.8
Meat specialties	10
Canned soups and baby foods	22
Tomato–cheese–starch combinations	29
Sauced vegetables	85
Syrups, jams, and jellies	2.4
Chinese and Mexican foods	14
Breaded frozen products	48

Source: http://www.p2pays.org/ref/26/25046.pdf. Wastewater Characterization for the specialty Food Industry. EPA.

in water may result in a salt deposition on the walls of the process equipment, which in turn would affect the heat transfer process in addition to spoiling the taste of the syrups.

FPIs generate wastewater that has distinctive characteristics with a high concentration of biochemical oxygen demand (BOD) and suspended solids (SS). Most of the components may be biodegradable and nontoxic, but they are often complex to predict due to the differences in the BOD and the pH. Some of the changes in the effluents from processing vegetables, fruits, and meat products are also due to the seasonal nature and the extent of maturity at the time of processing. High-grade water in large volumes is used in food processing. High loads of particulates, organic compounds, and surfactants are present in vegetable wash waters. The wash water from a slaughtering house contains very strong organic waste, antibiotics, growth hormones, and a variety of pesticides. Salts, flavors, coloring material, acids or alkalis, and oil or fats are present in significant quantities.

Another distinct feature of FPIs is that the effluents coming out of the various process industries are not similar in composition because the processing methods that are employed vary in nature; these processing methods are tabulated in Table 10.3. Various FPIs (shown in Table 10.4) employ one or more of these operations, resulting in the discharge of effluents that need to be characterized.

10.1.2 Characterization of Effluents

Some of the important parameters of water based on which effluents are characterized are

- pH
- Temperature
- Volume/quantity of wastewaters coming out
- Physical parameters such as color, odor, and turbidity
- Biological oxygen demand in 5 days (BOD_5)
- Chemical oxygen demand (COD)
- Total soluble solids (TSS)
- SS
- Volatile soluble solids (VSS)
- Fats
- Total Kjeldahl nitrogen (TKN) (mg/L)
- Total phosphorous (TP)
- Oils and greases

TABLE 10.3

Various Processing Methods Used for Food Preservation

S. No.	Processing Method	Possible Products	Remarks
1	Salting	Drying of fish, fruits, and vegetables	Less effluent waters. Used only for cleaning raw materials and equipment
2	Pickling in • Acid • Oil • Brine	Pickles made out of fruits, vegetables, roots, fish, meat, chicken, etc.	Less effluent waters. Used only for cleaning raw materials and equipment
3	Fermentation	Fermentation of fruits, vegetables, marine products, meat products, cereal dough and batters, some pulse products, and dairy products	Less effluent waters. Used only for cleaning raw materials and equipment
4	Roasting and grinding	Cereals and pulse products; mostly convenience foods	Minimum quantity of effluent waters
5	Drying and dehydration	Drying of fish, fruits and vegetables, animal products, cereals, and pulse products	Less effluent waters, except in animal products. Used only for cleaning raw materials and equipment
6	Freezing	Marine and animal products, fruits and vegetables, and dairy products	Less effluent waters, except in animal products. Used only for cleaning raw materials and equipment
7	Cold storage	Marine and animal products, fruits and vegetables, and dairy products	Less effluent waters, except in animal products. Used only for cleaning raw materials and equipment
8	Canning	Marine and animal products, fruits and vegetables, and dairy products	Effluent waters are mostly from cleaning equipment
9	Concentration and evaporation	Fruit juices, cane sugar, and sugar candies	Effluent waters are mostly from cleaning equipment
10	Packaging: • Packing and wrapping • Aseptic packaging • Modified atmospheric packaging (MAP) • Controlled atmospheric packaging (CAP)	Fresh fruits and vegetables, fresh meat, pulps, and beverages	Effluent waters are mostly from cleaning raw materials
11	Chemical preservation in • Potassium meta bisulfite • Benzoate • Acids	Pulps and beverages	Effluent waters are mostly from cleaning raw materials and equipment

Note: The animal product processing results in large quantities of effluent waters that come from slaughtering and cleaning the raw materials and are contaminated with blood, gut fluids, offal, and vaccines.

TABLE 10.4

Classification of Food Processing Industries Based on Raw Materials

S. No.	Classification of Industry Based on Raw Materials	Various Product Industries
1	Animal-based industries	• Fresh meat processing and packaging • Fresh poultry meat processing • Egg powder plants • Dehydration of fish and meat • IQF units • Canning of fish and meat • Convenience foods • RTE foods
2	Bakery products	• Bread • Biscuits • Cakes and pastries
3	Beverage products	• Alcoholic • Nonalcoholic • Carbonated • Noncarbonated
4	Cereal- and pulse-based products	• Breakfast foods • Flours • Pasta products • Convenience foods • Ready mixes • RTE foods • HAE foods • Puffed products • Edible wrappers
5	Convenience foods and confectioneries	• RTE foods • RTS foods • HAE foods • Chocolates and chiclets • Éclairs • Ready mixes
6	Dairy products	• Chilled milk • Pasteurized milk • Packed milk • Condensed milk • Ice creams • Fermented foods • Cheese • Ghee • Milk powder
7	Fruit- and vegetable-based products	• Dehydrated foods • Canning • Fruit pulps • Freezing and chilling • Modified atmosphere packaging • Fruit bars and toffees • Fruit beverages

TABLE 10.4 (Continued)

Classification of Food Processing Industries Based on Raw Materials

S. No.	Classification of Industry Based on Raw Materials	Various Product Industries
8	Microbiology and fermented foods	• Food enzymes • Alcoholic beverages • Fermented foods
9	Spices and condiments	• Spice powders • Spice mixes • Spice oils • Spice oleoresins
10	Protein and proteinaceous products	• Baby foods • Weaning foods • Breakfast foods • Energy foods • Health foods

Notes: IQF units, individual quick freezing units; RTE foods, ready-to-eat foods; RTS foods, ready-to-serve foods; HAE foods, heat-and-eat foods.

BOD for all food processing wastewaters is relatively high compared with that of the wastewaters from other industries and is used as a gauge to measure the level of treatment needed. Table 10.5 highlights some of the typical values for BOD and SS for some specialty food industries (EPTS 1974). The COD values are approximately double the BOD values. Greases and oil

TABLE 10.5

BOD, COD, and SS Values (mg/L) for Some Food Industries

Industry Category	BOD	COD	SS
Prepared dinners	1900		1500
Frozen bakery products	3200	7000	2200
Dressings, sauces, and spreads	2600		1200
Meat specialties	820	900	460
Meat packing	1433	2746	
Canned soups and baby foods	560		320
Tomato–cheese–starch combinations	370		220
Sauced vegetables	310		250
Syrups, jams, and jellies	2400	4000	400
Chinese and Mexican foods	570		200
Breaded frozen products	2400		3700
Poultry processing	1306	1581	
Dairy	2700	4700	

Source: http://www.p2pays.org/ref/26/25046.pdf. Wastewater Characterization for the specialty Food Industry. EPA.

concentrations are of the order of 0–2000 ppm. The wide variations in the composition of the effluent waters could be primarily due to

- The moisture content of the raw materials, other ingredients, and products
- The various unit operations and processes used in the production process
- The variation and richness of the product ingredients
- The quality of the raw materials and the intermediaries processed elsewhere
- The production capacity, the type of equipment and machinery used, and the sophistication of the equipment used to process the goods
- The management's commitment to ensure quality

Thus, the quality and the quantity of the wastewater generated are dependent on the sophistication of the processing plants and the plant sizes. Generally, if the plant capacity is high, the quantity of the effluent per ton of production will be less. The effluent waters can be conveniently classified as

- Toxic
- Nontoxic
- Nonhazardous

With the exception of a few products, most of the effluents are organic in nature and hence can be treated by biological methods. However, the effluent waters from the animal-based processing industries, namely, meat processing, poultry processing, and fish processing industries (freshwater fish and seafood processing/marine food processing industries), contain high amounts of pathogenic organisms because they contain most of the body fluids, such as blood, gut contents, a number of vaccines (used during rearing and breeding), feathers, hairs, and hoofs. Such waters need a high dosage of treatment before being discharged.

10.1.3 Treatment Methods

The impurities present in the wastewaters are varied in nature, namely,

- SS
- Microorganisms
- Minerals such as iron or manganese in dissolved form
- Organic matter that imparts color or odor or taste
- Dissolved gases, if any

Hence, they require different types of treatment methods, which are classified as

- Physical methods
- Chemical methods
- Biological methods
- Incineration

The previously mentioned methods used for treating the wastewaters consist of a heterogeneous group of ingredients and are shown in Table 10.6 (Rao 2010). Most of the SS materials and the dissolved minerals are treated by *physical methods,* namely, sedimentation, flocculation, filtration, and

TABLE 10.6

Various Treatment Methods for the Purification of Wastewaters

Impurity	Treatment Method	Remarks
(i) Suspended solids/matter	Settling coagulation filtration through sand beds	Done by gravity sedimentation in large tanks Fine suspensions are agglomerated by adding flocculating agents, namely, aluminum sulfate, sodium aluminate, chitosan, etc. Water is passed through a filter bed consisting of fine sand supported on pebbles
(ii) Dissolved minerals such as Fe and Mn in the form of bicarbonates	Aeration, settling, and filtration	Sodium hexametaphosphate is added as a sequestering agent. Mn is oxidized by aeration. The insoluble oxides are removed by filtration
(iii) Dissolved salts in the form of sulfates, chlorides, and bicarbonates of Na, Mg, and Ca	Precipitation process	$Ca(OH)_2$ and Na_2CO_3 precipitate the salts in the form of insoluble hydroxides or sulfates $Ca(HCO)_3 + Ca(OH)_2 \rightarrow 2CaCO_3\downarrow + 2H_2O$ $CaSO_4 + Na_2CO_3 \rightarrow CaCO_3\downarrow + Na_2SO_4$
	Zeolite processes	The cation-exchanging ability of zeolites is exploited to remove the hardness of water $Ca(HCO_3)_2 + Na_2Z \rightarrow 2NaHCO_3 + CaZ$ The zeolite can be regenerated by treatment with NaCl
(iv) Organic matter that imparts color due to the presence of colloidal suspensions of ultramicroscopic particles	Coagulation, settling, and filtration/decanting	Coagulants such as activated silica are used. Aeration and superchlorination are also employed. Adsorption of activated carbon is also recommended by passing through a fixed bed of activated carbon

(*continued*)

TABLE 10.6 (Continued)

Various Treatment Methods for the Purification of Wastewaters

Impurity	Treatment Method	Remarks
(v) Microorganisms in surface waters	Chlorination/ treatment with chlorine derivatives	Common microorganisms are diatoms, fungi, algae protozoa, rotifera, nematodes, and some pathogenic bacteria. Chlorination under optimum pH, temperature, contact time, and sufficient quantity would remove the microorganisms
(vi) Dissolved gases	Coagulation, settling, filtration, softening, chlorination, and degassing	The dissolved gases are usually CO_2, N_2, O_2, and H_2S (in sulfur waters). Sodium sulfite and hydrazine solutions are used for oxygen scavenging

Source: Reproduced with permission from Rao, D.G., *Fundamentals of Food Engineering*, PHI Learning Pvt. Ltd., New Delhi, 2010. Copyright (2010), Prentice-Hall of India, New Delhi.

centrifugation, or by air flotation methods. Aeration or air flotation is an effective method for treating wastewaters, especially to reduce the BOD.

Various treatment methods are summarized in Table 10.7. By using these methods, the contaminated water effluents coming from FPIs are treated. The purpose of the treatment is to make the treated water either potable or reusable in processing. On many occasions, the treated water may not be fit for potable purpose, but it should at least be useful for reuse or it may be used for agricultural purposes. The following steps may have to be considered before choosing the treatment operation (Crittenden et al. 2005):

- The initial effluent water quality and the purpose of its use
- The predesign studies and the design criteria
- The detailed design of the alternative based on the criteria of usage
- The construction, operation, and maintenance of the facilities

In the following sections, we discuss the various treatment methods.

10.2 Classical Treatment Processes

Food processing is a traditional sector with relatively less returns on investment; hence, by and large, less sophisticated techniques are used in processing. Thus, most of the water treatment techniques employed in FPIs are classical physical separation methods (unit operations), such as

TABLE 10.7

Summary of Traditional Methods Used for Wastewater Treatment[a]

S. No.	Treatment Operation	Treatment Process
1	Mechanical separation	• Sedimentation • By screening screens, scrubbers, and filters • Centrifugation • Filtration
2	Coagulation	• By the action of coagulants, the suspended solids are drawn together into groups. Later, they are removed by mechanical separation but without any significant chemical change in the water
3	Chemical purification	• Adjustment of pH • Softening—by using lime • Iron removal • Disinfecting by chlorination • Ozone • Addition of copper sulfate
5	Biological processes	• Oxidation of organic matter to reduce BOD • Microbial treatment for killing objectionable organisms • Simple biomethanation to degrade harmful organic matter
6	Aeration	• Aeration of the effluent for the purification of certain chemicals • Aeration also supports the growth of water purifying organisms • Helps reduce BOD and COD of the effluents
7	Boiling	• For purification from soluble solids, multistage flash distillation is used

[a] Some information is based on Crittenden (2005).

sedimentation, filtration, aeration, and some biological methods, which we will briefly discuss.

10.2.1 Sedimentation

Sedimentation is the process of separating SS and liquid using gravitational force. Hence, the efficiency of separation depends on

- Gravitational force
- Particle size
- Liquid properties

The efficiency of sedimentation can be improved by increasing the particle size, as the other two parameters (particle size and liquid properties) cannot be changed independently. Even though the definition of sedimentation includes the separation of solid suspensions from gases or liquids by gravitational force, here we restrict our discussion to the separation by gravitational settling of solids in liquids. Sedimentation, in general, is used for solid–liquid separation or liquid–liquid separation.

In solid–liquid separation, the main purpose could be the removal of the solids to clarify the liquids. Rarely are the solids the desired products. In some cases, the primary purpose could be to produce the solids in a highly concentrated form; the process is called thickening. If the purpose is to clarify the liquids, the process is called clarification. In the FPI, for wastewater treatment, the thickeners and the clarifiers are used essentially for separating the SS from the liquid effluents. If the settling of the solids is not fast, flocculants are used to increase the particles and enhance the process.

In the sedimentation process, during the clarification or settling of the solids, three layers will form, namely, (i) the topmost clarification layer, (ii) the zone-settling layer, and (iii) the compression layer at the bottom, as shown in Figure 10.1 (Mc Cabe et al. 2005). Each of the three layers requires a certain area. Hence, the design of a thickener is based on the largest of these areas, and that area is designed as the area of the thickener. As the clarification proceeds, the middle layer will fade out and ultimately we have only two layers, consisting of (i) a clear liquid and (ii) a concentrated cake (or solids).

The conventional sedimentation tanks are designed and scaled-up based on thumb-rules of experience. However, a rigorous procedure would be to find out the critical solids flux, G_c, which determines the minimum design area of the thickener. The tank area should be such that the separation of the solids and the liquids is clearly made. A faulty design would lead to the

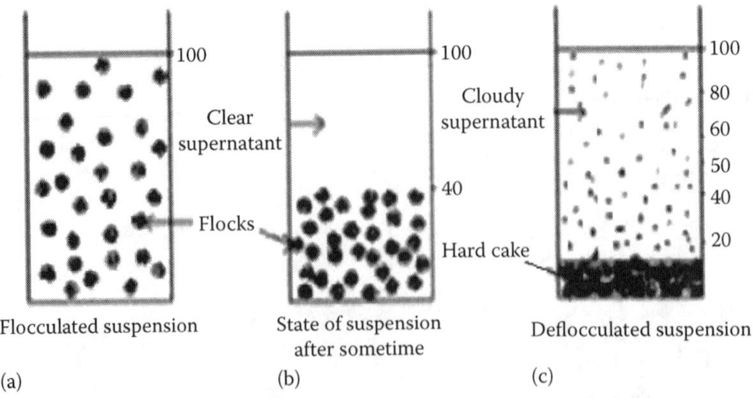

FIGURE 10.1
A schematic representation of batch sedimentation process: (a) initially at the start, (b) at the intermediate stage, and (c) finally, when a concentrated hard cake forms at the bottom.

intermixing of the solids and the liquids at the time of discharge. The area of the thickener (A) is calculated from the critical solids flux (G_c), the feed flow rate (Q), and the concentration (C_f), using a simple mass balance that assumes a complete separation of all the feed solids. Generally, for the design of large-scale continuous sedimentation thickeners, the data are obtained on the zone settling in a laboratory batch sedimenter by experimentation to evaluate the critical solid flux. Such design procedures are available in the literature (Rao 2010; Foust et al. 1980).

10.2.1.1 Equipment Used for Sedimentation

The sedimentation equipment is classified as (i) batch-operated settling tanks and (ii) continuously operated thickeners or clarifiers. The batch-operated sedimentation process is used for the treatment of small quantities of liquids, as in the case of the separation of fat globules from the process waters. Most sedimentation processes are operated in a continuous manner in view of their operational efficiency and economy. Here, they are described as clarifiers and thickeners.

10.2.1.1.1 Clarifiers

As has been mentioned, clarifiers are used for clarifying dilute suspensions in the water-treatment industry. In one-pass clarifiers, the feedwaters are fed horizontally into circular (or rectangular) vessels with the feed at one end and the overflow at the other end. Depending on the feed concentration and the nature of the solid suspensions in the feed, it may be preflocculated in a flocculator. Settled solids are pushed into a discharge trench by paddles or blades on a chain mechanism or suspended from a traveling bridge.

In most of the circular basin clarifiers, the feedwater is usually fed through a centrally located feed pipe. The overflow is led into a trough around the periphery of the basin. The bottom of the tank slopes to the center so that the settled solids are pushed down the slope by slowly moving motor-driven scraper blades attached to the central shaft (Figure 10.2). In some of the newer designs, a vertical flow combined with a flocculation unit is introduced. These units achieve higher overflow rates and are known as rapid-settling clarifiers or high-rate clarifiers. Most of the wastewaters coming from the fruit processing industries (after cleaning the fruits) contain a large amount of SS and are treated in this way. Flocculation accelerates the settling velocity of the solids in the feed.

10.2.1.1.2 Sludge-Blanket Clarifiers

The sludge-blanket clarifier has the specific advantage that the feedwater is fed from the bottom of the clarifier through a blanket of SS that acts as a filter. The inverted cone shape in the clarifier increases the cross-sectional area

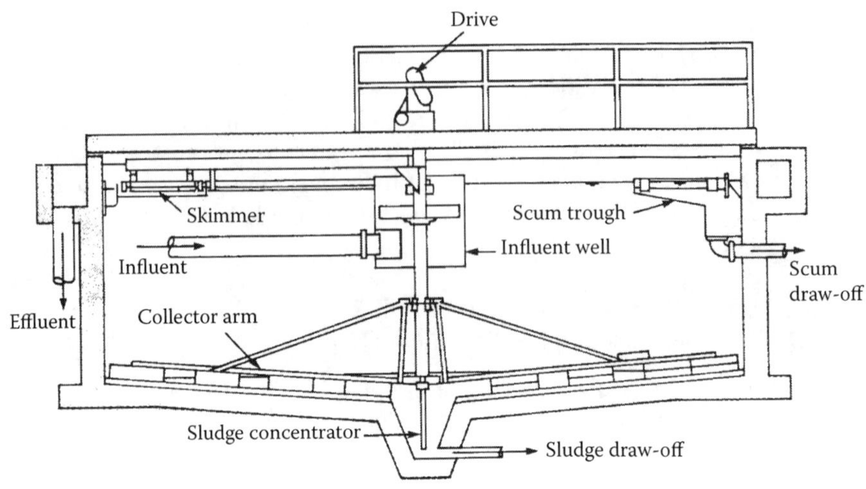

FIGURE 10.2
A wastewater clarifier.

as the feed moves from the bottom of the clarifier to the top (Figure 10.3). This helps decrease the upward velocity of the water as it approaches the top. At some point, the upward velocity of the water exactly balances the downward velocity of a solid particle. Thus, the particles are in a state of suspension, with the heavier particles suspended closer to the bottom,

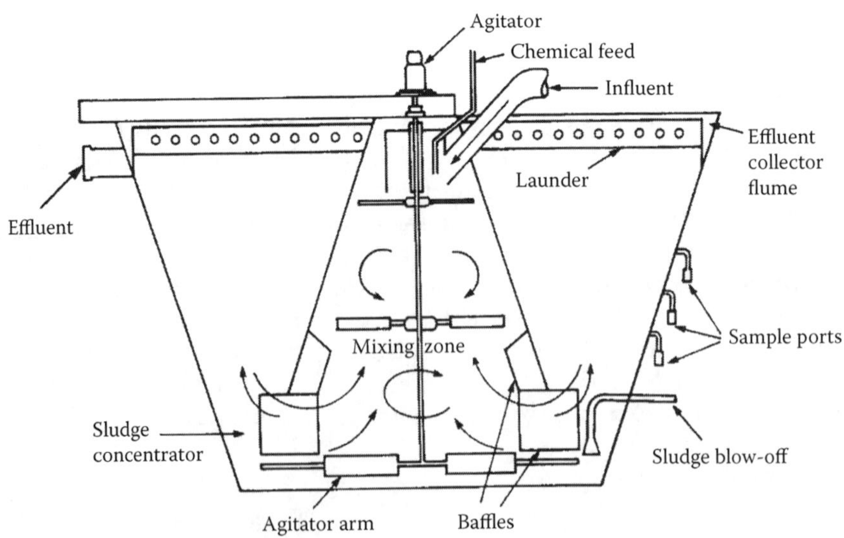

FIGURE 10.3
A sludge-blanket clarifier.

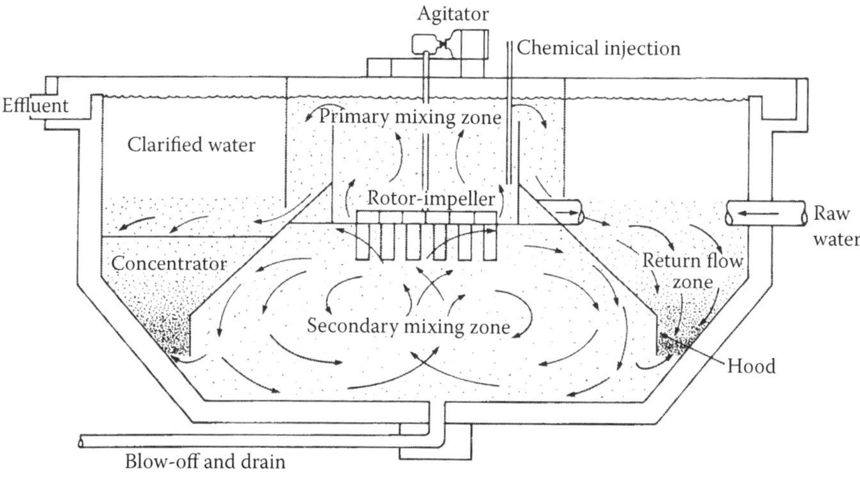

FIGURE 10.4
Solids contact clarifier.

forming a blanket. As the water containing the flocculated solids passes up through this blanket, the particles adhere to the larger flocs, thereby increasing the floc size. The larger flocculated particles then drop to a lower level. They finally settle at the bottom of the clarifier and may be further recirculated or drawn off.

10.2.1.1.3 Solids Contact Clarifier

The solids contact clarifier (Figure 10.4) is another type of clarifier that provides two mixing zones for the solids. The feedwater is initially drawn into the primary mixing zone in which coagulation and flocculation take place. In the secondary mixing zone, the particles will collide with each other so that the smaller particles are entrained in the larger floc. Water passes out of the inverted cone into the settling zone. The solids settle to the bottom of the clarifier, and the clarified water flows over the weir. Subsequently, the solids are drawn back into the primary mixing zone, causing recirculation of the large floc, and the process continues. The solids in the secondary mixing zone are occasionally or continuously blown down, which helps maintain the concentration of the solids in the sludge.

10.2.1.1.4 Thickeners

The most common thickener is the circular basin type. The wastewater is treated with flocculating agents, if necessary, before feeding it to the center of the well of the thickener. A typical thickener has three operating layers: clarification, zone settling, and compression (Figure 10.1). A very popular

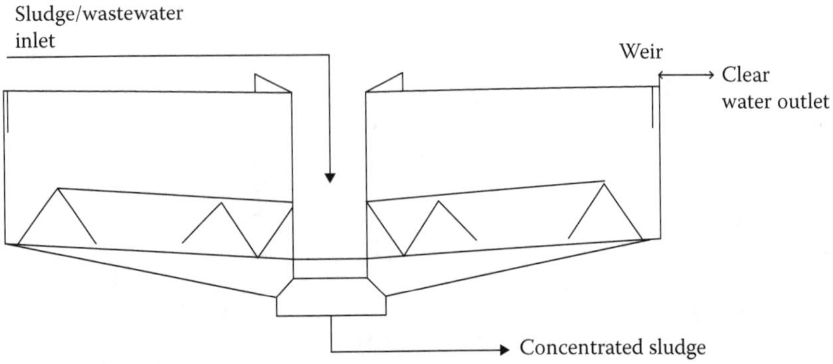

FIGURE 10.5
A gravity thickener.

thickener is a gravity thickener (Figure 10.5). It consists of a circular basin with a central raking arrangement and a slowly moving mechanical raker. The slow motion of the raker assists the solids to separate out and settle. It uses the physical process to concentrate the sludge by removing the water portion to increase the amount of solid percentage. It is used on both the untreated wastewaters and the treated activated sludge. The sludge is fed from an aeration system to the center well. Adequate time is given so that good settling takes place. The waste material collects at the bottom of the tank and settles, and the clarified water is disposed of through a weir in the form of an overflow. The process can be improved by decreasing the feed rate.

10.2.2 Filtration

Filtration is a mechanical operation that is used to separate the SS from a liquid, based on the fact that the solids will have a larger particle size and hence can be used in the separation process by passing the suspension through a filter cloth or medium that retains the bigger solid particles on the filter cloth and percolates the liquid through. Thus, it is a physical separation process. Some slurries separate well by sedimentation over a long period of time. If the solid particles are very small in size, they will flocculate using a chemical flocculant. However, the sedimentation process takes quite a long time, and it is generally used if the concentration of the solids is low and the volume of the effluent is large. Filtration is used for colloidal suspensions. One of the oldest product applications of filtration is in the purification of wine and water, practiced by the ancient Greeks and Romans. Cake filtration, such as the rotary vacuum filter press, was developed much later from the necessity to filter sewage.

The filtration operation can be classified into two categories: (i) constant pressure filtration and (ii) constant volume filtration. In industrial practice, the constant pressure filtration is usually employed in view of its operational advantages.

Some important factors to be considered before selecting a filtration operation are

1. The type of filter to be used
2. The filtration rate
3. The filtrate driving force
4. The number and the size of the filter units
5. The backwash water requirement

In a filtration system, a porous filter medium is housed in a container with a flow of suspension onto the filter medium, and the clear liquid leaves after filtration. A driving force, usually in the form of a static pressure difference, is applied to achieve the flow through the filter medium on the upstream side. Alternatively, vacuum may also be applied on the downstream side to suck the liquid through the porous medium. Fundamentally, it is immaterial how the pressure difference is generated. There are four main types of driving force:

1. Gravity
2. Vacuum
3. Pressure
4. Centrifugal

As the filtration proceeds, the solids build up on the filter medium in the form of a cake. Hence, such filters are known as cake filters. In this process, the liquid flows continuously, but the cake is disposed of intermittently in what is called a batch process. The liquid filtering out is known as filtrate. When the accumulation of the cake is excessive, the process is stopped and the cake is removed. Again, the process is continued until further buildup of the cake. The successive layers of the batch operation can be discarded using continuous filters.

The three types of filters that are very popular in the industry are

- Plate-and-frame filter press
- Leaf filter
- Rotary vacuum filter

The first two filters are semicontinuous in operation, whereas the rotary vacuum filter is a continuous filter and is used in wastewater treatment.

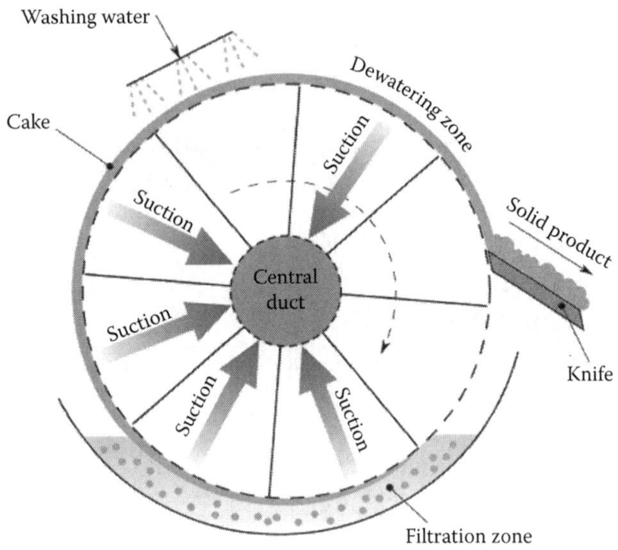

FIGURE 10.6
A rotary vacuum filter.

10.2.2.1 Rotary Vacuum Filter

It is a continuous filtration unit that is frequently used in process operations because it is simple to use and versatile. It consists of a slowly rotating cylindrical drum (at 1–2 rpm). The rotating drum is partly dipped (approximately up to one-third of its circumference) into the feed, which is a suspension of solids in a liquid (Figure 10.6). The feed slurry is continuously fed into the tank. The drum has a metallic ring firmly attached to the rim (circumference) with a metallic screen fixed to it, which acts as the filter medium. A suitable filter fabric is fixed to the screen. Vacuum is maintained inside the drum so that the slurry is sucked into the drum when it is brought in contact with the filter screen. Depending on the solid concentration, approximately 33–66 kPa of vacuum is applied. Filtration takes place, the solids adhere to the filter cloth, and the liquid is sucked into the drum, which is subsequently discharged. When the solid cake emerges from the slurry, it is washed with a spray of water or the solvent and scraped continuously by means of an arrangement known as doctor's blade. The filter cake is collected in the product tray. Various types of arrangements are available for the discharge of the cake as well as the filtrate.

10.2.2.2 Filter Aids

As has been mentioned, as filtration progresses, the cake will accumulate on the filter medium, which in turn makes for better filtration by offering resistance to the free flow of the suspensions through the pores of the filter medium. In the initial stages of filtration, there will not be any cake formation and hence the resistance for flow will be least. This (no cake formation) affects the efficiency of filtration. The problem is more acute if the slurry is dilute (dilute suspensions of less than 0.1% by volume). Very dilute or very fine and slimy suspensions are too difficult to filter by cake filtration because of the fast percolation of the slurry. In such cases, filter aids are used, which are rigid, porous, and highly permeable powders. They are added to the feed suspensions to extend the efficiency of the cake filtration. The filter aid helps in the initial pressure buildup. The filter aids are used either as a precoat on the filter medium as a coarse support material, or they are mixed with the feed suspensions to increase the permeability of the resulting cake. In the precoat mode, the filtrates allow the filtration of very fine compressible solids from suspensions of 5% or lower of the solid concentration on a rotary drum precoat filter. If the filter aid is mixed with the slurry, during filtration, the filter aid first forms a precoat on the filter medium. This is followed by filtration of the free liquids, which may have the filtrate mixed with it as the body feed, in order to improve the permeability of the resulting cake. The proportion of the filtrate to be added as the body feed is determined by the nature of the slurry and by the nature of the filter aid. The recovery and the regeneration of the filter aids from the cakes are not normally practiced, except in very large insulations where it might be economical to do so. In fact, the addition of filter aids is discouraged if the cake is the desired product.

10.2.3 Coagulation and Flocculation

If the particle size of the SS is very small, of the order of a few microns, it will take quite a long time for the solids to separate out during sedimentation. The purpose of the flocculants is to agglomerate the small individual particles to increase their size. Thus, the particle diameter increases, which helps increase the settling velocity (u_t); hence, the separation is faster.

$$u_t = d_p^2 g \left(\rho_p - \rho_f\right)/18\mu_f, \tag{10.1}$$

where d_p is the diameter of the particle; g is the acceleration due to gravity; ρ_p and ρ_f are the densities of the particle and the fluid, respectively; and μ_f is the viscosity of the fluid.

Coagulation and flocculation help increase the effective particle size, which results in additional benefits, such as higher settling or flotation rates, higher permeability of the filtration cakes, and better particle retention in deep bed filters. Coagulation brings particles into contact to form agglomerates. Flocculation is done by adding chemicals such as hydrolyzing coagulants, for example, alum or ferric salts or lime, and the subsequent agglomeration can produce particles up to 1 mm in size. The process of flocculation is facilitated by destabilizing the surface forces of the particles. Some of the coagulants simply neutralize the surface charges on the primary particles. Alum salt is an effective and frequently used flocculating agent in water treatment operations. Flocculating agents, usually in the form of natural or synthetic polyelectrolytes of higher molecular weights, interconnect and enmesh the colloidal particles into giant froths up to 10 mm in size. They are relatively expensive and hence the correct dosage is critical but essential. The optimum flocculant type and the dosage depend on factors such as the solid concentration, particle size distribution, surface chemistry, electrolyte content, and pH values; the effect of these is very complex. Overdosing with flocculant is not economical and may inhibit the flocculation process by coating the particles completely, with the subsequent destabilization of the suspension, or it may cause operating problems such as blinding of the filter media. Flocculent selection and dosage optimization require extensive experimentation with only a general guidance as to the ionic charge or the molecular weight (or chain length) required.

Subsequent to flocculation or coagulation, the separation of the solids is achieved either by sedimentation and/or centrifugation or by filtration. The type of flocculant required depends on the subsequent separation process, for example, the rotary vacuum filtration requires evenly sized, small, strong flocs that capture ultrafines to prevent cloth blinding and cloudy filtrates. In gravity thickening, large and relatively fragile flocs are needed to allow high settling rates and fast collapse in the compression zone.

10.2.4 Centrifugation

Centrifugal separation is a mechanical means of separating the components of a mixture of liquids or of liquids and solid particles. It is akin to sedimentation, except that the material is accelerated in a centrifugal field, which acts on the mixture in the same manner as a gravitational field. In Equation 10.1, the acceleration due to gravity (g) term is replaced by the acceleration created by the centrifugal force, and hence the settling is very fast as compared with simple sedimentation. The centrifugal field can, however, be varied by changes in the rotational speed and the equipment dimensions, whereas gravity is essentially constant. Commercial centrifugal equipment can reach an acceleration of 20,000 times gravity (i.e., 20,000 g); laboratory equipment can reach a much higher acceleration of up to 360,000 g. Thus, most of the centrifugation equipment is designed to separate immiscible or insoluble components from a liquid medium with very low-density differences among the components. For example, in Equation 10.1, if ($\rho_p - \rho f$) is small, the settling velocity will be lower in the usual gravitational operations, such as the sedimentation or the flotation of solids in liquids. This is compensated for by increasing the acceleration term. Thus, the drainage or the squeezing of the liquids according to the density is accomplished more effectively in a centrifugal field. The centrifugal force separates the solid particles from the fluids by throwing them away from the center of rotation.

Centrifugation is used in water treatment for the following operations:

- The separation of solid substances from highly concentrated suspensions
- The separation of oil and/or fat suspensions (globules)
- The separation of concentrated sludge

The three most frequently used centrifuges are

- Tubular bowl centrifuge—used for a small concentration of solids
- Disk bowl centrifuge—used for moderate quantities of solids
- Basket centrifuge—used like a thickener where the major product is the clarified liquid

These centrifuges are only briefly described here, as a detailed description of their treatment is available in any standard textbook of unit operations (Mc Cabe et al. 2005; Foust et al. 1980; Rao 2010).

10.2.4.1 Tubular Bowl Centrifuge

It is a tall and narrow centrifuge rotating at a very high speed of the order of 15,000 rpm. The feed is admitted from the bottom into the rotating narrow bowl at the center (Figure 10.7) and is centrifuged. It separates into two phases, the lighter phase moving toward the center and the heavier phase moving away from the center. The two phases slowly rise up. An adjustable ring at the top makes the streams to discharge through the respective outlets. If the feed contains only SS, they settle at the bottom in the form of a cake (Figure 10.7) and are discharged intermittently.

10.2.4.2 Disc Bowl Centrifuge

It consists of a number of coaxial conical discs housed in a bowl (Figure 10.8). The discs are mounted on a central axis and are rotated at very high speeds. The holes on either side of each disc align to form a path to allow the heavier liquid to pass to the outlet. The lighter phase of the feed accumulates at the center and rises up to be discharged through its designated outlet. The solids, if any, accumulate at the bottom and are occasionally discharged by flushing with a jet of water. Thus, the centrifuge operates continuously and is also useful for separating a liquid–liquid–solid mixture. We may not come across such operations in the treatment of wastewaters from FPIs, but they are extensively used in the dairy industry.

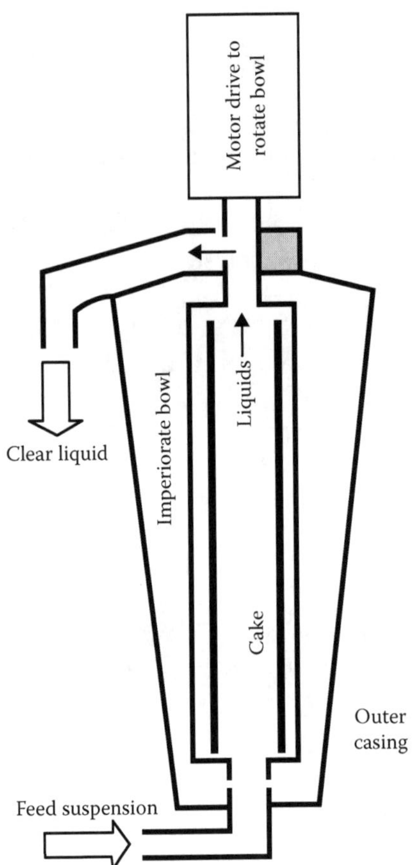

FIGURE 10.7
A tubular bowl centrifuge.

10.2.4.3 Basket Centrifuge

A typical basket centrifuge usually contains a perforated drum with holes to allow the water to pass through (Figure 10.9). The feed enters the drum (basket) from the top (or bottom, depending on the feeding arrangement) and then, under centrifugal force, the water will move away toward the outside of the drum, where it will be collected and drained out. Some basket centrifuges do not have perforations in the basket, and they are known as imperforate basket centrifuges. They are used with feeds containing a high solid content and consist of a simple drum-shaped basket or bowl, usually rotating around a vertical axis. The solids accumulate due to the centrifugal force, but they are not dewatered. The filtrate is drained after stopping the centrifuge.

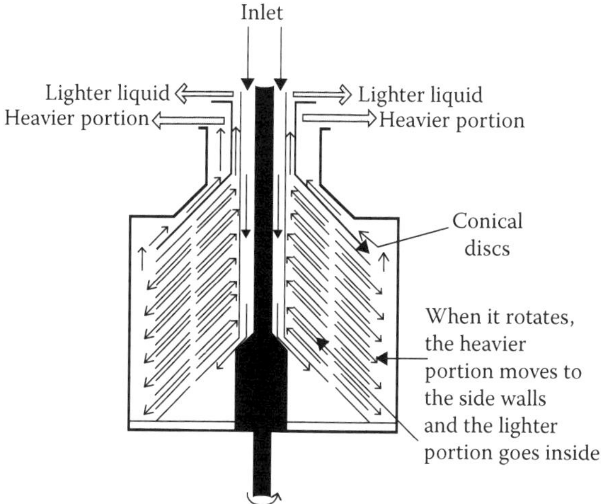

FIGURE 10.8
A disc bowl centrifuge.

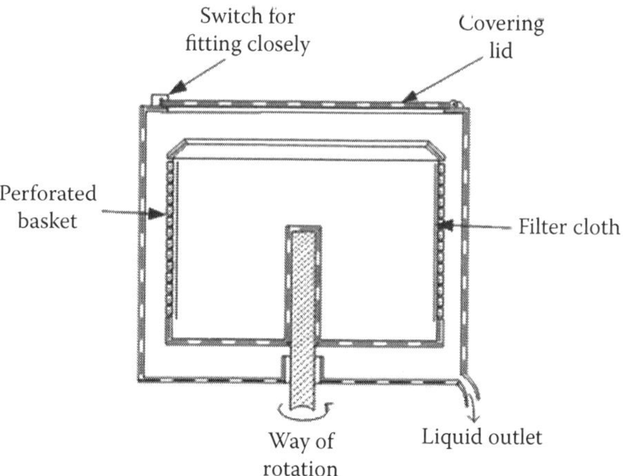

FIGURE 10.9
A perforated basket centrifuge.

10.2.5 Air Flotation Methods

Air flotation is a frequently used industrial operation to reduce the BOD content in the wastewater by aerating the system. Air is pumped into the wastewater under pressure and then released at atmospheric pressure into a flotation tank or basin to separate the suspended matter in the wastewater. The released air forms tiny bubbles that combine with the suspended

FIGURE 10.10 (See color insert)
An activated sludge treatment plant. (Courtesy ARTES Ingegneria, Cannon, Italy.)

matter, causing the suspended matter to float on the water surface in the form of froth. It can be removed subsequently by a skimming device. Air sparging helps meet the oxygen requirement of the organic matter, which in turn forms sludge (popularly known as biomass). An air flotation system that forms the activated sludge is shown in Figure 10.10.

10.2.6 Chemical Methods

The chemical methods of treatment include precipitation and the use of sanitizing chemicals, such as chlorine-based sanitizers, quaternary ammonium compounds, iodophores, and amphoteric bactericides. While using sanitizers, it is essential to note the purpose for which they are being used, for instance, whether to remove:

- Organic matter such as fats, proteins, vegetable matter, or animal-based materials
- Microorganisms associated with the organic matter
- Bacteria, fungi, or any other pathogenic organisms

10.2.6.1 Chlorine-Based Sanitizers

They are used more for sanitizing the water than for treating the wastewaters. If the waters are not too contaminated, particularly those waters coming out of vegetable washing units, they are treated with sodium hypochlorite solutions that contain about 9%–12% chlorine. Chlorine can be used effectively against both Gram-positive and Gram-negative bacteria. Hence, these chemicals are effective for bactericidal purposes.

10.2.6.2 Quaternary Ammonium Compounds

Quaternary ammonium compounds are used for sanitizing purpose. They are a combination of ammonium radicals with chloride or bromide anions. Their major drawback is that they foam, which is sometimes undesirable for reusing the wastewater after treatment.

10.2.7 Biological Methods

Biological methods are mostly used to reduce the BOD of the wastewater. The BOD is an important measure of the pollutant organic materials in the wastewater. It is a measure of the amount of oxygen required to decompose the putrescible organic matter present in water. Therefore, a low BOD is an indicator of good quality water, while a high BOD indicates polluted water. The BOD also measures the chemical oxidation of inorganic matter (i.e., the use of oxygen from water via a chemical reaction). The oxygen present in water, known as dissolved oxygen (DO), is consumed by bacteria when large amounts of organic matter from sewage or other discharges are present in the water. Thus, the actual amount of DO in the water is consumed by microorganisms. Hence, when such waters are let into water streams, the life-forms in those waters would be unable to continue at a normal rate. The BOD specifies the strength of the sewage. Most wastewaters contain BOD of the order of 150 ppm, which needs to be brought down to 20–30 ppm before being discharged.

Thus, the purpose of the biological treatments is to reduce the BOD by bringing the cells into close contact with oxygen, thereby satisfying their oxygen demand. This is achieved by two methods, namely, mechanical agitation (the activated sludge method) and nonmechanical means of mixing.

10.2.7.1 Activated Sludge Method

It is a very popular method of wastewater treatment. This method consists of two tanks (basins). In the first tank, air is pumped from the bottom into the feedwater by means of a good sparging arrangement. Air rises to the top in the form of bubbles, creating good agitation and also meeting the desired oxygen demand of the microorganisms (bacteria). These microorganisms eat the organic matter in the wastewater in the presence of oxygen and produce a cellular biomass. This biomass, known as sludge, will be transferred into the next tank, where the absence of agitation allows the biomass to settle. The clear water is decanted from the top and the biomass is discharged from the bottom. A part of the biomass is recycled to the first tank to act as an inoculum for further biomass to grow. A typical activated sludge processing tank is shown in Figure 10.10.

10.2.7.2 Trickling Bed Filters

The trickling bed filters are a type of packed bed arrangement in which the packing is some type of pebbles or rocky stones on which the wastewater is sprayed. Air is blown from the bottom. The feedwater is sprayed from the top and makes contact with the packing materials on which the slime (fungi and algae) develops. The water comes in contact with air in the presence of the slime and develops more biomass. The packing materials help establish a better and close contact between the water and the slime. When the biomass generated is considerable, it slips to the bottom of the bed, or the process is stopped and the biomass is scraped. There are some arrangements in which the outlet waters are collected separately, the biomass (sludge) is separated, and a part of the effluent water is recycled (Figure 10.11). Care should be taken in this process to ensure that good contact is established between the feed and the packing material without the formation of channeling.

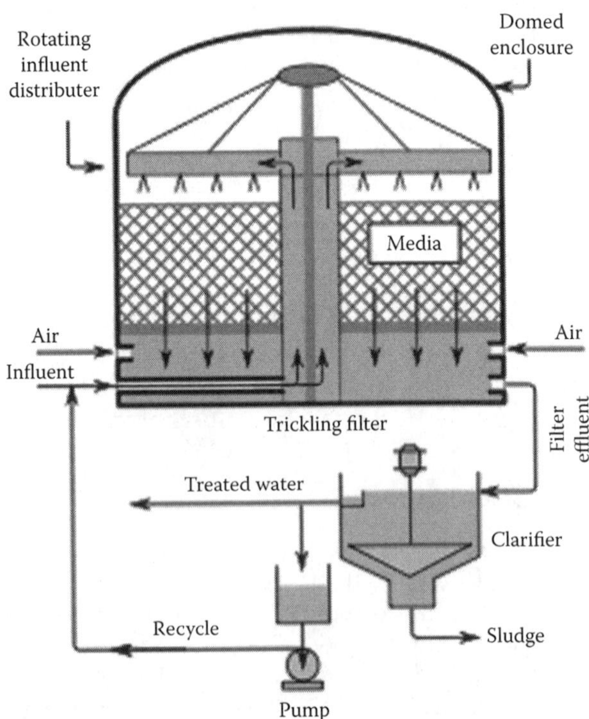

FIGURE 10.11 (See color insert)
A trickling filter bed.

10.3 Advanced Treatment Processes

Some specific and advanced treatments, such as an activated sludge combined with a fluidized media, a sequenced batch reactor system, a multipurpose activated sludge process environment control system, and an upflow anaerobic sludge bed system, are used to treat the complex nature of the wastewaters from FPIs. The majority of the clean technology advances in research have been in reducing the volume of the wastewaters, generated in the food processing operations.

10.3.1 Ozonation

Ozone, first discovered in 1840 by Schonbein, was initially utilized as a disinfecting agent in the production of potable water in France in the early 1900s. The majority of its early development was limited to Europe, where it became more widely used in drinking water treatment. The potential utility of ozone in the food industry lies in the fact that ozone is 52% stronger than chlorine and has been shown to be effective over a much wider spectrum of microorganisms than chlorine and other disinfectants, leaves no chemical residue, and degrades to molecular oxygen on reaction or natural degradation. The treatment of the wash water used in the processing of carrots reduces bacteria by 3 log, and a combination of screening, diatomaceous earth filtration, and ozonation yields the highest quality of water with the total microbial loads (total coliforms, *Escherichia coli*, and salmonella) reduced by 99%. Ozone and ozone/hydrogen peroxide are very effective in removing color from dye-processing wastewater. The advances in the ozone generation and applications technologies have continued to make the process more economical and reliable. Ozone has 1.5 times the oxidizing potential of chlorine and 3000 times the potential of hypochlorous acid (HOCl). The contact times for antimicrobial action are typically four to five times less than chlorine. Ozone rapidly attacks the bacterial cell walls and is more effective against the thick-walled spores of plant pathogens and animal parasites than chlorine, at practical and safe concentrations.

10.3.2 Anaerobic Treatment

Anaerobic treatment is a recognized, well-established, and proven technology for the treatment of various industrial effluents. Currently, over 1000 anaerobic treatment plants have been constructed worldwide for the treatment of industrial effluents.

10.3.3 Dissolved Air Flotation System

The dissolved air flotation (DAF) systems can be used to treat many types of industrial effluent streams and are ideal for use in FPIs. The unit has a countercurrent surface-scraping design, which ensures outstanding results in terms of solid capture and overall performance. The standard unit capacities range from 1 to 1000 kL/h. Polymer chemicals may be introduced prior to the DAF, in order to promote flocculation of the wastewater removal of SS in this process. The DAF causes the suspended material in the wastewater to separate, with the contaminants floating to the top while the cleaner water flows under the bottom baffle and overflows into a holding tank, known as the treated water storage tank.

Raw water is automatically delivered to the DAF from the retort waste tank via a pump whenever the upstream tank has water and the treated water storage tank needs water. A control valve is located on the raw water delivery pipe to control the flow to the DAF. A polymer may be introduced into the raw water to cause a floc to form. Polymers are used to flocculate the suspended material in the water; however, they are not used in every installation. Air is introduced into the raw water via the aeration pump, which pushes the water from the side of the DAF through the aeration vessel and back into the bottom of the DAF joining the raw water inlet. Air is dissolved in the water in the aeration vessel by the system air pressure on the vessel. The bottom half of the aeration vessel has water and the top half has air. The air introduced into the DAF brings the waste oil and the suspended material to the surface, where they removed by the skimmer arms. The skimmer arms operate on an interval basis. The treated water is allowed to overflow in a weir from behind a baffle into the treated water storage tank. The weir height may be adjusted so that the water level is maintained in the DAF for proper floc removal (dewatering) ramp operation. The bottom of the DAF is allowed to drain periodically when operating, and this period is set by the microprocessor system. The DAF should be drained manually whenever the system has been shut down, after the skimmer arms have been allowed to run for a period of time.

The DAF systems have some serious limitations. While the small bubbles used in such systems yield better contaminant removal efficiencies, the rise time of the particles attached to the bubbles is in minutes, which results in a longer water residence time inside the flotation tanks and a larger footprint-tank size. The solubility of the air in water and the necessity of recycling instead of the full-flow treatment limit the number of bubbles that can be produced in such systems. One of the recent developments in flotation technology circumvented some of these problems. In particular, the air-sparged hydrocyclone (ASH) couples a porous cylindrical membrane with the design features of a hydrocyclone. Gas is introduced through the porous membrane while wastewater is pumped through the hydrocyclone. Such a device is not dependent on the gas solubility and can introduce air-to-water ratios as high as 100:1; however, the removed particulates are forced through an overflow device, known as the vortex finder. This results in a separate stream of contaminated water with a low concentration of solids. This deficiency results in a sludge with low particulate concentrations and a larger volume of waste. To overcome this deficiency, bubble accelerated flotation (BAF) technology evolved to address the operational limitations resulting from the traditional stream-splitting characteristics of the hydrocyclones. Removing the underflow restriction that forces the froth and the contaminants to be ejected through a vortex finder improves the consistency and ease of operation in the BAF. At the point when the stream exits the BAF hydrocyclone, the bubble/particle aggregates have already formed, and coagulation and flocculation are complete before the froth particles are ejected with the cleaned water through the underflow.

10.3.4 Pressure Filtration

Pressure filtration refers to a pressure vessel and a filter media. The vessels range from a 54 inch steel housing to a plastic cartridge housing. Filter media typically used are anthracite, silica sand, bag filters, and cartridges. The multimedia filter is operated automatically. It is designed to remove particles to the 10 μ range. The term "multimedia" refers to the two layers of media contained in the filter. In a usual application, the bottom layer of the media (or the lower working media) is 18″ in depth and is made up of 20 mesh silica filter sand. The upper layer of the media (or the upper working media) is 18″ in depth and is made up of anthracite. This material enables the penetration of the larger suspended particles, thereby allowing a greater period of time between the backwash cycles than a normal sand filter. The average media life is 3–5 years in constant service. The filter is fully automated through the use of air control valves and the systems microprocessor control center. The filter may be set up to backwash either on a set pressure differential or on a periodic basis. The microprocessor control center, sensing the inlet and the outlet pressures, will take the filter off-line, air-scour and backwash the media, and then put the filter back online. The filter could also backwash through a timer, for example, once every 24 h of operation.

An industry standard bag filter pressure vessel is utilized as a prefilter for the larger nanofiltration units. The 5 μ bag filter is typically placed after the multimedia filter and is charged by the same pump. The bags will need to be replaced on a periodic basis whenever they have been loaded. The system can be isolated with valves and drained to allow for ease of changing. Bag filters have advantages over cartridge filters in that they offer a greater flow rate, a longer life, and less downtime. Cartridge filters are used as prefiltration for a smaller membrane system as well as a wash filter on all the membrane systems. The most common type of cartridge filters used is polypropylene fiber–wound.

10.3.5 Membrane Separations

Membrane filtration refers to reverse osmosis (R/O) and, in particular, to low pressure membranes, known as nanofilters. Nanofilters were originally developed in the late 1970s by several membrane manufacturers as softening membranes designed to replace ion exchange water softening with high recovery, low-pressure membranes. Nanofiltration is a low-pressure (100 psi) form of R/O using thin film elements. It removes four-fifths of the hardness, two-thirds of the salt, and 80%–90% of the dissolved organics. Nanofiltration effectively removes all pathogens including viruses, most of the bivalent ions such as calcium and magnesium, as well as large quantities of monovalent ions such as sodium and chloride. Furthermore, the cost of nanofiltered water is not much different from the cost of conventionally treated drinking water.

Each membrane system is specifically designed to meet the individual user's need. The systems operate at a relativity low pressure with a normal recovery rate of 80%; this provides the user with a much lower operating cost and a higher payback potential than the traditional "high pressure" R/O systems. A polymeric material is introduced prior to the prefilter unit, in order to reduce the formation of scale in the elements. The membrane unit is pressurized using a multistage centrifugal pump. The unit is equipped with an automatic shut-off in the event that the pressure should drop below or above the specified level, indicating that there are problems that need to be corrected.

The permeate or the product water is split into two different qualities; a higher level and a lower level. The higher-quality product water comes from the first stage pass, which has higher pressure/higher reject elements. The lower-quality product water is produced from the second stage, which has lower pressure/lower reject elements, sent to a tank for return to the retort cooling system. The higher-quality product water passes into a separate holding tank for boiler makeup. Flow meters are installed to track the actual production of the product water. Conductivity meters are also used to monitor the quality of the product water produced. The concentrate or the reject water is delivered to the drain. A booster pump is utilized to recirculate a portion of the reject water back to the second stage. As the membranes are operational, their flow rate will decrease, and periodic cleaning of the membranes will restore production. This technology will allow today's food processing plants to realize substantial savings in water, sewerage, chemicals, and heat (energy). Additionally, the quality of the "product water" produced will exceed the quality of the water from the municipal source.

10.4 Case Studies

The wastewaters generated from FPIs have some distinct features that need special treatment. In this section, we describe some of the case studies indicating successful industrial processes.* The examples are mostly taken from the animal product processing industries because they produce a lot of slaughter and processing products that usually contain body fluids, such as blood and gut contents. Such waters are also highly contaminated with antibiotics and growth hormones and sometimes even with a variety of pesticides. Thus, they have high BOD and SS.

10.4.1 Beef Processing

A beef processing plant slaughtering 1200 cattle/day produces 80,000 gpd (gallons per day) of wastewater with a BOD of 3000 ppm and a TSS of 1800 ppm. Integrated Engineers, Inc.† renovated the existing DAF unit by installing a 1.5 m belt press to dewater the solids initially, and adding a dosage of 216 ppm of Floccin KP+ to the DAF unit. The treatment reduced the BOD to 400 ppm and the TSS to 250 ppm. In another beef processing plant, Floccin E was the flocculating agent used to reduce the BOD and TSS from 2600 and 2000 ppm to 310 and 150 ppm, respectively.

Pine Valley Meats (Norwalk, WI) is a progressive slaughtering and meat packaging company with a capacity to slaughter 800 heads of cattle per day. The blood and the water from floor washing and skin and body washing of the slaughtered animals were the wastewaters generated from the processing, which were to be treated in a treatment plant. The wastewater treatment plant consisted of an anaerobic digester and aeration and clarification units. The excess fat in the water was removed prior to it (water) reaching an anaerobic digester. Aeromix Systems Inc.‡ installed a 2 hp ZEPHYR Induced Air Flotation System to remove about 5000 lb of fat per day.

10.4.2 Poultry Processing

The major contaminants in the poultry processing industry are fats, feathers, and biological solids. A chicken processing facility in California has availed of the consultancy services of Integrated Engineers to reduce the BOD and the COD by utilizing Floccin 1105 as the flocculating agent, followed by cationic and anionic flocculants to treat 180,000 gpd of wastewater to bring down the BOD to 2000 ppm and the TSS to 500 ppm. The wastewater was initially pumped through a rotary screen into the DAF unit. The pH was adjusted to 7–8. Later, the water was treated with a coagulant followed by cationic and anionic flocculants. The dissolved air in the flotation unit raised the solids to the surface, which could be skimmed later. Integrated Engineers reported that the new treatment had saved the food processing unit $1400/day and

* The information provided here is not exhaustive, it is only indicative.
† http://wecleanwater.com/html/success/food_processing/beef_processing.htm.
‡ http://www.aeromix/documents/PinaValleyMeats-000.pdf.

had facilitated an increase in the production capacity while disposing of the wastewaters into the municipal treatment system.

Shade Foods (Kansas, TX) had a similar problem with disposing of 20,000 gpd of wastewater into the municipal treatment with the BOD as low as 200 ppm. Aeromix Systems Inc.* recommended and installed a 3 hp ZEPHYR Induced Air Flotation System. This formed a 3″ thick jelly-like foamy material that could be removed through skimming. This helped reduce 50% of the fat with a detention time of 3 h in the tank.

10.4.3 Fishery and Seafood Processing

As in most processing industries, the seafood processing operations produce wastewater containing a substantial amount of contaminants in soluble, colloidal, and particulate forms, depending on the processing operations involved. Usually, the wastewater is very high in dissolved and suspended organic materials, which results in BOD and COD; fats, oil, and grease (FOG); SS; and nutrients, such as nitrogen and phosphate. One of the major concerns with the seafood processing units is the unpleasant odor and the high concentrations of sodium chloride from boat unloading, processing water, and brine solutions. The processing wastewaters usually contain blood, offal products, viscera, fins, fish heads, shells, skins, and meat fines. The major processing operations that generate a large amount of wastewaters are (i) product receiving; (ii) unloading from boats; and (iii) product processing, such as pickling in brine, and further processing (canning and bottling). Organic materials in the wastewater are produced in the butchering process, which generally include blood and gut materials. The degree of pollution of the wastewater depends on several parameters, and hence they need careful consideration, including good manufacturing practices. Clean Water Technologies Inc.† follows a stepwise approach to wastewater treatment to yield the best results in the most economical way. The treatment steps are flow equalization, screening, sedimentation, pH adjustment, flocculation, flotation, and microfiltration. A hybrid centrifugal–dissolved air flocculation–flotation system is the key component of the wastewater treatment plant for treating 500 gpm of wastewater requiring only an 8′–16′ area. The TSS and FOGs are almost completely removed to less than 20 and 1 mg/L, respectively (Colic et al.), allowing for successful breakpoint chlorination–dechlorination and fecal coliform removal (99.995%) at the Ocean Gold plant in the United States.

High Liner Fishery Products (Danvers, MA) is a leading supplier of seafood in the United States, supplying a variety of fishes, shrimps, lobsters, and their processed foods. The company was treating 72,000 gpd of wastewater, utilizing

* http://www. aeromix.com/documents/ShadeFoods_000.pdf.
† http://www.cleanwater tech.com.

the conventional trickling filter technology; it was facing the problems of plugging bad odors, flies, etc. The BOD and the TSS were of the order of 1500 and 1000 ppm, which needed to be brought down to <250 and <100 ppm, respectively. The company works for 350 days a year with only a 2-week gap for maintenance, in which time the installation needs to be done. With these constraints, Headworks Bio Inc.* (Houston, TX) took up the work of installing the moving bed biofilm reactor (MBBR). The wastewater was made to flow from the plant through a rotary screen into a collection tank of 40,000 gal capacity, where it was continuously agitated with a blower and an air-sparging system. Later, the water was pumped at a predetermined rate into a DAF unit, where the TSS and FOG were flocculated and coagulated. The DAF unit was specially designed by Headworks Bio. The clarified water was then sent to two active cell biofilm reactors (1000 gal capacity) for biodegradation. The active cell process (designed by Headworks Bio) biodegraded the wastewater using thousands of suspended biofilm carriers that operated in a continuously mixed, aerated environment. The biofilms work by carrying an active surface area per each biofilm to sustain heterotrophic and autotrophic bacteria within the protected cells.

IBC Water Pvt Ltd. (Mansfield, Australia)† provided ozone treatment technology for a seafood processing unit. Ozone offers the advantages of (i) killing microorganisms on floors and in drains and in making ice, etc.; (ii) BOD and COD limits in the wastewaters are lower; (iii) cleaning in process (CIP) is effective; and (iv) plant sanitary conditions are ensured. The ozone-treated waters ensure better microbiological control of the raw material, plant, and equipment.

10.4.4 Dairy Processing

Mid-American Dairy (Winthrop, MA) is a leading dairy industry in the United States, processing milk and milk products. In the process, it also generates 20,000–60,000 gal/day of wastewater with BOD and TSS loadings of 850–4200 ppm and 532–1900 ppm, respectively. The fat loading is 76–1370 ppm. The range is high because the load will vary in a day, depending on the process conditions and the processed products. AEROMIX Systems Inc. (Minneapolis, MN)‡ suggested installing a 2 hp ZEPHYR Induced Air Flotation System to remove a portion of the fats and solids from the process water before it reached the ditch. The ZEPHYR was installed in a $10' \times 6'$ tank with a 6′ water depth and a top skimmer. The present ZEPHYR system was effective and put an end to the need for two 30 hp surface aerators, and it could remove 20%–35% of the BOD.

* http://www.headworksusa.com/useriles/Fishery%20products%20Intl.pdf.
† http://www.ibcwater.com.au.
‡ http://www.aeromix.com/documents/Mid-American Dairy.pdf.

10.4.5 Fruits and Vegetable Processing

Amrutech, Inc., (Moraga, CA)* is a US-based company providing a complete system installation of waste treatment systems and processes using membrane technology for FPIs. The wastewaters are treated for high BOD, COD, TSS, and FOG to significantly reduce the contaminant levels in the wastewater produced from washing and peeling fruits and vegetables, rinse water, and cooking equipment wash water. The technology allows the recycling of the treated water back to the process for full or partial reuse by removing the contaminants to satisfactory limits.

CRS Industrial Water Treatment Systems† is an Australian-based engineering company offering consultancy in the areas of water and wastewater treatment to FPIs. They have successfully installed a 480 kL/day water treatment plant for a vegetable processing unit in Lidcombe, NSW. The technology is based on a DAF unit, where coagulated and flocculated solids, oil, and grease are separated in the form of a surface sludge. This sludge is slowly scraped from the surface and collected on a sludge hopper from where it is dewatered through a belt filter press and collected separately as refuse which will be discharged after adjusting the pH if needed.

10.4.6 Food and Beverage Processing

Most food and beverage processing units release wastewaters that have a high BOD mainly because they contain dissolved sugars, fibers, and carbohydrates. Thermoenergy Corp. (Castion Corp., Worcester, MA)‡ belongs to the group of ThermoEnergy's CASTion systems and is one company with special expertise in the area of wastewater treatment for the food and beverage industries, which release effluents with high sugars that come from various process operations and pollute the municipal sewage lines. The technology involves the removal of the BOD and the subsequent recovery of the sugars, proteins, starches, and yeast as salable products and reusable water, by using their special CASTion systems (proprietary products CAST and RCAST). The sugars can be used for various purposes such animal feed, etc. The details of CAST and RCAST are not available.

10.4.7 Miscellaneous Industries

Amrutech, Inc. developed and tested a two-stage membrane system to treat the wastewaters coming from a yeast manufacturing facility. An effluent water stream coming from a beer-manufacturing unit had BOD and COD as high as 75,000–85,000 and 100,000–120,000 ppm, respectively. By utilizing the

* http://www.amrutech.com/en_water_foodproc.html.
† http://www.watertreatment.net.au/.
‡ http://www.castion.com/Industrial Solutions/food-and-beverage.aspx.

two-stage membrane system, the BOD and COD levels were brought down by 90%.

A California-based pizza dough manufacturing company has faced the problem of excess BOD and TSS in their effluent water. Integrated Engineers* designed a 20 gal/min continuous flow flocculation unit that utilizes their proprietary flocculant Floccin J. This resulted in the reduction of the BOD and TSS from 210 and 550 ppm to 33 and 26 ppm, respectively, which saved the company $70,000 pa.

10.5 Conclusions

The importance of treating the effluent waters from FPIs need not be overemphasized in view of its obvious relevance in not only conserving the water utility but also in combating the environmental pollution. The very fact that large FPIs would consume as much as 1 million gallons of potable water per day, out of which a good amount would be discharged as effluent waters, speaks volumes of the need for the treatment of wastewaters, which has resulted in incessant and intensive research in this area. This has resulted in a number of treatment methods that are classified based on

- Physical methods
- Chemical methods
- Biological methods

In view of the obvious importance and ease of operation, the physical separation methods are usually resorted to at an industrial scale unless otherwise warranted. Next, are the chemical and biological methods. In almost all of the methods, the approach is more toward protecting the environment rather than conserving and reusing the treated water. Hence, most of the water that is treated by the chemical and biological methods is used for activities such as irrigation or is discharged into the environment, while the water treated by the physical methods usually goes for reuse in the same industry. Hence, most of the process industries show an increased interest in resorting to the physical methods of treatment, as in the case of fruit and vegetable processing industries. However, in the case of animal-based industries, the effluent waters, which are contaminated with body fluids such as blood, gut contents, and milk, will be treated by chemical or biological methods.

Wherever the treated waters are used for irrigation purposes, efforts are being made to study the soil profile also. Depending on the field conditions,

* http://wecleanwater.com/html/success/food_processing/pizza_dough.htm.

the treated waters are used; highly saline waters cause osmotic stress on the plants, whereas low pH waters reduce the fertility of the land. Combining all of these soil management studies, the water treatment strategies are made. Another classical approach followed by most of the industries is to wash or cleanse the raw materials before processing so that the quantity of effluent water is minimized. As in the case of fruit and vegetable processing and animal product processing, some of the processing is done at field level so that the load on the effluent waters in the industry is considerably reduced.

Several R&D activities have been implemented in the food processing industries to reduce the environmental pollution and the discharge of effluent waters by resorting to a number of clean technologies. The industry will continue to strive to implement advanced innovative methods to combat the environmental impact of the wastewaters being discharged from the food processing industries.

References

Colic, M., W. Morse, J. Hicks, A. Lechter, and J. Miller. 2007. Case study: Fish processing plant waste water treatment. Proceedings of the Water Environment Federation: Industrial Water Quality, 2007, pp 1–27. Clean Water Technologies, Inc., http://www.cleanwatertech.com.

Crittenden, J.C., R.R. Trussell, D.W. Hand, K.J. Howe, and G. Tchobanoglous. 2005. *Water Treatment: Principles and Design.* Hoboken, NJ: John Wiley.

EPTS (Environmental Protection Technology Series). 1974. Water characterization for the specialty food industry. EPA-660/2-74-075. Portland, OR: US Environmental Protection Agency. http://www.p2pays.org/ref/26/25046.pdf.

Foust, A.S., L.A. Wenzel, C.W. Clump, L. Maus, and L.B. Andersen. 1980. Particulate solids flow and separation through fluid mechanics. *Principles of Unit Operations*, 2nd edn., New York: John Wiley, pp. 622–637.

Mc Cabe, W.L., J.C. Smith, and P. Harriott. 2005. *Unit Operations of Chemical Engineering*, 7th edn. New York: McGraw-Hill.

Rao, D.G. 2010. *Fundamentals of Food Engineering.* New Delhi: PHI Learning Pvt. Ltd.

11

Removal of Lower-Molecular-Weight Substances from Water and Wastewater: Challenges and Solutions

V. Jegatheesan, J. Virkutyte, L. Shu, J. Allen, Y. Wang, E. Searston, Z. P. Xu, J. Naylor, S. Pinchon, C. Teil, D. Navaratna, and H. K. Shon

CONTENTS

11.1 Introduction

There are many different sources of water accessible to humans. It is this variety that allows humans to inhabit almost every corner of the planet. These sources include lakes, rivers, creeks, natural springs, aquifers, man-made dams, and oceans. Each source of water, due to its different location, has different concentrations of substances including contaminants. Water can gather these substances at various stages of its cycle due to various factors including the characteristics of surrounding water bodies (freshwater running into the ocean), the catchments and land that the rain falls within, and the type and frequency of interactions with biological processes and humans. Figure 11.1 presents a "life cycle" for contaminants, contamination sources, treatment methods, and potential health concerns.

There are a variety of different sources of contamination to water supplies. These include naturally occurring substances (not biologically), biological materials, and synthetically produced substances. From a human health perspective, we consider these materials as contaminants due to the fact that we are unable to consume water in which they are present without some side effects on health or aesthetic displeasure. Despite this, most of these contaminants are totally natural and play important ecological roles that help maintain the conditions of the water. Natural organic matter (NOM) and its derivatives are the cause of many of the aforementioned water sources having a significant quantity of organic matter within them. These organic compounds play important biological roles within

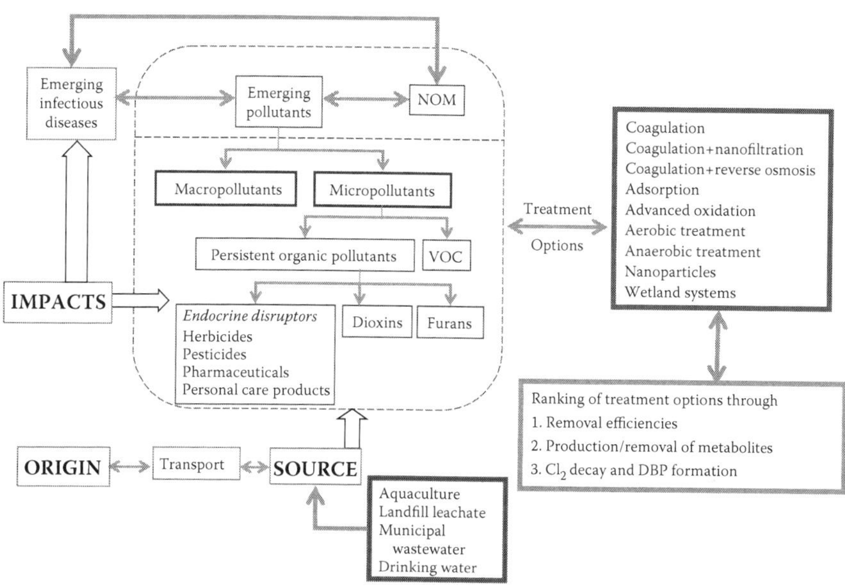

FIGURE 11.1
Life cycle of contaminants, their sources, treatment options, and potential health concerns.

their environmental system, but when used for drinking water supplies, NOM is seen as a contaminant. NOM is probably the second most common contaminant found in the earth's water bodies, with the first being the high concentration of sodium chloride (table salt) found in the sea. This concentration of organic matter is due to the abundance of life on earth, especially in the vicinity of plentiful water environments. Many chemicals in their natural form are completely harmless to humans, including many of the organic chemicals found in NOM. It is when these substances react with other chemicals such as those added during the purification process that a lot of potentially harmful substances are produced. Substances that react with disinfectants are known as disinfection by-product precursors (DBPPs). Of these substances, those that react with chlorine during disinfection are known as chlorination by-product precursors (CBPPs). The end products are called disinfection by-products (DBPs) and chlorination by-products (CBPs), respectively.

In addition, human activities are increasingly affecting these water bodies through pollution due to inputs from various contaminants. A large number of these contaminants are produced by lower-molecular-weight organic compounds (LMWOCs), mainly in the form of pesticides and herbicides (Dsikowitzky et al. 2004). LMWOCs are defined as the organic compounds with molecular weight less than 1000 Da (Perry and Linder 1989), including bioorganic acids, vitamins, chlorophyll, and other toxic substances such as persistent organic pollutants (POPs), endocrine-disrupting chemicals (EDCs), and pharmaceuticals and personal care products (PPCPs) (Pempkowiak and Obarska-Pempkowiak 2002; Stull et al. 1996; Wang et al. 2009). It is reported that even in trace levels, they might deteriorate the quality of water, such as its color, taste, and odor (Wang et al. 2009). Contamination of water bodies causes long-term adverse effects in the aquatic environment, for example, bleaching of corals in the Great Barrier Reef (GBR) World Heritage Area (Negri et al. 2005). Little is known about what long-term effects these herbicides can have on the GBR and its ecosystems.

The widespread use of pesticides promotes leaching into surface water and groundwater, and these contaminants are present in drinking water (Younes and Galal-Gorchev 2000). Conventional water treatment processes include coagulation/flocculation, sedimentation, and traditional filtration, and these conventional water treatment processes are not effective in the removal of pesticides and herbicides (Navaratna et al. 2010). In recent years, Australia has upgraded many of its water treatment plants to meet higher water quality standards and for the removal of micropollutants (Navaratna et al. 2010). These upgrades include advanced water treatment processes such as powdered activated carbon (PAC) filtration, granular activated carbon (GAC) filtration, and high-pressure membrane filtration such as reverse osmosis (RO). These advanced water treatment processes are considered to be efficient in the removal of micropollutants.

LMWOCs are found mainly in the form of pesticides and herbicides (Younes and Galal-Gorchev 2000). Annually, it is estimated that 4 million tons of pesticides are applied to crops worldwide, but less than 1% of the total applied pesticides gets to the target pests (Gavrilescu 2005). As herbicides are frequently used near water bodies, they have often been found in surface waters. The contamination of surface waters generally occurs during the wet season due to the discharge from agricultural lands. Many herbicides are also fairly mobile in soil and can easily migrate into the groundwater. Pesticides and herbicides are considered to be POPs, which persist in the environment for a long period of time. POPs bioaccumulate through the food web. They are capable of long-range transport and pose a risk of causing harmful effects to human health and to the environment, in general (Navaratna et al. 2010).

11.1.1 Water Quality Guidelines

The World Health Organization (WHO) is the directing and coordinating authority for health in the world. Its main responsibility is to provide leadership to the international community on global health matters. By doing this, it helps shape the health research agenda. The WHO is also responsible for setting norms and standards, articulating evidence-based policy options, providing technical support to countries, and monitoring and assessing health trends (WHO 2006). These guidelines or norms are known as the *Guidelines for Drinking-Water Quality,* Volume 1 of which was published in

1996. These guidelines include the recommendation of a minimum chlorine level (including residual level) to deactivate the microbial activity in water, creating pathogen-free water supplies. Other relevant information in these guidelines includes contaminant levels within water supplies for DBPPs, organic substances, and inorganic substances.

Although the WHO is the international body for providing standards and guidelines, they are unable to enforce such compliance and emphasize that the values provided in the guidelines are not mandatory limits. The guidelines state that such limits should be set by national authorities, using a risk–benefit approach and taking into consideration the local environmental, social, economic, and cultural conditions. The Australian potable water quality guidelines are known as the 2004 *Australian Drinking Water Guidelines* (*ADWG*) (NHMRC 2004). The values within these guidelines take their point of reference from those recommended by the WHO's *Guidelines for Drinking-Water Quality*. The differences in these values usually arise from the Australian guidelines use 70 kg as the average adult weight, whereas the WHO guidelines use 60 kg. The differences could also be due to different risk assessments; for example, the Australian guidelines use a risk of one additional cancer per 1 million people, whereas the WHO guidelines give values for 1 in 1000 (NHMRC 2004).

11.1.2 Health Concerns

Of all the health interventions in modern history, the disinfection of public drinking water has had one of the largest if not the most profound and beneficial effects on human health. Disinfection by chlorine has been successful in the deactivation of pathogenic organisms and hence the prevention of waterborne disease outbreaks (Clark 1998). With so many different contaminants possible, it is highly unlikely that water from a water body that is collected for human consumption will lack all of these contaminants. As stated earlier, organic compounds are one of the most common contaminants in any water source. It is not predominantly the organic substances themselves that create health risks to humans, but the derivatives formed when the organic substances react with disinfection chemicals such as chlorine. There have been various epidemiological studies that have suggested an association between DBPs, in particular CBPs, and various cancers. Cancers of the bladder and rectum have been the most common cancers in this association, but specific concentrations at which CBPs might cause an increased risk to human health cannot be determined due to insufficient data (NHMRC 2004).

There have been experiments on mice that show no increase in the incidence of skin tumors when CBPs in solution are applied to the skin compared with unchlorinated samples. There was also no increase in tumors due to oral administration of chlorinated humic acids in drinking water compared with animals receiving unchlorinated humic acids or saline-treated controls. Studies of concentrated chlorinated drinking water supplies have shown the supplies to be mutagenic to some strains of test bacteria. These results were found consistently when the surface water samples used had a high content of NOM at the time of chlorination. A large proportion of this increased mutagenicity has been attributed to a chlorinated furanone. A review of the available data was carried out by the International Agency for Research on Cancer, and it was concluded that there is inadequate evidence to determine the carcinogenicity of chlorinated drinking water to humans. Action to reduce this concentration of DBPs is encouraged worldwide on the presumption that disinfection itself is not compromised. This is due to the relative risks associated with the lack of each treatment, with the risk posed by DBPs being considerably smaller than the risk posed by pathogenic microorganisms (NHMRC 2004). Epidemiological studies often examine DBPs as a generic group. This can be useful in determining the overall effects since there is the possibility of combined health effects of different DBPs, some known (without health data) and some unknown.

Furthermore, acetic acid, which can be found in water systems due to its use in herbicides, can cause irritation to the eyes and skin as well as burning. It can also irritate the stomach lining and cause kidney failure (Government 2007). Citric acid, which is also used in herbicides, can erode tooth enamel. Diuron is a herbicide that is used to kill weeds by preventing photosynthesis. This organic substance has been tested on animals to find that at levels of continual exposure, blood, liver, and spleen abnormalities can occur. It could also result in weight loss (Giacomazzi and Cochet 2004). The effects on humans are not as distinguishable; however, there is no evidence that states that cancer-causing attributes do exist in this chemical.

11.1.3 Water Treatment Processes

There are a vast variety of water treatment processes for preparing water for drinking. A lot of these processes have similar treatment mechanisms, differing only slightly for various applications and water sources. The treatment mechanisms include coagulation and the related mechanisms of flocculation and agglomeration, softening, addition of PAC, filtration and adsorption through GAC, the use of nanoparticles, deionization, membrane filtration, and disinfection. Conventional water treatment processes usually entailed coagulation, flocculation, sedimentation, filtration, and disinfection (Jegatheesan et al. 2009).

11.1.3.1 Membrane Filtration Processes

Membrane technology is becoming increasingly popular and more viable as an alternative treatment technology to the traditional treatment process for drinking water. This is due to the anticipation of more stringent water quality regulations, a decrease in availability of adequate water resources, and an emphasis on water for reuse (Davis and Cornwell 2008). A membrane is a thin barrier between two phases that allows preferential transport of certain species (Shu 2000). In combination with a pressure source, a solvent can be transported through the membrane via a pressure imbalance between the feed and the permeate solution streams. Membranes have the capability of separating materials based on their physical and chemical properties. They are often described by the membrane pore size, molecular weight cut-off (MWCO), material and geometry, targeted materials to be removed, type of water to be treated, and treated water quality (Davis and Cornwell 2008). Membrane filtration can be categorized into pressure-driven and electrically driven processes. For water treatment, pressure-driven membranes are the only relevant processes. Pressure-driven membranes can be categorized from largest particle removal to smallest particle removal as microfiltration (MF), ultrafiltration (UF), nanofiltration (NF), and RO. Each membrane utilizes various separation mechanisms. Table 11.1 shows the membrane types and their rejection efficiency.

NF has gained momentum in recent years as having an important role to play in the drinking water purification processes. Its purpose to date has mainly been for water softening, decoloring, and micropollutant removal. The viability of NF becoming a major water purification process is based

TABLE 11.1

Membrane Types and Their Rejection Efficiency

Membrane Filtration Process	Pore Size (μm)	MWCO (Da[a])	Operation Pressures (kPa)	Primary Applications
Microfiltration (MF)	≥0.1			Particulate and microbial contamination removal
Ultrafiltration (UF)—tight	0.001–0.1	1000	70–700	Removal of some organic materials from freshwater
UF—loose		>50,000	70–200	Liquid/solid separation, that is, particle and microbial removal
Nanofiltration (NF)	<0.001	<1000	500–1000	Membrane softening; removes ions contributing to hardness: calcium and magnesium; very effective in removal of color and DBP precursors
Reverse osmosis (RO)	<0.001	<200	1000–8000	Removal of salts from brackish water and seawater

Source: Davis, M.L. and Cornwell, D.A., *Introduction to Environmental Engineerig*, 4th edn. In ed. I.A. Dubuque, McGraw Hill Companies, 2008. With permission.

[a] Dalton—a unit mass equal to 1/16 the mass of the lightest and most abundant isotope of oxygen (1 amu).

on its performance in removing low-molecular-weight organic compounds versus economic, health, and environmental benefits that could result.

11.1.3.2 Disinfection

There are various types of disinfectants used in water supply treatment. These include ozonation, ultraviolet (UV) light, and chlorine in its various forms. The disinfection process relies on the ability of the disinfectant to deactivate pathogenic organisms. Chlorine is one of the most effective and commonly used disinfectants (Clark 1998). Ozone, on the other hand, has powerful oxidation properties but does not provide a residual past the point of treatment. One part per million (ppm) of ozone destroys all bacteria within 10 min but is more costly than chlorine. Ozone has the limitation of having to be manufactured on site. Like ozone, UV has no residual effect and, therefore, is limited in use to plants close to the point of use. Because of chlorine's all-round performance as a disinfectant in terms of cost, ease of handling, and having a lasting residual, it has been the most widely used of all disinfectants (Gray 2005). Currently, there are no other large-scale, economically viable disinfection methods that can compete with chlorination, especially for developing countries. With chlorine continuing to be the most widely used disinfectant in the future, it is essential to improve the chlorination process in all possible ways, especially safety, by understanding how to make the chlorine residual last longer while decreasing the chance of DBP formation and the quantity of DBPs formed.

11.1.3.3 Coagulation/Flocculation

Coagulation alters the colloids so that they can adhere to each other by reducing the surface charge of the colloids. This is done by adding a coagulant (chemical), which in turn adds a positive ion to the water (Davis and Cornwell 2008). Flocculation, which is synonymous with agglomeration and coagulation, is defined as a process of contact and adhesion whereby the particles of dispersion form larger-size clusters (Calvert 1990). Rapid mixing is the most important factor affecting the efficiency of coagulation, whereas flocculation is the most important factor affecting particle-removal efficiency (Davis and Cornwell 2008). Coagulation not only has importance in the treatment of water itself, but without it, many membranes would not be able to filter raw water effectively due to flux declination by fouling of the membrane.

11.1.3.4 Use of Nanoparticles

A nanoparticle is defined as a small particle in the size range of 1 to 50 nm (Zhou et al. 2009). Due to their extremely small particle size, nanoparticles have relatively large specific surface areas. The high usage and manipulation of nanoparticles for adsorbents, catalysts, and sensors are a consequence of high reactivity and surface area properties of those particles (Li et al. 2008). Nanoparticles are extremely useful for a number of applications attributable to the electronic, optical, magnetic, and catalytic properties of those particles. The use of nanoparticles is developing into a highly desirable new technology for treating drinking water sources. A nanoparticle dispersed in water is referred to as a hydrosol, and in organic solvents (organic compounds capable of dissolving other substances), a nanoparticle is called an organosol (Zhou et al. 2009).

Over the years, low-molecular-weight organic materials have caused issues regarding the complete removal of organic matter contained in a water source. Filtration has been the main choice when it comes to the removal of suspended organic matter from drinking water sources. However, there are much smaller molecules that pass through filtering systems. At present, technology is evolving to develop processes in which nanoparticles are used to adsorb low-molecular-weight organic matter from drinking water sources.

11.1.3.5 Adsorption onto Activated Carbon

Activated carbon is a popular adsorbent due to its large porous surface area, controllable pore structure, thermostability, and low acid/base reactivity (Foo and Hameed 2009). Activated carbon has an advanced ability for removing a range of organic and inorganic pollutants that are dissolved in aqueous media and even from gaseous environments (Foo and Hameed 2009). Activated carbon is produced by heating organic matter to make it extremely porous. Adsorption processes in drinking water treatments can successfully control issues relating to trace organic substances such as taste- and odor-causing compounds, volatile organic compounds, and organic compounds including pesticides (Knappe et al. 1998).

Activated carbon adsorption has been designated by the US Environmental Protection Agency as the "best available technology" for the treatment of LMWOC, that is, herbicides in drinking water (Adams and Watson 1996). There are different forms of activated carbon that can be used. These forms include powdered, granular, and fiber or cloth. The most commonly used adsorbents for the removal of herbicides are PAC and GAC. Activated carbon cloth has gained increasing attention in recent times with its high specific surface area, adsorption capacity, and mechanical strength (Ayranci and Hoda 2005).

11.1.3.6 Membrane Bioreactor

Membrane bioreactor (MBR) is a new improvement of the conventional activated sludge treatment processes, where the secondary clarifier is replaced by a membrane unit for the separation of treated water from the mixed solution in the bioreactor (Xing et al. 2000). The main advantages of the technique are associated with the absolute retention of all microorganisms, which in turn ensures an increase in the sludge concentration and results in complete disinfection of the treated water. Also, it allows to control the hydraulic retention time (HRT) and the sludge retention time (SRT) independently (Brindle and Stephenson 1996; Nagano et al. 1992). Thus, due to the absence of the secondary clarifier, the overall size of the treatment plant can be significantly reduced (Manem and Sanderson 1996).

11.2 Application of NF

In recent years, there has been a lot of interest in a variety of areas relating to water resources. These areas include water usage and allocation, where to source primary supplies from, and what treatments should raw water undergo before delivering it as potable water. These are very important questions that will be investigated and debated for many years.

The history of potable water and its link to the transmission of waterborne diseases provides the answers as to why significant time and efforts have gone into water treatment research. Since water is the life-providing substance, it is not hard to understand why it may be inhabited by microbes and other living organisms. Some of these microscopic organisms are pathogens such as Cryptosporidium and Giardia. These two pathogens are relatively easy to remove with MF membranes (Gray and Bolto 2003). Unfortunately, these are not the only waterborne pathogens, and even if they are almost completely removed at the treatment plant, there is the chance of them reemerging within a distribution system due to factors such as biofilm within the distribution network (Gray and Bolto 2003). This is the very reason why static treatments (at the point of the treatment plant) alone cannot control microbial growth, and residual treatment (providing a source of control right through the distribution system) such as disinfection by chlorination has been used so extensively. Chlorination of potable water has occurred since the beginning of the twentieth century. Its use in effectively inactivating pathogenic microbial organisms and almost fully deterring any serious waterborne disease outbreaks is one of humanity's most successful public health interventions (Clark 1998). Until the last 30 years, there has been relatively no concern for what chlorine itself might be doing to human health since there was no known reason for concern. In 1976, a humic fraction within NOM was identified as a major precursor for DBP trihalomethane (THM). Since then, there has been extensive research into chlorine decay and the related DBP formation and DBP toxicities (Gang et al. 2003). International standards arose, putting limits on the allowable DBPs in potable water. This has led to research and technological advancements in water treatment processes to achieve the benchmarks set by the guidelines.

The development of membrane filtration, in particular, NF, has opened the doors to the possibility of DBP-free water. Due to its low-molecular-weight rejection size, NF may be able to remove the low-molecular-weight organic substances that are the main contributors to DBP formation. Most research has been focused on performing investigations on natural water sources. This is often done on the basis of simulating real-world conditions (Visvanathan et al. 1998). For a more detailed understanding of the performance of NF in removing specific low-molecular-weight organic substances that are DBP precursors, studies on individual species need to be conducted. This should determine the mechanisms behind the interactions of NF membranes with DBP precursors and their associated organic functional groups.

11.2.1 Membrane Filtration Mechanisms

Some of the parameters that affect the performance of a membrane include the fixed charge density, the sign of the charge, and the hydrophobicity/hydrophilicity of the substances (Linder et al. 2006). Even though a membrane can be described by a certain pore size or MWCO, other rejection mechanisms may be present, retaining particles of size smaller than the specified pore size or MWCO (Sterlitech 2002). There are three general categories of membrane rejection mechanisms. These include:

- Sieve effect—rejection based on size differences of species in solution
- Solution–diffusion—rejection based on differences in solubility and diffusivity of species in solution through the membrane
- Electrochemical effect—rejection due to charge differences of species in solution

The general rejection mechanisms governing the rejection of solute through commercial polymeric NF membranes are the size of the solute (sieve effect) and the Donnan effect (electrochemical effect) (Peeters et al. 1998).

11.2.1.1 Fouling

Fouling can be described as the alteration of the filtration mechanisms such that a flux decline occurs or the driving-pressure force is required to be increased to maintain constant flux. Therefore, fouling decreases the efficiency of a membrane. Fouling can be categorized into

- Reversible fouling—can be removed by cleaning and washing processes
- Irreversible fouling—cannot be removed by cleaning and washing processes

11.2.1.2 Resistance Mechanisms

11.2.1.2.1 Concentration Polarization

During NF, the solute is retained within the feed with less concentration in the permeate solution. The retained solute can accumulate on the surface until the surface concentration of the solute is greater than the feed solution concentration. This phenomenon is known as concentration polarization. Once this polarized solute distribution is achieved, a diffusive flow of solute from high concentration on the membrane surface to lower concentration in the bulk feed will occur until a steady state is established for solute flow in both directions. This is important as flux is dependent on the solute concentration at the membrane surface and in the feed solution.

11.2.1.2.2 Gel Layer and Boundary Layer Resistance

Important to this study is the gel layer resistance model since it is often associated with organic applications where adsorption on the surface of the membrane occurs. This adsorbed layer forms a "gel layer," causing a resistance to solvent transport through the membrane. This results in flux decline and reduced membrane efficiency (Denyer 2005). Over the filtration period, an increase in the solute concentration in the feed results in the boundary layer exerting a resistance on the passage of the solvent molecules (Denyer et al. 2007).

11.2.1.2.3 Osmotic Pressure

Osmotic pressure is the pressure opposing the driving pressure due to the concentration gradient between the feed and the permeate solution. This results in flux decline as the concentration increases in the feed. Osmotic pressure has less of an effect in NF, as a greater concentration of the solute exists in the permeate solution. This has a flux-increasing effect, balancing out, at least partially, some of the osmotic pressure resisting the flow direction (Denyer 2005).

11.2.1.3 Surface Forces and Rejection Mechanisms

11.2.1.3.1 London–van der Waal's Force

The London–van der Waal's force is generated by the instantaneous dipole moments generated by the temporary asymmetrical distribution of electrons around the atomic nuclei. This force is important in particle adhesion on membrane surfaces (Jegatheesan and Vigneswaran 2005).

11.2.1.3.2 Electric Double Layer (EDL) Interaction

Surfaces charges will be acquired in aqueous environments due to the adsorption of ions or the dissociation of surface groups. A double layer of charge is established by the surface charge and the counter charged ions present in the solution. This double layer of charge will cause attractions and repulsions depending on the magnitude of the charge (Jegatheesan and Vigneswaran 2005).

11.2.1.3.3 Born Repulsion Force

The Born repulsion force is a strong repulsive force arising from electron cloud overlapping of atoms. This force is also known as hard-core repulsion and determines how close two atoms or molecules can come (Jegatheesan and Vigneswaran 2005).

11.2.1.3.4 Hydration

This force is due to the disruption or change in the order of liquid molecules surrounding a surface during the approach of a second surface. Hydration forces are believed to result from the strong hydrogen-bonding surface groups (Jegatheesan and Vigneswaran 2005).

11.2.1.3.5 Donnan Exclusion

Donnan exclusion may play a significant role in the rejection of organic acids. This is because an ionic solution can undergo an ion shift if the membrane has a surface charge. This results in the exclusion/rejection of ions in solution with the same charge as the membrane surface. Depictions of how membrane charge affects rejection can be seen in Figure 11.2. Figure 11.3 depicts a decrease in rejection with an increase in ionic concentration due

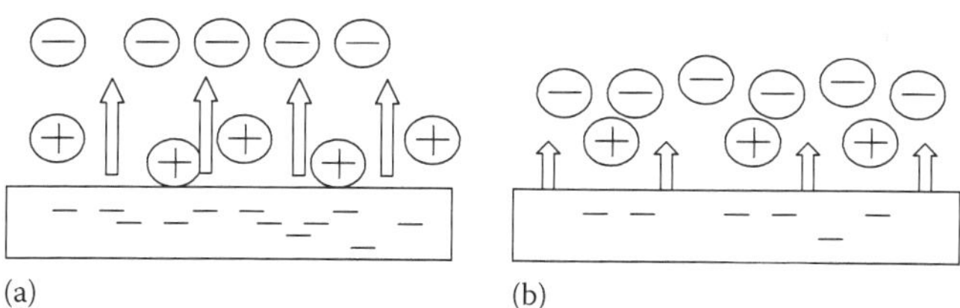

FIGURE 11.2
(a) Membrane with a strong negative charge rejects anions more. (b) Membrane with a weak negative charge rejects anions less.

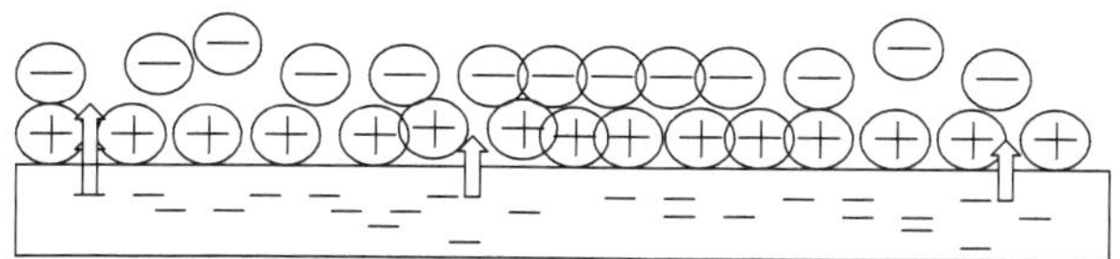

FIGURE 11.3
Concentrated ionic solutions (monovalent cations) neutralize the membrane surface charges (negative) and thus provide small rejection of anions.

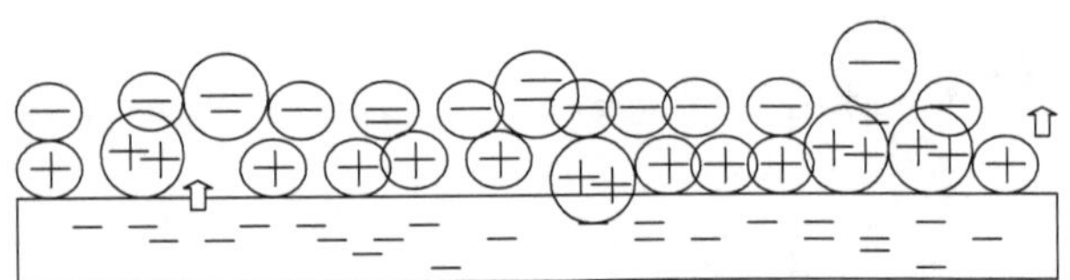

FIGURE 11.4
Concentrated ionic solutions with monovalent and multivalent ions (cations) will provide even smaller rejection of anions (compare with Figure 11.3).

to shielding, as found by Bartels et al. (2005). Multivalent ions of the same charge as the membrane are rejected to a greater extent, although multivalent ions of opposite charge to the membrane surface are better at shielding the membrane charge as seen in Figure 11.4. Bartels et al. (2005) also found that Donnan exclusion is negligible for solutions of <300 mg/L solute concentration.

Visvanathan et al. (1998) found that rejection of organic substances decreases substantially with increasing concentration of divalent cations. This also agrees with the findings of Bartels et al. (2005).

11.2.2 NF Efficiency and Performance

A membrane's efficiency and performance are determined by its selectivity and the applied flow (Davis and Cornwell 2008). The efficiency of a membrane is quantified by the flux through it, which is defined as the volume flowing through the membrane per unit area of the membrane in unit time, and usually has the units of L/m²/h. The equation (Equation 11.1) defining flux (J) is given by:

$$J = \frac{(m_{t_2} - m_{t_1})}{\rho A(t_2 - t_1)}, \tag{11.1}$$

where m_{ti} is the mass measurement taken at time t_i (kg), ρ is the density of feed solution (kg/L), A is the surface area of the membrane, and t_i is the time since the commencement of the membrane filtration.

Flux declination is given by the relative efficiencies of a membrane after one or more factors (time, feed solution concentrations, temperature, or pH) have changed. The performance (removal capacity) (Equation 11.2) is based on the selectivity of a membrane toward a contaminant mixture or substance and is expressed by its retention (or rejection), R, which is given by

$$R = \frac{c_f - c_p}{c_f} \times 100\% \tag{11.2}$$

where c_f is the contaminant concentration in the feed (retentate) and c_p is the contaminant concentration in the filtrate (permeate) (Davis and Cornwell 2008).

11.2.3 Membrane Materials and Structures

The membrane material along with its structure is an important factor in the performance of a membrane for rejecting the compounds from a solution. These two factors determine the rejection mechanisms that may apply during filtration of a solution. Membranes are produced from a variety of materials, including silver, cellulose acetate, ceramic, glass fiber, nitrocellulose mixed esters (MCE), nylon, polycarbonate (PCTE), polyethersulfone (PES), polyester (PETE), polypropylene, and polytetrafluoroethylene (PTFE—Teflon)-laminated and PTFE-unlaminated membranes (Sterlitech Corporation, 2002). There is a far greater variety to this list mentioned, but for the purpose of this study, the variety is not of concern. The general types and the rejection mechanisms that membranes use are of more importance. The selection of NF membranes requires care as chemical reactions can take place between the membrane and the dissolved compounds in solution, leading to membrane damage. Two examples of this are polycarbonate membranes, where *N*-methyl-2-pyrrolidone, methylene chloride, chloroform, and acetone can dissolve the membrane; and polyester membranes, which can be dissolved by *m*-cresol, *o*-chlorophenol, and hexafluoroisopropanol (Sterlitech 2002).

The structure of a membrane can vary by its thickness, degree of crosslinking and morphology, charge density, porosity, and electrokinetic properties. Some of these properties are due to the membrane's base material, while others are due to graphed material on the membrane.

11.2.4 Summary of Main Observations

Jarusutthirak et al. (2007) found that an increase in NOM concentration from 0 to 25 mg/L caused higher NOM rejection, membrane fouling, and greater flux decline. This was assumed to be a result of NOM accumulation on the membrane. It showed a correlation/relationship between the initial concentration of the NOM or DBPPs with the flux and rejection. This is an important finding for this study, as an overall view of performance will be based on rejection of DBPPs and flux decline through the membrane. The fact that fouling leads to a better rejection may require the concentration of the model compounds to be used to be selected carefully. They may possibly have to be selected on the basis of simulating the real world so that the results represent something more quantifiable. Also, fouling patterns and mechanisms can change with a changing NOM concentration in the feed. From low to high NOM concentrations, the pattern changes from pore blocking and constricting to cake formation (Jarusutthirak et al. 2007). These mechanisms should be considered in the selection of model compounds in terms of some that will foul differently as compared with others. Importantly, an increase in pH showed significant flux decline due to increased salt concentration on the surface and/or within the pores of the membrane, fouling the membrane. This result is significant as pH will need to be considered during the experimentation before and after NF.

Gray and Bolto (2003) found that the flux decline for strongly hydrophobic acid (SHPA), weakly hydrophobic acid (WHPA), and charged hydrophilic fractions (CHPI) reached a steady state if liquid backwashing is used (leading to a minor flux recovery), which is consistent with fouling due to filling of all adsorption sites and blocking of the pores. This means that the hydrophobicity/hydrophilicity and acidic/basic nature of the model compounds to be chosen are crucially important. It was also found that associations between SHPA and WHPA may increase the extent of the flux decline. This may open an area for further research by testing a number of combinations of model compounds and noting if this behavior is consistent.

According to Visvanathan et al. (1998), precompaction of membranes can bring about better structure of the membrane. This structure is then capable of producing a higher rejection capacity and a more stable performance. While this is the case, the overall flux is not likely to be affected due to precompaction. The initial concentration of the model compounds is significant when considering the temporary compaction caused by short-term applied pressure. The removal efficiency of low concentrations of humic substances increases with an increase in pressure (and temporary compaction) (Visvanathan et al. 1998). Such compaction is far from being as good as a precompacted membrane. However, for both types of compaction, the rejection was found to increase with an increase in the total organic carbon (TOC) in the feed. These factors should possibly be investigated further; otherwise, factors such as the initial concentration and temporary compaction (pressure) should remain constant. A pH between 7 and 9 should be used for NF. The initial model compound solutions may be within this range, solving any need for adjustment. In the presence of divalent ions, the flux decreases slightly, whereas the rejection decreases substantially with the increase in the concentration of divalent ions suspected to be due to concentration polarization.

When NF of model compounds was conducted using pressurized stirred cell with a nanomax-50 membrane (500 kPa of operating pressure), the lower concentrations (20 mg/L) of citric acid, ascorbic acid, and lactic acid achieved 51%, 39%, and 43% of rejections, respectively. NF of the higher-concentration solutions (200 mg/L) achieved rejections of 19%, 29%, and 24%, respectively. The mechanism behind the rejection is the Donnan effect, which could be justified by analyzing various proportions of the anionic species in different solutions (Table 11.2). The undissociated model compound diuron was found to have a rejection up to 26%. The sieve effect is

TABLE 11.2

Initial Species Distribution of Organic Acids in Investigated Solutions

Model Compound	Concentration (mg/L)	Measured pH	Species Distribution (%)		
			H_3A	H_2A^{-1}	HA^{-2}
Citric acid	200	3.16	45	53	2
	20	4.13	6	74	20
Ascorbic acid	200	3.17	89	11	—
	20	4.51	29	71	—
Lactic acid	200	3.47	74	26	—
	20	4.14	31	69	—

Note: All acids will have the following species: H_3A, H_2A^{-1}, and HA^{-2} at different percentage distribution depending on the pH of the solution.

thought to be the main rejection mechanism for diuron due to its larger molecular weight (Allen 2008). The flux obtained during the experiments is shown in Figure 11.5.

11.3 Application of Coagulation and Adsorption in the Presence of NF Membranes

The presence of organic compounds, especially LMWOCs, in the treated water can have adverse influence on the quality of those effluents. One of the ways to remove these substances is to adsorb them onto the surface of the activated carbon and achieve more than 80%–90% of removal of the target contaminants (Wang et al. 2009). Despite the effectiveness of the treatment, the operating costs of activated carbon columns are very high. To overcome such a problem, hybrid techniques such as coagulation and/or adsorption in the presence of NF membrane are used (Wang et al. 2009, 2010).

Membrane filtration processes are currently gaining increased popularity and viability as alternative treatment methods for the removal of LMWOCs from water sources, as there is no formation of by-products in membrane filtration processes. Moreover, reuse of organic substances such as pesticides present in a concentrate is also possible. NF and RO are two membrane processes, and NF is becoming widespread because it could produce higher permeate flux at lower operating pressures relative to RO (Wang et al. 2009). The membranes used for NF have an MWCO in the range of 200–1000 Da. In order to obtain a higher removal of organic compounds while improving the fouling of the NF membrane, a hybrid system composed of coagulation, adsorption, and NF has been introduced in the field of water treatment (Carroll et al. 2000; Tran et al. 2006; Wang et al. 2009, 2010).

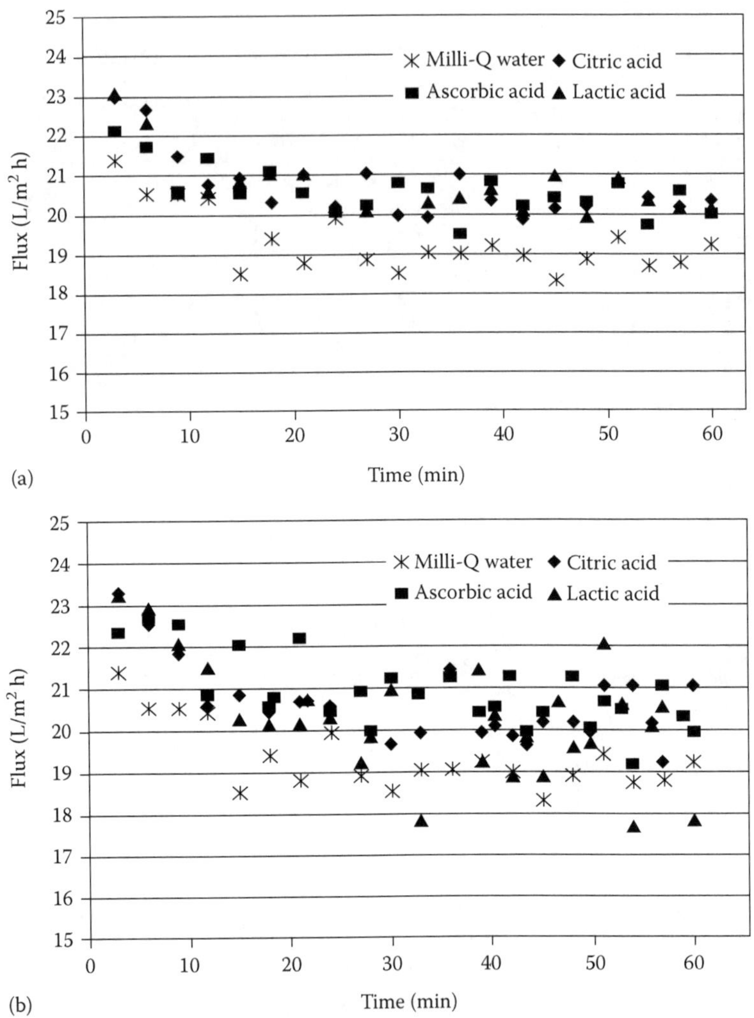

FIGURE 11.5
Flux of model compounds during nanofiltration: (a) 20 mg/L; (b) 200 mg/L.

In a recent study by Wang et al. (2009), it was found that the removal of herbicide (diuron) was influenced by the nature of the membrane (hydrophobicity, charge density, porosity, and morphology), by the operating conditions (temperature and operating pressure), and by the characteristics of the tested solution (ionic strength and pH). Other studies (Plakas and Karabelas 2008; Zhang et al. 2004) established the significance of the inorganic ions present in the system.

In a study by Wang et al. (2009), experiments conducted with oxalic acid concentrations of 100, 200, and 500 mg/L clearly indicated that NF in

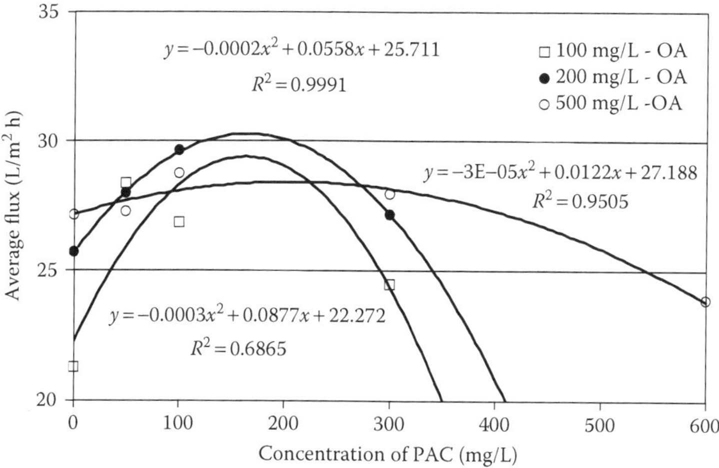

FIGURE 11.6
Average flux of 100, 200, and 500 mg/L oxalic acid (OA) solutions at different poly aluminum chloride (PAC) doses.

combination with coagulation was quite an effective treatment to remove the lower-molecular-weight (LMW) organic acid from water (Figure 11.6). During the NF treatment, coagulation by poly aluminum chloride (PAC) was very helpful for the removal of oxalic acid especially at higher concentrations. However, membrane fouling became more serious if optimum doses of PAC were used. Therefore, PAC doses that are lower than the optimum but still would provide a better TOC removal could be used to achieve less fouling of the membrane. This consideration will reduce the operating costs while maintaining higher removal efficiency.

For the tested diuron solutions, removal efficiencies around 23% were obtained by using the DK-1812 NF membrane even with the addition of PAC. Nevertheless, the use of NaCl in the NF experiments enhanced the removal efficiency up to 60%, with no additional fouling of the membrane. The RO membrane has a good capacity to remove diuron if its relatively high operating cost compared with NF is ignored (Table 11.3).

11.4 Use of Nanoparticles

Nanoparticles are an extremely new and advanced technology. These particles are on the scale of one billionth of a meter and are excellent candidates for the removal of very low-molecular-weight organic matter contained in most drinking water sources. These particles can be biologically or

TABLE 11.3

TOC Removals and pH for Diuron after Nanofiltration (NF) and Reverse Osmosis (RO)

	NF	NF	NF Coagulation (PAC = 800 mg/L)	NF Ionic Solution (0.02 *M* NaCl)	RO
Operating pressure (bar)	5	10	5	5	35
Initial concentration of diuron (mg/L)	7.59	8.07	7.77	9.24	4.28
pH	7.16	7.54	4.75	6.91	6.76
TOC removal (%)	22.78	23.52	23.74	60.02	95.47

Source: Wang, Y., Shu, L., Jegatheesan, V., and Gao, B., *Desalin. Water Treat.*, 11, 23–31, 2009. With permission.

chemically produced and are naturally formed in the environment, although this is extremely rare. Nanoparticles are available in a number of different forms and are used for a number of different processes in medicine, science, and engineering (Poole and Owens 2003).

The use of nanoparticles to remove low-molecular-weight organic matter in water sources is the most recent technology emerging in the water treatment industry. Thus, it is an interesting and developing area in which there is continual research. It has been previously found that nanoparticles are difficult to contain in large water bodies, and as these are potentially harmful to human health, no use of these particles has occurred in large water sources to date (Li et al. 2008). The limits to which nanoparticles can be ingested into the human body before impacting upon human health are not yet known. The large-scale use of nanoparticles will increase once safe limits are established for nanoparticle concentrations in water sources, as this can then be monitored to avoid any detrimental effects. A system of nanoparticle retention also needs to be developed and used in conjunction with health limits to decrease the chances of nanoparticle release. Using a retention system and monitoring to maintain safe health levels will make large-scale use of nanoparticles even more achievable in the near future. Retaining nanoparticles is an important process not only to prevent human health effects but also to decrease costs that would be related to the loss of nanoparticles. Nanoparticles are beneficial for use in water treatment as they can be reused and recycled; therefore, any loss of these materials would be seen as an unnecessary expense (Li et al. 2008). After application in a water system, nanoparticles can be reused and recycled by implementing an efficient filtration process to separate and remove the nanoparticles from a water source downstream of the application point.

Two processes that can be used to keep nanomaterials in a stationary situation are the use of filters impregnated with nanomaterials and the use of nanomaterials as a surface coating. When added to a water body, these materials will remain in a stationary position, making it impossible to lose nanoparticles. A comparison of the cost-effectiveness of nanomaterial use needs to be made with the cost-effectiveness of other methods of wastewater treatment. Limitations of use and lack of study leave us with a process the cost estimation of which has not yet been established (Li et al. 2008). Therefore, the cost-effectiveness has not yet been complied for this type of water treatment. Nevertheless, it is envisaged to provide a much greater benefit for the water treatment industry with respect to human and environmental health.

Chemical treatment of wastewater can also cause human and environmental health problems. Conventional treatment in the wastewater treatment industry can involve chemical processes for the removal of certain matter; these chemicals can produce harmful by-products. To reduce the production of DBPs in water, a number of nanomaterials have been investigated to treat wastewater sources. Due to the fact that these materials are inert when contained in water bodies, the success of these particles is extremely high in removing harmful matter and not producing DBPs. These materials include chitosan, silver nanoparticles, photocatalytic TiO_2, aqueous fullerene nanoparticles, and carbon nanotubes. Potential replacement or enhancement of current water treatment processes by nanomaterial use could be effected because of the benefit of reduced creation of harmful DBPs (Li et al. 2008).

11.4.1 Layered Double Hydroxide

Layered double hydroxides (LDHs) are used extensively due to their high capability for ion exchange (Violante et al. 2009). These molecules have been known to the chemistry world since 1842 (Zaneva and Stanimirova 2004). LDHs, as the name suggests, have a layered structure containing outer layers that are brucite-like and an interlayer containing exchangeable anions (Xu et al. 2008).

The outer layer is referred to as brucite-like, as it has a similar form to brucite, which is the mineral form of magnesium hydroxide, $Mg(OH)_2$. The molecular formula of the outer layer in the LDH structure is in the form $M^{2+}_{1-x}M^{3+}_{x}(OH)_2$ (Newman and Jones 1998). This shows the similarity with brucite. In this formula, M^{2+} and M^{3+} are divalent and trivalent cations, respectively, and x is the ratio $M^{3+}/(M^{2+} + M^{3+})$. From experimental results, the purest LDH form is said to be produced with the x value being between 0.2 and 0.33 (Braterman et al. 2004). It has been said that LDH exists for x values ranging from 0.1 to 0.5. However, anything outside of the values 0.2–0.33 results in an increased chance of formation of hydroxides and other compounds (Braterman et al. 2004).

The molecules used for M^{2+} include magnesium, nickel, zinc, calcium, copper, cobalt, iron, and manganese. The molecules used for M^{3+} include aluminum, iron, chromium, manganese, cobalt, and vanadium (Zaneva and Stanimirova 2004). The ionic radius needs to be similar to that of magnesium so that the cations are able to fit between the gaps in the closely packed hydroxide groups. Overall, the structure of LDH is hydrotalcite-like. Partial substitution occurs between the trivalent and the divalent ions, which then causes an overall positive charge of the outer layers. The interlayer as shown in Figure 11.7 contains exchangeable anions with the molecular formula $A^{n-}_{x/n}$, where n is the valence of the anion. The molecules of the anions in this layer include bicarbonate, sulfate, and chloride (Zaneva and Stanimirova 2004). The choice of anions for the interlayer is more versatile than the choice of cations for the outer layers. There are no limits to which an anion can be used, the only stipulation being that the anion must not obstruct the metal ion and its interaction with the hydroxide layer (Braterman et al. 2004). Sufficient charge density must also be provided by the anion for a cross section. This is usually no less than 3 e/nm^2. Anions in the interlayer are generally orientated in a way that will maximize the interactions that can occur with the surroundings. This layer also contains water molecules to occupy the remaining space in the layer. Water molecules are usually either bonded to the anions in the layer or the hydroxide layers. The overall positive charge created by the outer layers is neutralized by the interlayer. Overall, the formula of an LDH is as follows (Newman and Jones 1998):

$$\left[M^{2+}_{1-x}M^{3+}_{x}(OH)_2\right]A^{n-}_{x/n}\cdot mH_2O,$$

where the expression contained within brackets represents the outer layer and the expression outside the brackets is the interlayer.

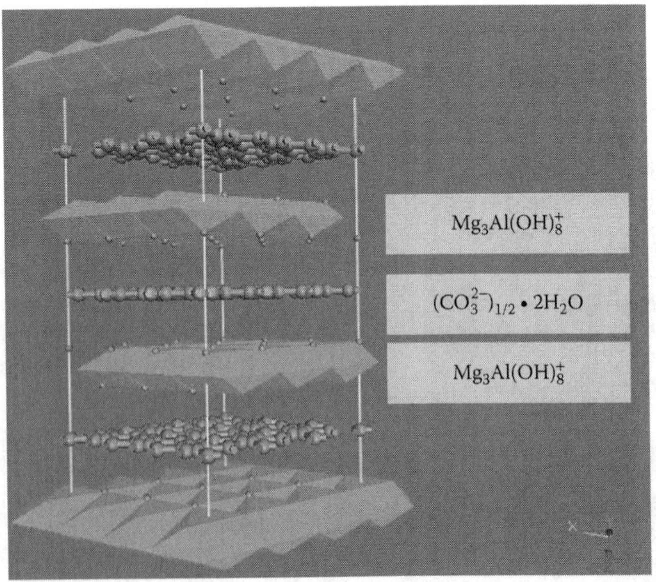

FIGURE 11.7 (See color insert)
Schematized interlayer structure of an LDH.

Previous studies have determined that calcined LDH structures are effective in the adsorption of phenols from water sources (Chen et al. 2009). In solutions, LDH structure is reformed by taking up new exchangeable anions in the interlayer. Through this process, a number of organic materials in wastewater can be removed to eliminate the effects on human and environmental health.

LDH of MgAl can be used for a number of different applications. Major applications in which LDH is used are as adsorbents for pollutants in air or water, in fuel cell technology, and in paper coatings (Auxilio et al. 2009).

11.4.1.1 Properties of LDH Materials

The properties that are displayed by LDH materials are related to the formation processes and treatment before synthesis. Therefore, the properties of LDH are highly dependent upon the reaction time, temperature, concentration of reactants, treatment before synthesis, and reaction solvent used in the synthesis process (Braterman et al. 2004). Greater crystallinity and increased particle size are attributed to materials that are gradually formed through a dissolution or reformation process. Crystallinity is said to also increase when a one-step process is used for the formation of LDH particles. The level of saturation of ions that is used in the preparation of solutions also influences crystallinity. A more concentrated solution will produce a poorer crystalline structure.

Porosity is another property that is determined through the preparation processes of the LDH particles. The porosity of a particle is the pore size and distribution within an LDH particle. The method that is used for the preparation and interconnection within the LDH is the process that determines the pore size (Braterman et al. 2004). The distribution of pores is more dependent upon the method of formation of LDH and the ions involved. N_2 adsorption/desorption and pore size distribution analysis are the methods used to determine the porosity of an LDH material. Through the investigation of porosity, a material can be classified as either microporous or mesoporous, depending upon the pore size. Microporous materials are related to a pore size of 0–2 nm, and mesoporous materials are related to a pore size of 2–10 nm.

11.4.1.2 Synthesis of Nanomaterials

The synthesis of nanoparticles is the creation process. Some nanoparticles are made through chemical processes, while others are created through natural processes. The most widely used synthesis process is direct synthesis. This process involves the combination of a solution of two metal salts and a solution of the desired anion and a base to produce a metal hydroxide layer (Braterman et al. 2004). Previous studies by Chen et al. (2009) involved the use of MgAl–CO_3 LDHs. In this process, a solution containing Al_2Cl_3 and

$MgCl_2$ was added to a second solution containing NaOH and Na_2CO_3. This solution was then centrifuged and heated for 4 h at 100°C to remove as much moisture as possible. Through this process, individual metal hydroxides also have the potential to form (Braterman et al. 2004).

Limitations are also encountered through the use of this process. The anion that is to be used must be as tightly held as the counter ion in the metal salt to maintain the structural value. Due to this factor, the use of sulfate is avoided and metal chlorides and nitrates are used. Another important consideration is that the LDH anion must not easily produce insoluble salts with the cations in the LDH (Braterman et al. 2004). The formation of LDH materials is influenced by the pH at which the process occurs and is said to begin at pH values much less than that required for the formation of soluble hydroxides. The pH titration curves can show this attribute with two plateau points, the first occurring when soluble hydroxides are formed and the second occurring when LDH is formed (Braterman et al. 2004). In the LDH structure, the most strongly held anion is carbonate. This can easily form in the structure through the introduction of carbon dioxide from the atmosphere. It is, therefore, extremely important when trying to intercalate another anion that the experiment is conducted in a way to exclude carbon dioxide. It is also extremely important to exclude oxygen from an experiment where oxidation of cations in the metal hydroxide can easily occur.

11.4.1.3 Characterization of Nanoparticles

Atomic force spectroscopy, transition electron microscopy (TEM), small-angle x-ray scattering, and x-ray diffraction (XRD) are all used for the characterization of the size and shape of nanoparticles (Zhou et al. 2009). Infrared (IR) spectroscopy can be used to determine the physical and electronic factors of a molecule structure (Kirkland and Hutchison 2007). Fourier transformation infrared spectroscopy (FT-IR) can be used to determine the alkyl chains present in the structure and their quality, as this can be an important factor for adsorption (Zhou et al. 2009). Photon correlation spectroscopy is used in the characterization process to determine the size of the particles in the LDH structure. The characterization processes used in most of the studies include XRD, TEM, and FT-IR.

11.4.2 Forces Involved in LDH Structure and Adsorption

11.4.2.1 Electrostatic Forces

Electrostatic forces are involved in the adsorption onto LDHs (Duan and Evans 2006). This means that molecules with a particular charge are attracted to other molecules with the opposing charge, while molecules with like charges are repelled (Silberberg 2006). Electrostatic forces also exist within the LDH structure between the cations and the hydroxide ions within the outer layers. This type of force is also important in the combination of the layers within the LDH material and its overall structure.

11.4.2.2 Hydrogen Bonding

Hydrogen bonding exists within the interlayer of the LDH structure and also in the complex combination of the interlayer and outer layers (Duan and Evans 2006). This bonding is also important in the adsorption of organic material onto nanoparticles. Hydrogen bonding occurs when electron density is removed from a hydrogen molecule, leaving a positive hydrogen atom. This hydrogen atom is then strongly attracted to a potentially negative lone pair that is provided by the removal of electron density from another molecule: $A\delta^- H\delta^+ - B\delta^-$, where A and B are nitrogen, oxygen, or fluoride.

11.4.2.3 Ion Exchange

Adsorption using a calcined LDH material involves two processes. First, the LDH is reconstructed from a mixed oxide. After this, the anion is then intercalated into the interlayer. This can also happen in one simultaneous step rather than two defined processes (Braterman et al. 2004). Adsorption can also occur starting with an LDH structure where ion exchange is the only step in the adsorption process. Figures 11.8 and 11.9 represent these processes.

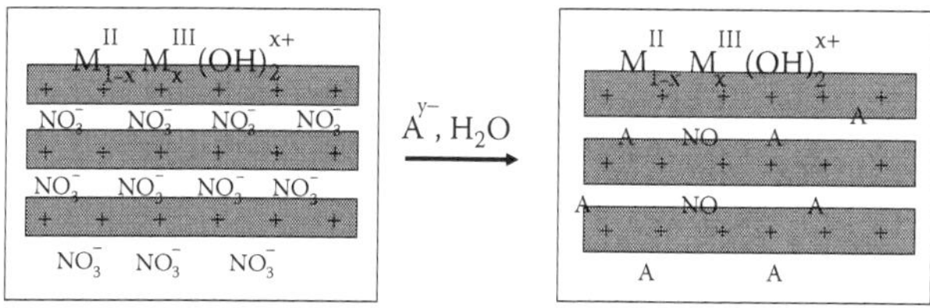

FIGURE 11.8 (See color insert)
Anion exchange of LDH where the anion to be intercalated is denoted as A^{y-}.

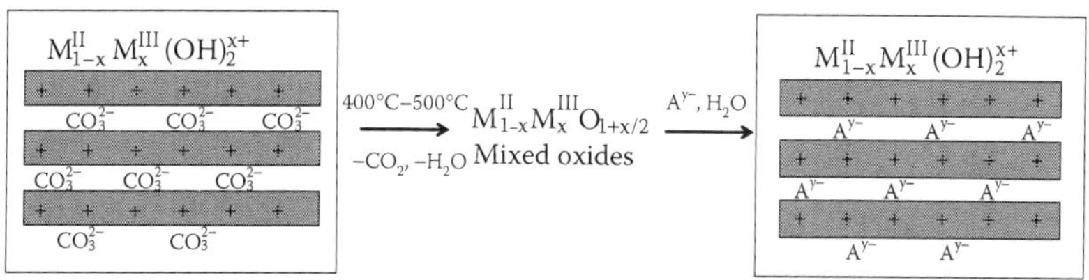

FIGURE 11.9
Reconstruction and intercalation processes. LDH materials are first converted to a mixed oxide; then reconstruction occurs with the intercalation of the anion denoted as A^{y-}.

11.4.2.4 Organic Materials and Reaction with LDH

Two nanoparticles were used to investigate the adsorption of the aforementioned three organic chemicals, two of which are predominately found in herbicides and pesticides and one in herbicides (Searston 2009). These three chemicals are citric acid, acetic acid, and diuron (Table 11.4). The nanoparticle types chosen for investigation were LDHs in the form of calcined $Mg_2Al–CO_3$ LDH and calcined $Mg_3Al–CO_3$ LDH.

The FT-IR plots for both $Mg_2Al–CO_3$ LDH and $Mg_3Al–CO_3$ LDH are shown in Figure 11.10. For $Mg_3Al–CO_3$ LDH, the band at 3400–3500 cm is attributed to the stretching vibrations of both O–H bonds in the lattice and water. The peak at around 1350 cm is attributed to the vibrations of CO_3^{2-}. A broad shoulder at 3000–3400 cm is associated with the presence of CO_3^{2-} in the interlayer due to vibrations of the O–H hydrogen bonded to CO_3^{2-}. Peaks in the range of 400–1000 cm are attributed to the M–O bonds in the LDH phase. $Mg_2Al–CO_3$ LDH showed similar IR spectra as shown in the FT-IR plot. The calcined product of $Mg_3Al–CO_3$ LDH shows less peaking from 3000 to 3400 cm, which

TABLE 11.4

Representation of the Organic Materials Used and the Interaction with the LDH Structure

Chemical Used	Chemical Formula	Chemical Structure	Interaction with LDH
Citric acid (pH 8–9)	$C_6H_5Na_3O_7$	HO, CH_2—C(=O)—O^-; $^-O_2CCH_2$—C—CO_2^-; Na^{3+} + (McMurry, 2003)	1. Reconstruction of LDH structure from calcined metal oxide 2. Intercalation into the interlayer of structure
Citric acid (pH 3)	$C_6H_8O_7$	COOH; HO—C—CH_2COOH; CH_2; COOH (McMurry, 2003)	1. Reconstruction of LDH structure from calcined metal oxide 2. Intercalation into the interlayer of structure
Diuron	$C_9H_{10}Cl_2N_2O$	NH—CO—N(CH_3)(CH_3); Cl; Cl (Giacomazzi and Cochet, 2004)	1. Reconstruction of LDH structure from calcined metal oxide 2. Intercalation into the interlayer of structure
Acetic acid (pH 8–9)	$C_2H_4O_2$	O; H—C—C—O—H; H H (McMurry, 2003)	1. Reconstruction of LDH structure from calcined metal oxide 2. Intercalation into the interlayer of structure

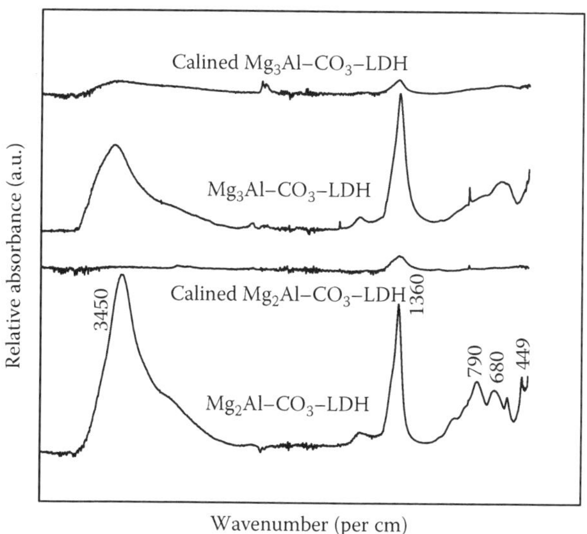

FIGURE 11.10
FT-IR of LDH samples.

is attributed to the removal of CO_3^{2-} bonds in the interlayer through calcinations. This is also similar for calcined $Mg_2Al–CO_3$ LDH.

The adsorption of citric acid was much greater at higher pH levels compared with the adsorption of all other chemicals tested (Table 11.5). The removal of organic material in this case was around 90%. The LDH material was also regenerated to determine the capability of reuse. The regenerated material provided about half the adsorption that could be achieved by a newly created LDH material.

11.5 Adsorption

The adsorption process is a surface phenomenon in which a multicomponent fluid (gas or liquid) mixture is attracted to the surface of a solid adsorbent and forms an attachment as either physical or chemical bonds (Foo and Hameed 2009). The adsorption capability depends on the porosity, specific surface area of the adsorbent, and number of sites available. Another important aspect of the adsorption mechanism between the herbicide and the activated carbon is the structure of the herbicide. The process of adsorption of herbicides onto activated carbon can be classified through kinetic studies as first-order and second-order mechanisms.

TABLE 11.5

Summary of Dynamic Testing Results Including Maximum Adsorption Level and Related Equilibrium Concentration and Isotherm Equation for Each Experiment

Experiment No.	Organic Solution	Concentration (mg/L)	pH	LDH Type	Equilibrium Concentration (C_e)	Max Equilibrium Adsorption (Q_e)	Isotherm Equation	R^2 Value
1	Citric acid	200	8.62	Mg_2Al–CO_3 LDH	53	70	Freundlich $\log(q_e) = \log(8.16) + \frac{1}{2.05}\log C_e$	0.8143
2	Citric acid	200	8.62	Mg_3Al–CO_3 LDH	48	33	Langmuir $\frac{1}{q_e} = \frac{1}{47.61} + \frac{1}{2.51}\left(\frac{1}{C_e}\right)$	0.8332
3	Citric acid	200	8.62	Mg_2Al–CO_3 LDH	61	31	Langmuir $\frac{1}{q_e} = -\frac{1}{14.71} + \frac{1}{0.1671}\left(\frac{1}{C_e}\right)$	0.9018
4	Citric acid	200	3.28	Mg_3Al–CO_3 LDH	49	70	N/A	N/A

5	Citric acid	200	3.28	Mg_2Al–CO_3 LDH	23	10	Freundlich $\log(q_e) = \log(165.42) - \frac{1}{1.1345} \log C_e$	0.5532
6	Diuron	10	7.10	Mg_2Al–CO_3 LDH	1.2	4.2	N/A	N/A
7	Diuron	10	7.00	Mg_3Al–CO_3 LDH	1.1	4.2	N/A	N/A
8	Acetic acid	200	9.20	Mg_2Al–CO_3 LDH	59	78	Freundlich $\log(q_e) = -\log(10^{54.273}) + \frac{1}{0.0316} \log C_e$	0.7322
9	Acetic acid	200	8.80	Mg_3Al–CO_3 LDH	74	12	Langmuir $\frac{1}{q_e} = -\frac{1}{0.5865} + \frac{1}{0.007565}\left(\frac{1}{C_e}\right)$	0.9126

11.5.1 Issues Related to LMWO Compounds: Herbicides and Pesticides

As herbicides and pesticides accumulate in the fatty tissues of living organisms, animals, and humans, a large number of studies have been conducted to determine the effect of pesticide and herbicide residuals (Navaratna et al. 2010). Through transport to surface water and leaching into groundwater, herbicides are present in drinking water. Literature review has demonstrated that the toxicity to humans from herbicides includes acute as well as long-term effects. In terms of acute effects, a large number of pesticides display their toxicity in the central and peripheral nervous systems (Younes and Galal-Gorchev 2000). There is also evidence that associates the long-term exposure of some pesticides with chronic diseases including cancer (Younes and Galal-Gorchev 2000), reproductive effects, fetal damage, delayed neurologic manifestations, and possible immunological disorders (Ahmad et al. 2008).

Herbicides are usually transported to waterways by entering through surface water and groundwater, which results in contamination of, for instance, the Great Barrier Reef (GBR) World Heritage Area in Queensland. Herbicides are also transferred to the GBR through antifouling paints that are used to prevent algae growth on marine boats and ships. Relatively low levels of herbicide residues can reduce the productivity of marine plants and corals, especially sensitive species. It has been determined that these residues may cause a change in the community structures of mangrove, seagrass, and coral reef ecosystems (Lewis et al. 2009). This may be due to atrazine, which has been known to affect the reproduction of aquatic flora and fauna and impact on the community as a whole (Graymore et al. 2001). Diuron has also been found to decline the reproductive output of corals following an exposure of over 50 days (Lewis et al. 2009). At high concentrations and over a long period of time, it has been found that the effects of diuron are nonreversible and result in coral bleaching (Negri et al. 2005). Chronic exposures of diuron and ametryn residues have implicated the dieback of mangroves in the Mackay region (Lewis et al. 2009). The properties of the herbicides are given in Table 11.6.

11.5.2 Activated Carbon Applications

Recently, Baup et al. (2002) completed an investigation into the removal of atrazine and diuron using GAC as the adsorbent and concluded that the crushing up of the GAC to PAC improved the accessibility of the adsorption sites. PAC is thought to be an effective method for the treatment of pesticides as it can be used in temporal and emergent cases, for example, seasonal contamination of waterways during the wet season.

Moreover, Ormad et al. (2008) studied the removal of numerous pesticides, such as diuron, atrazine, and ametryn, in the process of drinking water production. The techniques used included preoxidation by either chlorine

TABLE 11.6

Properties of Herbicides Used in This Investigation

Properties	Diuron	Atrazine	Ametryn
Molecular weight (g)	233.10	215.69	227.33
Molecular formula	$C_9H_{10}Cl_2N_2O$	$C_8H_{14}ClN_5$	$C_9H_{17}N_5S$
Chemical structure	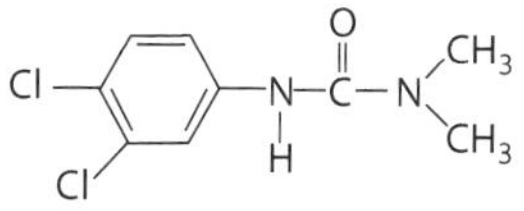	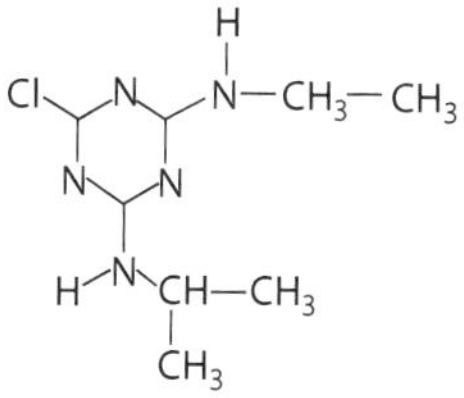	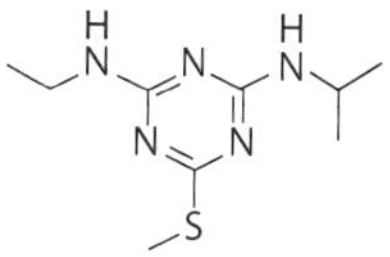
Melting point (°C)	158–159	173–175	84–85
Appearance	White crystalline solids	Colorless crystals	White powder
Solubility	36–42 mg/L in water (25°C)	34.7 mg/L in water (22°C) and 31 g/L in acetone (25°C)	185 mg/L in water (20°C) and readily dissolves in solvents (acetone)
Purpose	Phenylurea herbicide to enhance grass killing	Chlorotriazine herbicide to control broad leaf weeds	Methylthiotriazine herbicide to control grass

or ozone, chemical precipitation with aluminum sulfate, and adsorption through PAC. It was determined that activated carbon (coupled with oxidation by ozone) removed 90% of the pesticides tested. It was also concluded that this technique was the most efficient in the removal of the majority of the pesticides studied. Overall, this study tested 22 pesticides. Another study carried out on the removal of atrazine by adsorption on GAC filters inoculated with bacterial culture resulted in a degradation of approximately 70% (Jones et al. 1998).

Ayranci and Hoda (2005) removed ametryn and diuron utilizing a commercial activated carbon cloth. Batch studies were completed with varying weights of carbon-cloth pieces with pesticide solutions of constant initial concentration for 48 h. They found that the adsorption process followed both pseudo-first-order and pseudo-second-order kinetics over a period of 125 min. The isotherms were found to fit almost equally well to Langmuir and Freundlich isotherms. There was 85% and 50% removal of ametryn and diuron, respectively. Importantly, the use of activated carbon cloth also eliminated competition between NOM and the herbicides.

There are numerous studies on the pore blockage effect of NOM on the adsorption of pesticides (Figure 11.11). NOM in natural water has been found to have a negative effect due to the direct competition for sites and pore blockages (Li et al. 2003).

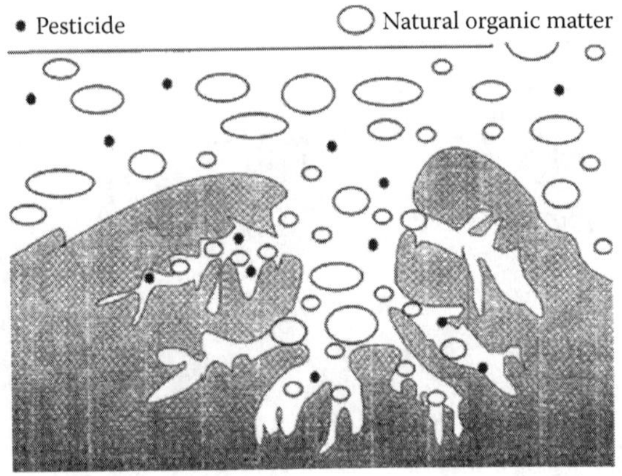

FIGURE 11.11
A schematic of the impact of NOM on pesticide adsorption. (From Heijman, S.G.J. and Hopman, R., *Colloids Surf. A Physicochem. Eng. Asp.*, 1999. With permission.)

11.5.3 Adsorption Isotherms

Adsorption isotherms are usually obtained by examining the batch reactions at a fixed temperature either in a water bath shaker or with jar test apparatus. Sorption isotherms are described by the equilibrium relationships between sorbents and sorbates; they result in the capacity of a sorbent (Ho 2006). To determine the best-fitting isotherms, linear regression is frequently used. The method of least squares has widely been used to confirm the experimental data and isotherms using the coefficients of determination. Although linear analysis is not quite possible for the three parameter models, a trial-and-error process can be used to determine the parameters for these models.

There may also be a bias in the results obtained through linear analysis, presenting difficulties in determining which model produces the best fit to the experimental data. However, from the literature researched, the constants determined by nonlinear regression are consistent and quite similar to those of linear transformation values (Ho et al. 2002). Table 11.7 shows isotherms and their linear forms.

11.5.3.1 Langmuir Isotherm

The Langmuir isotherm can be linearized into four different types (Table 11.7); the theory is based on the assumption that there is a saturated monolayer of adsorbate over the homogeneous adsorbent surface, with no lateral interaction between the sorbed molecules (Hameed et al. 2009). The constants q_m (mg/g) and K_a (L/mg) are related to the loading and energy of adsorption (Chen et al. 2009) and can be found by plotting the linear form of the isotherm.

TABLE 11.7

Isotherms and Their Linear Forms

Isotherm		Linear Form	Plot
Freundlich	$q_e = K_F C_e^{1/n}$	$\log(q_e) = \log(K_F) + 1/n \log(C_e)$	$\log(q_e)$ versus $\log(C_e)$
Langmuir 1	$q_e = \frac{q_m K_a C_e}{1 + K_a C_e}$	$\frac{C_e}{q_e} = \frac{1}{q_m} C_e + \frac{1}{K_a q_m}$	$\frac{C_e}{q_e}$ versus C_e
Langmuir 2	$q_e = \frac{q_m K_a C_e}{1 + K_a C_e}$	$\frac{1}{q_e} = \left(\frac{1}{K_a q_m}\right)\frac{1}{C_e} + \frac{1}{q_m}$	$\frac{1}{q_e}$ versus $\frac{1}{C_e}$
Redlich–Peterson	$q_e = \frac{A C_e}{1 + B C_e^g}$	$\ln\left(A\frac{C_e}{q_e} - 1\right) = g\ln(C_e) + \ln(B)$	$\ln\left(A\frac{C_e}{q_e} - 1\right)$ versus $\ln(C_e)$
Sips	$q_e = \frac{q_m b C_e^{1/n}}{1 + b C_e^{1/n}}$		$\left(\frac{1}{C_e}\right)^{1/n}$ versus $\frac{1}{q_e}$
Temkin	$q_e = \frac{RT}{b}(\ln A C_e)$	$q_e = \frac{RT}{b}\ln A + \frac{RT}{b}\ln C_e$	q_e versus $\ln C_e$

11.5.3.2 Freundlich Isotherm

The Freundlich isotherm model is an experimental equation used to describe a heterogeneous system. K_F and n_F are Freundlich's temperature-dependent constants and can be determined by plotting the linear form of the isotherm. It has also been suggested that these constants are associated with the adsorption capacity (K_F) and intensity (n_F) (Mead 1981). Values of $n_F > 1$ represent a favorable adsorption condition, and the smaller the value of $1/n_F$, the stronger is the adsorption bond. The greater values of K_F correspond to a greater capacity of adsorbent (Jiang and Adams 2006).

11.5.3.3 Redlich–Peterson Isotherm

The Redlich–Peterson model contains three parameters and incorporates the aspects of both the Freundlich and Langmuir isotherms into one single equation. At low surface coverage, the equation is reduced to a linear isotherm; at high surface coverage, it is reduced to Freundlich isotherm; and when $B = 1$, it is reduced to Langmuir isotherm (Allen et al. 2003). The three constants can be determined by the pseudo linear plot seen in Table 11.7 and using a trial-and-error method to determine the best value of A that yields an optimal value of r^2.

11.5.3.4 Temkin Isotherm

The Temkin isotherm model assumes a linear fall in the heat sorption, rather than logarithmic as Freundlich's model implies. The equation incorporates

b_T (J/mol), a constant related to the heat of sorption; A_T (L/g), an isotherm constant; the gas constant R; and the absolute temperature T. The two constants can be found by plotting the linear form of the isotherm.

11.5.3.5 Sips Isotherm

Also known as the Langmuir–Freundlich isotherm, the equation reduced to Freundlich isotherm at low sorbate concentration and at high concentrations, it predicts Langmuir isotherm of a monolayer adsorption capacity (Ho et al. 2002). The isotherm incorporates three constants: the maximum adsorption capacity, q_m (mg/g); Sips model isotherm constant b (L/mg) related to the adsorption strength; and Sips model exponent, n. These constants can be found by plotting the linear form of the isotherm.

11.5.4 Adsorption Kinetics

The modeling of the kinetics of adsorption can be investigated by two common models: pseudo-first-order model and pseudo-second-order model. The following equations are used to determine the rate constants of the adsorption. The Lagergren and Svenska's expression can be used to determine the pseudo-first-order equation (Equation 11.3):

$$\log(q_e - q_t) = \log q_e - \frac{k_1 t}{2.303}, \tag{11.3}$$

where q_e and q_t are the amounts of adsorbate adsorbed onto the adsorbent in (mg/g) at time t (h). A linear plot of $\ln(q_e - q_t)$ against time allows you to determine the rate constant k_1(1/h). If the plot is found to be linear with a correlation coefficient close to 1, it can be said that Lagergren's equation is an appropriate description of the adsorption kinetics. The Ho and McKay's expression was used to determine the pseudo-second-order equation (Equation 11.4):

$$\frac{t}{q_t} = \frac{1}{k_2 q_e^2} + \frac{1}{q_e} t, \tag{11.4}$$

where q_e and q_t have the same definitions as previously stated and k_2 (g/mg/h) is the rate constant, which can be determined by a linear plot of t/q_t against time. Again, if the plot is found to be linear with a good correlation coefficient, it can be assumed that the adsorption follows pseudo-second-order kinetics.

The adsorption equilibrium and the kinetics of atrazine, ametryn, and diuron using PHO 12/30 and GI 1000 12/30 GAC types were studied using a jar test system by Naylor (2010). Both types of activated carbon were supplied by Haycarb Limited, Sri Lanka. The experimental data were analyzed by using linear forms of the following isotherm models: Langmuir 1 and 2, Freundlich, Sips, Redlich–Peterson, and Temkin. The results indicated that the equilibrium data fitted well with all models for each herbicide, with Langmuir 2 and Sips showing the best fit with high correlation values. The results define GAC as an effective adsorbent for the removal of herbicides atrazine, ametryn, and diuron. The isotherm parameters confirm this by the results stating that the adsorbent shows a high affinity, suggesting a strong bonding between the herbicides and the GAC surface. The maximum adsorption capacities calculated were 156.3 mg/g, 98.7 mg/g, and 114.2 mg/L for atrazine, ametryn, and diuron, respectively, and were all observed with the Langmuir 2 isotherm model. The pseudo-first-order and pseudo-second-order kinetics were used to test the adsorption kinetics, with the pseudo-first-order kinetics resulting as the best suited to each herbicide. The effect of the solution pH was studied to determine the best solution pH for removal, with a pH of 10 resulting in the highest removal.

11.5.5 Relationship with Chlorine

Chlorination is used worldwide as a method for final disinfection of treated water to eliminate microorganisms such as viruses and bacteria. It is, therefore, useful to determine the relationship between chlorine and the herbicides, to determine the effectiveness of chlorine in removing the herbicides. Jiang and Adams (2006) determined in a study that a free chlorine dosage of 2 mg/L with a 30 min contact time led to no significant removal of atrazine. A recent study determined that ametryn had a high reactivity with chlorine compared with that of diuron and that ametryn inclined to react with chlorine and exerted fast consumption of hypochlorous acid (Xu et al. 2009). It was, however, found that the chlorination of ametryn is accelerated by the presence of bromide.

In another study, citric acid, ascorbic acid, and diuron were chlorinated to determine their chlorine demand and DBP formation potential (Allen 2008). Chlorine decay was observed for all three model compounds. Diuron was found to consume chlorine at a faster rate after NF but had a much lower chlorine demand.

11.6 Applications of MBR

The MBR process, which is a combination of biological treatment and membrane filtration for the separation of biomass, is one of the most novel wastewater treatment processes available at present. Bioreactor and membrane filtration cannot be considered as individual unit operations in MBRs, as these processes interact in many different ways. For the past two decades, many MBR plants have been installed for the treatment of domestic and industrial wastewaters worldwide. MBR technology is now becoming very popular with an approximate market value of US$217 million and a growth rate of 10.9% in 2008 (Judd 2008), due to its wide array of advantages over conventional treatment technologies, such as the production of superior quality of treated effluent, confinement to smaller footprints, higher efficiency in the removal of micropollutants and POPs, and its ability to produce higher-quality effluents even when the sludge is bulked.

The demand for MBR systems is increasing steadily because they are now becoming more cost-effective due to the continuous fall in the costs of membrane modules and related accessories that could be associated with high competition and advances in technology as well as the imposition of more stringent environmental laws and regulations in every state and region in the world. Due to fast-growing industrial applications of MBR technology in wastewater treatment, the number of related research studies continue to increase for finding solutions to the currently identified drawbacks of MBR systems (mainly fouling of membrane) and for optimizing their performance (especially in nutrient removal, the treatment of micropollutants such as pesticides, herbicides, and pharmaceuticals, etc.) in order to use them as a reliable treatment process.

MBRs mainly comprise either MF or UF as shown in Figure 11.12; in the submerged MBR systems, the membranes are placed inside (flat-sheet or hollow-fiber membranes) the bioreactor, and in the side-stream MBR systems, the membranes (multitube/tubular) are placed outside the bioreactor (Judd 2008; Le-Clech et al. 2006). Currently, most of the MBRs are operated aerobically (98%) and the rest function anaerobically (Mulligan and Gibbs 2003). In submerged MBRs, air is supplied for biodegradation and membrane cleaning (coarse bubbling).

Membrane fouling is caused by the restriction, occlusion, or blocking of the membrane pores (Judd 2008) at the surface of the membrane. This

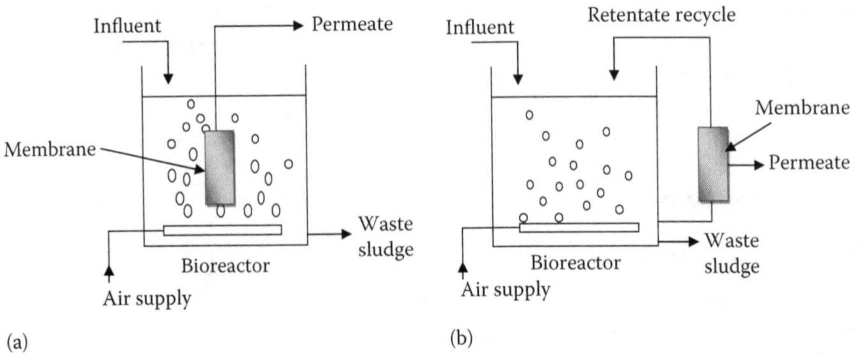

FIGURE 11.12
Configurations of MBR systems: (a) submerged MBR; (b) side-stream MBR.

reduces the permeate flux (volumetric flow rate per unit membrane area) through the membrane. Thus, fouling is considered as the main obstacle to the widespread application of MBRs. Fouling of membrane is mainly caused by the physical (nominal particle size of microbial flocs), chemical (hydrophobicity), and biological (extracellular polymeric substances (EPSs) and viscosity) factors related to biomass. According to Meng et al. (2009), the fouling mechanisms in an MBR are (1) adsorption of solutes and colloids within or on membrane surface; (2) deposition of sludge flocs onto the membrane surface; (3) formation of a cake layer on the membrane surface; (4) detachment of the foulants attributed mainly to shear forces; and (5) spatial and temporal changes of the foulant composition such as the change of microbial community and biopolymer components in the cake layer during a long-term operation (Meng et al. 2009). Most of the previous research work (Jang et al. 2006; Laspidou and Rittmann 2002; Le-Clech et al. 2006; Rosenberger et al. 2006) confirmed that soluble microbial products (SMPs), which are referred to as free EPSs, and bound EPS (eEPSs), which are secreted by microorganisms, are the main organic compounds that cause membrane fouling. Free and bound EPSs mainly consist of polysaccharides (carbohydrates) and proteins, and they play a major role in the formation of cake and gel layers on the membranes.

Operating MBRs at subcritical flux (below the "critical flux," where the flux starts to form the cake or gel layer on the membrane surface) is considered as one of the most practical strategies to control the fouling of membranes in MBR. In addition to this, subcritical flux operation reduces the consumption of energy and hence minimizes the operational cost of MBR. Field et al. (1995) originally introduced the concept of critical flux in MF using an empirical approach, and they defined "critical flux" as "the flux below which a decline of flux with time does not occur (that is, at subcritical flux, where $dTMP/dt = TMP' = 0$) and above which (supercritical flux) fouling is observed" (Field et al. 1995). However, subsequent to that, Le-Clech et al. (2003) showed that a zero rate of TMP increase may never be obtained ($TMP' \neq 0$) during the short-term (common flux step method) critical flux determination tests carried out for synthetic and real sewage (Le-Clech et al. 2003). Since then, different types of short-term critical flux determination and long-term subcritical flux operational studies have been carried out under different feed-wastewater characteristics, biomass/sludge conditions, and other operations (Defrance and Jaffrin 1999; Le-Clech et al. 2003; Torre et al. 2009; van der Marel et al. 2009).

In a recent study, the common flux step method was used to determine the critical flux value for a laboratory-scale MBR system, which was fed with synthetic wastewater consisting of ametryn (Navaratna and Jegatheesan 2010). During that study, the step height and the step duration were maintained at 3 $L/m^2/h$ and 40 min, respectively. Also, the concentration of mixed liquor suspended solids (MLSS) was recorded around 9 g/L and the chemical oxygen demand (COD) of the synthetic wastewater was kept around

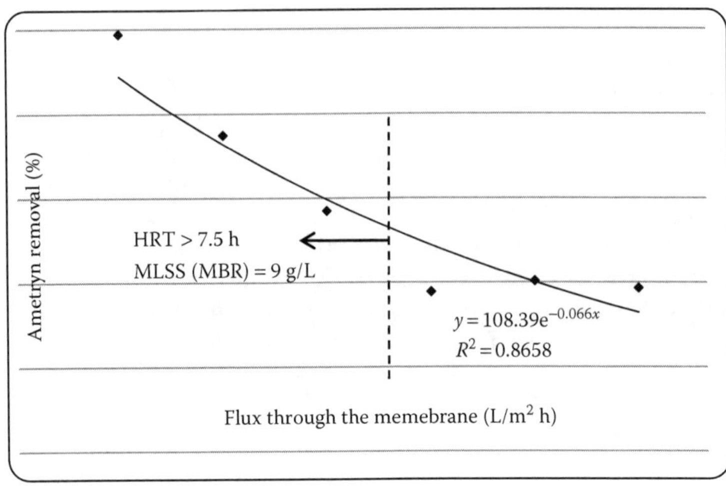

FIGURE 11.13
Variation in the percentage removal of ametryn with the flux through the membrane.

750 mg/L. The concentration of ametryn in the influent was maintained at around 1 mg/L. Figure 11.13 shows the variations in the percentage removal of ametryn with membrane flux during the critical flux determination experiment. The results revealed that the percentage of ametryn removal declined exponentially with the increase in membrane flux. It was also found that the removal of ametryn by the MBR was over 50%, when it was operated at an HRT of 7.5 h. The performance of the MBR could be further improved by operating it with an intermittent permeate suction mode. Removal of the remaining ametryn could be carried out by incorporating an adsorption process using activated carbon.

11.7 Conclusions

The removal of LMW substances from water poses a huge challenge, as most of these substances are persistent and thus do not degrade biologically and their molecules are too small to be captured by granular or membrane filtration processes. The following ranking could be made in terms of the removal efficiencies of the existing processes (in the order of highest to lowest removal of LMW substances): RO, hybrid coagulation and NF, adsorption, nanoparticles, MBRs. This chapter did not include processes such as advanced oxidation and wetland treatment. While the performance of advanced oxidation process is comparable to those of membrane filtration processes, the performance of wetland treatment could only be ranked below the performance of

MBR. However, when available, all these systems should be used appropriately to form a hybrid system that could provide efficient treatment of LMW substances.

References

Adams, C.D. and T.L. Watson. 1996. Treatability of [bold s]-triazine herbicide metabolites using powdered activated carbon. *J. Environ. Eng.* 122(4): 327–330.

Ahmad, A.L., L.S. Tan, and S.R. Abd. Shukor. 2008. The role of pH in nanofiltration of atrazine and dimethoate from aqueous solution. *J. Hazard. Mater.* 154(1–3): 633–638.

Allen, J. 2008. Nanofiltration performance in removing small molecular weight organics from water. A Honors thesis submitted to the School of Engineering, James Cook University, Townsville, QLD.

Allen, S., Q. Gan, R. Matthews, and P. Johnson. 2003. Comparison of optimised isotherm models for basic dye adsorption by kudzu. *Bioresour. Technol.* 88(2): 143–152.

Auxilio, A.R., P.C. Andrews, P.C. Junk, and L. Spiccia. 2009. The adsorption behavior of C.I. Acid Blue 9 onto calcined Mg–Al layered double hydroxides. *Dyes Pigm.* 81(2): 103–112.

Ayranci, E. and N. Hoda. 2005. Adsorption kinetics and isotherms of pesticides onto activated carbon-cloth. *Chemosphere* 60(11): 1600–1607.

Bartels, C., R. Franks, S. Rybar, M. Schierach, and M. Wilf. 2005. The effect of feed ionic strength on salt passage through reverse osmosis membranes. *Desalination* 184(1–3): 185–195.

Baup, S., D. Wolbert, and A. Laplanche. 2002. Importance of surface diffusivities in pesticide adsorption kinetics onto granular versus powdered activated carbon: Experimental determination and modeling. *Environ. Technol.* 23(10): 1107–1117.

Braterman, P.S., Z.P. Xu, and F. Yarberry. 2004. *Layered Double Hydroxides (LDHS).* New York: Marcel Dekker, Inc.

Brindle, K. and T. Stephenson. 1996. The application of membrane biological reactors for the treatment of wastewaters. *Biotechnol. Bioeng.* 49(6): 601–610.

Calvert, J.G. 1990. Glossary of atmospheric chemistry terms: Recommendations 1990. *Pure Appl. Chem.* 62(11): 2167–2219.

Carroll, T., S. King, S.R. Gray, B.A. Bolto, and N.A. Booker. 2000. The fouling of microfiltration membranes by NOM after coagulation treatment. *Water Res.* 34(11): 2861–2868.

Chen, S., Z.P. Xu, Q. Zhang, G.Q.M. Lu, Z.P. Hao, and S. Liu. 2009. Studies on adsorption of phenol and 4-nitrophenol on MgAl-mixed oxide derived from MgAl-layered double hydroxide. *Sep. Purif. Technol.* 67(2): 194–200.

Clark, R.M. 1998. Chlorine demand and TTHM formation kinetics: A second-order model. *J. Environ. Eng.* 124(1): 16–24.

Davis, M.L. and D.A. Cornwell. 2008. *Introduction to Environmental Engineerig*, 4th ed. Singapore: McGraw-Hill. 1008 p.

Defrance, L. and M.Y. Jaffrin. 1999. Comparison between filtrations at fixed transmembrane pressure and fixed permeate flux: Application to a membrane bioreactor used for wastewater treatment. *J. Memb. Sci.* 152(2): 203–210.

Denyer, P. 2005. Nanofiltration of textile wastewater: Treatment and reuse, James Cook University, Townsville, QLD.

Denyer, P., L. Shu, and V. Jegatheesan. 2007. Evidence of changes in membrane pore characteristics due to filtration of dye bath liquors. *Desalination* 204(1–3): 296–306.

Dsikowitzky, L., J. Schwarzbauer, A. Kronimus, and R. Littke. 2004. The anthropogenic contribution to the organic load of the Lippe River (Germany). Part I: Qualitative characterisation of low-molecular weight organic compounds. *Chemosphere* 57(10): 1275–1288.

Duan, X. and D.G. Evans, eds. 2006. "Layered Double Hydroxides." *Struc. & Bonding*, Vol. 119. Springer-Verlag: Berlin Heidelberg, Germany.

Field, R.W., D. Wu, J.A. Howell, and B.B. Gupta. 1995. Critical flux concept for microfiltration fouling. *J. Memb. Sci.* 100(3): 259–272.

Foo, K. and B. Hameed. 2009. Detoxification of pesticide waste via activated carbon adsorption process. *J. Hazard. Mater.* 175(1–3): 1–11.

Gang, D., T.E. Clevenger, and S.K. Banerji. 2003. Relationship of chlorine decay and THMs formation to NOM size. *J. Hazard. Mater.* 96(1): 1–12.

Gavrilescu, M. 2005. Fate of pesticides in the environment and its bioremediation. *Eng. Life Sci.* 5(6): 497–526.

Giacomazzi, S. and N. Cochet. 2004. Environmental impact of diuron transformation: A review. *Chemosphere* 56(11): 1021–1032.

Government. 2007. Acetic Acid: Fact Sheet. National Pollution Inventory, Australia. (http://www.npi.gov.au/substances/acetic-acid/health.html; last accessed 9 March 2012.)

Gray, N.F. 2005. *Water Technology: An Introduction for Environmental Scientists and Engineers*, 2nd edn. Oxford: Elsevier Butterworth-Heinemann.

Gray, S.R. and B.A. Bolto. 2003. Predicting NOM fouling of low pressure membranes. In *International Membrane Science and Technology*, Sydney, NSW, November 10–14, pp. 203.

Graymore, M., F. Stagnitti, and G. Allinson. 2001. Impacts of atrazine in aquatic ecosystems. *Environ. Int.* 26(7–8): 483–495.

Hameed, B.H., J.M. Salman, and A.L. Ahmad. 2009. Adsorption isotherm and kinetic modeling of 2,4-D pesticide on activated carbon derived from date stones. *J. Hazard. Mater.* 163(1): 121–126.

Heijman, S.G.J. and R. Hopman. 1999. Activated carbon filtration in drinking water production: Model prediction and new concepts. *Colloids Surf. A Physicochem. Eng. Asp.* 151(1–2): 303–310.

Ho, Y. 2006. Isotherms for the sorption of lead onto peat: Comparison of linear and non-linear methods. *Pol. J. Environ. Stud.* 15(1): 81–86.

Ho, Y., J. Porter, and G. McKay. 2002. Equilibrium isotherm studies for the sorption of divalent metal ions onto peat: Copper, nickel and lead single component systems. *Water Air Soil Pollut.* 141(1): 1–33.

Jang, N., H. Shon, X. Ren, S. Vigneswaran, and I.S. Kim. 2006. Characteristics of biofoulants in the membrane bioreactor. *Desalination* 200(1–3): 201–202.

Jarusutthirak, C., S. Mattaraj, and R. Jiraratananon. 2007. Factors affecting nanofiltration performances in natural organic matter rejection and flux decline. *Sep. Purif. Technol.* 58(1): 68–75.

Jegatheesan, V., S.H. Kim, C.K. Joo, and B. Gao. 2009. Evaluating the effects of granular and membrane filtrations on chlorine demand in drinking water. *J. Environ. Sci.* 21(1): 23–29.

Jegatheesan, V. and S. Vigneswaran. 2005. Deep bed filtration: Mathematical models and observations. *Crit. Rev. Environ. Sci. Technol.* 35(6): 515–569.

Jiang, H. and C. Adams. 2006. Treatability of chloro-s-triazines by conventional drinking water treatment technologies. *Water Res.* 40(8): 1657–1667.

Jones, L.R., S.A. Owen, P. Horrell, and R.G. Burns. 1998. Bacterial inoculation of granular activated carbon filters for the removal of atrazine from surface water. *Water Res.* 32(8): 2542–2549.

Judd, S. 2008. The status of membrane bioreactor technology. *Trends Biotechnol.* 26(2): 109–116.

Kirkland, A.I. and J.L. Hutchison. 2007. *Nanocharacterization*. Cambridge: The Royal Society of Chemistry.

Knappe, D.R.U., Y. Matsui, V.L. Snoeyink, P. Roche, M.J. Prados, and M.-M. Bourbigot. 1998. Predicting the capacity of powdered activated carbon for trace organic compounds in natural waters. *Environ. Sci. Technol.* 32(11): 1694–1698.

Laspidou, C.S. and B.E. Rittmann. 2002. A unified theory for extracellular polymeric substances, soluble microbial products, and active and inert biomass. *Water Res.* 36(11): 2711–2720.

Le-Clech, P., V. Chen, and T.A.G. Fane. 2006. Fouling in membrane bioreactors used in wastewater treatment. *J. Memb. Sci.* 284(1–2): 17–53.

Le-Clech, P., B. Jefferson, I.S. Chang, and S.J. Judd. 2003. Critical flux determination by the flux-step method in a submerged membrane bioreactor. *J. Memb. Sci.* 227(1–2): 81–93.

Lewis, S.E., J.E. Brodie, Z.T. Bainbridge, K.W. Rohde, A.M. Davis, B.L. Masters, M. Maughan, M.J. Devlin, J.F. Mueller, and B. Schaffelke. 2009. Herbicides: A new threat to the Great Barrier Reef. *Environ. Pollut.* 157(8–9): 2470–2484.

Li, Q., S. Mahendra, D.Y. Lyon, L. Brunet, M.V. Liga, D. Li, and P.J.J. Alvarez. 2008. Antimicrobial nanomaterials for water disinfection and microbial control: Potential applications and implications. *Water Res.* 42(18): 4591–4602.

Li, Q., V.L. Snoeyink, B.J. Mariñas, and C. Campos. 2003. Pore blockage effect of NOM on atrazine adsorption kinetics of PAC: The roles of PAC pore size distribution and NOM molecular weight. *Water Res.* 37(20): 4863–4872.

Linder, C., M. Cantor, and Y. Oren. 2006. Relationships between material parameters of nanofiltration membranes and the resultant membrane performance. *Desalination* 199(1–3): 256–257.

Manem, J. and R. Sanderson. 1996. *Membrane Bioreactors*. New York: McGraw-Hill.

McMurry, J. 2003. Fundamentals of Organic Chemistry (5th ed.). California: Thomson Learning, Inc.

Mead, J. 1981. A comparison of the Langmuir, Freundlich and Temkin equations to describe phosphate adsorption properties of soils. *Aust. J. Soil Res.* 19(33): 3–342.

Meng, F., S.-R. Chae, A. Drews, M. Kraume, H.-S. Shin, and F. Yang. 2009. Recent advances in membrane bioreactors (MBRs): Membrane fouling and membrane material. *Water Res.* 43(6): 1489–1512.

Mulligan, C.N. and B.F. Gibbs. 2003. Innovative biological treatment process for water in Canada. *Can. Water Qual. Res. J.* 38(2): 243–265.

Nagano, A., E. Arikawa, and H. Kobayashi. 1992. The treatment of liquor wastewater containing highstrength suspended solids by membrane bioreactor system. *Water Sci. Technol.* 26(3–4): 887–895.

Navaratna, D. and V. Jegatheesan. 2010. Removal of ametryn using membrane bioreactor process & its influence on critical flux. In *Proceedings of the International Conference on Sustainable Built Environment (ICSBE 2010)*, University of Peradeniya, December 13–14.

Navaratna, D., L. Shu, and V. Jegatheesan. 2010. Existance, impacts, transport and treatments of herbicides in the Great Barrier Reef catchments in Australia. In *Treatment of Micropollutants in Water and Wastewater*, eds. Virkutyte, J., R.S. Varma, and V. Jegatheesan. London: IWA Publishing.

Naylor, J. 2010. Modeling the adsorption of lower molecular weight organic compounds (LMWOC) by activated carbon. A Honors thesis submitted to the School of Engineering and Physical Sciences, James Cook University, Townsville, QLD.

Negri, A., C. Vollhardt, C. Humphrey, A. Heyward, R. Jones, G. Eaglesham, and K. Fabricius. 2005. Effects of the herbicide diuron on the early life history stages of coral. *Mar. Pollut. Bull.* 51(1–4): 370–383.

Newman, S.P. and W. Jones. 1998. Synthesis, characterisation and application of layered double hydroxides containing organic guests. *New J. Chem.* 105–115.

National Health and Medical Research Council (NHMRC). Australian Drinking Water Guidelines. 6 ed. 2004, Canberra, ACT.

Ormad, M.P., N. Miguel, A. Claver, J.M. Matesanz, and J.L. Ovelleiro. 2008. Pesticides removal in the process of drinking water production. *Chemosphere* 71(1): 97–106.

Peeters, J.M.M., J.P. Boom, M.H.V. Mulder, and H. Strathmann. 1998. Retention measurements of nanofiltration membranes with electrolyte solutions. *J. Memb. Sci.* 145(2): 199–209.

Pempkowiak, J. and H. Obarska-Pempkowiak. 2002. Long-term changes in sewage sludge stored in a reed bed. *Sci. Total Environ.* 297(1–3): 59–65.

Perry, M. and C. Linder. 1989. Intermediate reverse osmosis ultrafiltration (RO UF) membranes for concentration and desalting of low molecular weight organic solutes. *Desalination* 71(3): 233–245.

Plakas, K.V. and A.J. Karabelas. 2008. Membrane retention of herbicides from single and multi-solute media: The effect of ionic environment. *J. Memb. Sci.* 320(1–2): 325–334.

Poole, C.P. and F.J. Owens. 2003. *Introduction to Nanotechnology*. Hoboken, NJ: John Wiley and Sons, Inc.

Rosenberger, S., C. Laabs, B. Lesjean, R. Gnirss, G. Amy, M. Jekel, and J.C. Schrotter. 2006. Impact of colloidal and soluble organic material on membrane performance in membrane bioreactors for municipal wastewater treatment. *Water Res.* 40(4): 710–720.

Searston, E. 2009. Removal of small molecular weight organic matter through the use of nanoparticles. A Honors thesis submitted to the School of Engineering and Physical Sciences, James Cook University, Townsville, QLD.

Shu, L. 2000. Membrane processing of dye wastewater: Dye aggregation, membrane performance and mathematical modelling, PhD Dissertation, University of New South Wales, Sydney, NSW.

Silberberg, M. 2006. *Chemistry: The Molecular Nature of Matter and Change*, 4th edn. New York: McGraw Hill.

Sterlitech. 2002. Sterlitech Polycarbonate (PCTE) and Polyester (PETE) Track Etch Filter Membranes: Frequently asked questions, Vol. 2008.

Stull, J.K., D.J.P. Swift, and A.W. Niedoroda. 1996. Contaminant dispersal on the Palos Verdes continental margin: I. Sediments and biota near a major California wastewater discharge. *Sci. Total Environ.* 179: 73–90.

Torre, T.D., V. Iversen, A. Moreau, and J. Stüber. 2009. Filtration characterization methods in MBR systems: A practical comparison. *Desalin. Water Treat.* 9: 15–21.

Tran, T., S. Gray, R. Naughton, and B. Bolto. 2006. Polysilicato-iron for improved NOM removal and membrane performance. *J. Memb. Sci.* 280(1–2): 560–571.

van der Marel, P., A. Zwijnenburg, A. Kemperman, M. Wessling, H. Temmink, and W. van der Meer. 2009. An improved flux-step method to determine the critical flux and the critical flux for irreversibility in a membrane bioreactor. *J. Memb. Sci.* 332(1–2): 24–29.

Violante, A., M. Pucci, V. Cozzolino, J. Zhu, and M. Pigna. 2009. Sorption/desorption of arsenate on/from Mg–Al layered double hydroxides: Influence of phosphate. *J. Colloid Interface Sci.* 333(1): 63–70.

Visvanathan, C., B.D. Marsono, and B. Basu. 1998. Removal of THMP by nanofiltration: Effects of interference parameters. *Water Res.* 32(12): 3527–3538.

Wang, Y., L. Shu, V. Jegatheesan, and B. Gao. 2009. Coagulation and nano-filtration: A hybrid system for the removal of lower molecular weight organic compounds (LMWOC). *Desalin. Water Treat.* 11: 23–31.

Wang, Y., L. Shu, V. Jegatheesan, and B. Gao. 2010. Removal and adsorption of diuron through nanofiltration membrane: The effects of ionic environment and operating pressures. *Sep. Purif. Technol.* 74(2): 236–241.

WHO. 2006. *Guidelines for Drinking Water Quality*, third edition, incorporating first and second addenda, Volume 1 – Recommendations.

Xing, C.H., E. Tardieu, Y. Qian, and X.H. Wen. 2000. Ultrafiltration membrane bioreactor for urban wastewater reclamation. *J. Memb. Sci.* 177(1–2): 73–82.

Xu, B., N. Gao, H. Cheng, C. Hu, S. Xia, X. Sun, X. Wang, and S. Yang. 2009. Ametryn degradation by aqueous chlorine: Kinetics and reaction influences. *J. Hazard. Mater.* 169(1–3): 586–592.

Xu, Z.P., Y. Jin, S. Liu, Z.P. Hao, and G.Q. Lu. 2008. Surface charging of layered double hydroxides during dynamic interactions of anions at the interfaces. *J. Colloid Interface Sci.* 326(2): 522–529.

Younes, M. and H. Galal-Gorchev. 2000. Pesticides in drinking water – A case study. *Food Chem. Toxicol.* 38(Supplement 1): S87–S90.

Zaneva, S. and T. Stanimirova. 2004. Crystal chemistry, classification position and nomenclature of layered double hydroxides. Bulgarian Geological Society, Annual Scientific Conference "Geology 2004", 110–112.

Zhang, Y., B. Van der Bruggen, G.X. Chen, L. Braeken, and C. Vandecasteele. 2004. Removal of pesticides by nanofiltration: Effect of the water matrix. *Sep. Purif. Technol.* 38(2): 163–172.

Zhou, J., J. Ralston, R. Sedev, and D.A. Beattie. 2009. Functionalized gold nanoparticles: Synthesis, structure and colloid stability. *J. Colloid Interface Sci.* 331(2): 251–262.

12

Treatment and Reuse Potential of Graywater from Urban Households in Oman

Mushtaque Ahmed, Abdullah Al-Buloshi, and Ahmed Al-Maskary

CONTENTS

12.1 Introduction

Oman was an arid zone country in the early stages of development. The country is highly dependent on groundwater as its main source of freshwater for domestic use and agricultural and industrial purposes, except for some areas that are served by desalinated water. It is likely that the gap between the water supply and the demand will continue to increase rapidly. The average rainfall in Oman is only about 100 mm/year. The total water requirement is 1600 MCM (million cubic meter)/year, and the country faces a water deficit of about 400 MCM/year (Ibnouf and Abdel-Magid 1994). Therefore, the reuse of wastewater, in general, and graywater, in particular, is an important step to overcoming or at least partially alleviate the problem of water shortage.

The water that has been used for washing dishes, laundering clothes, or bathing is considered graywater. Essentially, any water other than toilet waste (blackwater) drained from a household is graywater. When properly managed, graywater can be a valuable resource, which can be reused by horticultural and agricultural growers as well as for home gardens. It can also be valuable to landscape planners, builders, and contractors because of the design and landscaping advantages of on-site graywater treatment and management. About 65% of the residential wastewater is graywater (Duttle 2001). Studies have shown that there are very significant distinctions between graywater and blackwater and, under most conditions, it would be possible to reuse graywater with little or no treatment. The recent increase in environmental concerns has made the reuse of graywater a more valuable alternative for reducing the amount of contaminated wastewater discharged to the environment by reducing the use of freshwater, reducing the strain on septic tanks and treatment plants, and reducing the use of chemicals. Graywater can also be used for groundwater recharge for potential reuse in the future. The presence of phosphorous, potassium, and nitrogen makes graywater an excellent nutrient source for vegetation when this particular form of wastewater is made available for irrigation.

The main objectives of this study were: (i) to assess the quality of the urban household graywater and its potential use, in Oman, for irrigation and other purposes, (ii) to assess the variation of the graywater quality and quantity within one site and between different sites, (iii) to estimate the quantity of graywater produced from typical Omani households, and (iv) to design and test simple laboratory experiments for graywater treatment process.

12.2 Methodology

12.2.1 Sample Collection

Graywater samples from two urban households, one from Al Batinah (Barka) and the other from the Capital area (Al Mawaleh), were collected. In each sampling, six water samples (Table 12.1) from each house were collected over a period of 6 months (one sample per month). Each household consisted of six people (three adults and three minors). Graywater samples were collected (from each source) in 20 L plastic containers. A subsample of about 2 L was

TABLE 12.1

Graywater Samples Collected from Different Sources in Each House

Tap water	1 sample (fresh sample)
Bathrooms	1 sample from wash basin and 1 from shower
Kitchen	1 sample
Washing machine	1 sample
Mix (of all sources)	1 sample

taken from each container. The subsample was subjected to physical, chemical, and biological analyses.

12.2.2 Graywater Quantity Measurement

Flow meters and sampling devices were installed in each house (kitchens and bathrooms), and water meter readings were taken for 1 month. The total water use (as graywater and other uses) was calculated on the basis of per person per day.

12.2.3 Sample Analysis

Tests were carried out in accordance with APHA (1995), whenever appropriate. Measurements of the key water quality parameters were made on all samples: physical, chemical, and biological parameters. These were electrical conductivity (EC), total suspended solids (TSS), total dissolved solids (TDS), turbidity, pH, dissolved oxygen (DO), total alkalinity, carbonate, bicarbonate (HCO_3), chloride (Cl), nitrate (NO_3–N), phosphate (PO_4), sulfate (SO_4), major cations, biological oxygen demand (BOD), total coliform, and *Escherichia coli* bacteria.

12.2.4 Graywater Treatment

The purpose of the treatment experiments was to assess the suitability of the graywater quality for irrigation before and after treatment. The treatment included storage, application of a filtration system, aeration, and chlorination.

12.2.4.1 Storage

The storage of the raw graywater is used to assess the changes in the water quality with increasing residence time. The storage of graywater for 24 h may significantly improve the water quality through the rapid settlement of the organic matter particles, but storage beyond 48 h leads to depleted DO levels (Dixon et al. 1999). In this study, graywater was stored in a closed plastic container. The graywater sample was taken after 3 and 7 days. It was then analyzed for physical, chemical, and biological parameters.

12.2.4.2 Filtration System

A multimedia filter system consisting of wood chips and different sizes of stones was used to remove some of the suspended and organic material from the graywater. The filtered graywater was then analyzed for the key water quality parameters, including the physical properties, inorganic nonmetallic constituents, major cations, and trace elements.

12.2.4.3 Aeration

In this process, air was pumped into the graywater using an air pump to ensure a continuous supply of air that can be used by the bacteria to break down the organic matter. The sample was taken after 3 and 7 days. It was then analyzed for physical, chemical, and biological parameters.

12.3 Results and Discussion

12.3.1 Graywater Production

Table 12.2 provides data from the two sites. For site one, the total input of freshwater for 1 week is about 1800 gal/week (1 gal = 3.78 L) and the total water used for 1 week (bathroom and kitchen) is about 1004 gal/week. The remaining water (796 gal/week) is mainly used as blackwater and for lawn cleaning and garden irrigation. For site two, the total freshwater input for 1 week is about 1865 gal/week and the total water used in the kitchen and bathroom is 799 gal/week. The remaining water is mainly

TABLE 12.2

Graywater Production from Barka (Site One) and Al Mawaleh (Site Two)

Site	Bathroom (gal/day)	Kitchen (gal/day)	Total Graywater Used (gal/day)	Graywater (gal/person day)	Total Freshwater Used (gal/person day)	Graywater Produced as Percentage of Total Consumed Water (%)
Barka	103 (72%)	40 (28%)	143.0	23.8	42.9	55.5
Al Mawaleh	75.4 (66%)	38.7 (34%)	114.1	19.0	44.4	42.8

Note: Site one kitchen graywater includes laundry, whereas site two does not. 1 gal = 3.78 L.

used as blackwater and for lawn cleaning, car washing, and garden irrigation (Table 12.2).

The data presented in Table 12.2 indicate that, on average, the total water consumed from the kitchen and bathroom only (that can be collected as graywater) accounts for about 50% of the total water used in the house. This percentage could change significantly from one house to another. Jamrah et al. (2004) reported that, on average, 81% of the total water consumed in a typical Omani house in Muscat Governorate could be collected as graywater. Graywater collected from the kitchen accounted for one-third and that collected from the bathroom accounted for two-thirds of the total graywater collected in a house.

12.3.2 Graywater Sample Analysis

12.3.2.1 Site One

The results of the analysis of the physical, chemical, and biological quality parameters for the different sources of graywater generated from site one (Barka) are shown in Table 12.3.

12.3.2.1.1 Physical Properties

The value obtained for the turbidity in laundry graywater (778 NTU) was significantly higher than that of the kitchen, hand basin, shower, and combined graywater (CGW) (270, 61.5, 139, and 208 NTU, respectively). The graywater from the shower exhibited the lowest value for EC, which was 1427 μS/cm, whereas the laundry and kitchen wastewaters had the highest values at 3043 and 1803 μS/cm, respectively. The EC value measured for tap water was 1396 μS/cm. The salinity of the samples was a little higher compared with the medium salinity in the baseline tap water, because of the addition of salts contained in food and soaps in the case of kitchen graywater and those contained in soaps and detergents in the case of bathroom and laundry graywater.

The values obtained for the TDS varied considerably between the sample sources. The highest value for the TDS originated from the laundry graywater (1704 mg/L), whereas the shower graywater had the lowest TDS (799 mg/L). With respect to the EC, except for laundry (3043 μS/cm), all other values obtained for the physical properties (EC and TDS) in all sources comply with the Omani Ministerial Decision 145/93 dated 13 June 1993—Regulations for Wastewater Reuse and Discharge.

12.3.2.1.2 Chemical Properties

Measurements of the key water quality parameters were made on all samples, including pH, BOD, DO, alkalinity, nitrate, chloride, and total hardness (as $CaCO_3$). Graywater that originated from laundry was alkaline and had pH values in the range of 7.7–8.79. Graywater from other sources generally

TABLE 12.3

Mean Values for Graywater Sample Analysis for Chemical, Physical, and Biological Parameters at Site One

		Graywater Reuse					Omani Standard for Reuse
Parameter	Tap Water	HB	SHB	WM	K	CGW	
pH	8.06	7.44	7.57	8.20	6.22	6.73	6–9
EC (μS/cm)	1396	1668	1427	3043	1803	1831	2000–2700
Turbidity (NTU)	0.6	61.5	139	778	270	208	—
Total hardness		408	413	384.8	327.8	360	—
TDS	782	934	799	1704	1010	1026	1500–2000
DO	6.2	2.17	2.87	1.27	1.34	1.50	—
BOD5	2.8	52	118	376	356	375	15–20
Sulfate (SO_4)	1.0	145	136	288	135	168	400
Total phosphate (PO_4)	<0.2	0.41	<0.2	16.8	<1.07	8.16	—
Total alkalinity, as $CaCO_3$	130	183	172	530	127	158	—
Calcium (Ca)	14.9	73.8	73.1	69.6	55.5	68.7	—
Magnesium (Mg)	10.6	54.3	55.8	51.2	45.9	50.9	150
Sodium (Na)	19.6	137	151	242	182	145	200–300
Potassium (K)	0.63	12.1	18.9	10.4	17.4	11.7	—
SAR	0.95	2.94	3.12	5.37	4.79	3.21	10
Chloride (Cl)	21.2	251	276	450	331	285	650
Nitrate (NO_3)	<0.15	2.99	<0.15	2.36	3.8	<0.15	50

Notes: Values are in milligrams per liter, unless otherwise specified.
HB: hand basin, SHB: shower bathroom; WM: washing machine; K: kitchen, CGW: combined graywater.
— No Omani standard is available.

had lower pH values, ranging from 5.5 to 8.79 (Table 12.3). The high pH in graywater depends largely on the pH and the alkalinity in the main water supply. However, the higher pH values observed in the graywater from laundry show that the use of chemical products is also of importance (Eriksson et al. 2002).

Low DO concentrations were obtained from laundry and kitchen graywater, 1.27 and 1.34 mg/L, respectively, when compared with other sources. This indicates that oxygen had been used by the microorganisms that were present in the wastewater as well as by the chemical processes that occur in the graywater.

The BOD_5 is measured to determine the relative oxygen requirements of the wastewater, effluents, and polluted water to biodegrade organic pollutants. The BOD_5 concentrations varied from one source to another. The bathroom graywater from the hand basin and shower contained the lowest concentrations at 0, 70.3, and 85 mg/L, respectively. However, the highest BOD_5 concentrations were found in the laundry graywater (average of 376 mg/L) and kitchen graywater (average of 356 mg/L). The CGW contains an average of 375 mg/L. The BOD_5 concentrations from all other sources (ranging from 276 to 473 mg/L) are still high compared with the Omani standards for wastewater reuse and discharge (15–20 mg/L), which makes the graywater from all sources not suitable for reuse without treatment as per the existing Omani guideline.

Total hardness (H_T), expressed in milligrams per liter as calcium carbonate ($CaCO_3$), is an indication of the capacity of the water to precipitate soap. For irrigation purposes, soft water is preferred over hard water (the lower the concentration of $CaCO_3$, the softer is the water). Eriksson et al. (2002) have also reported that the total hardness found in the bathroom graywater ranges from 192 to 236 mg/L; laundry graywater, 282 mg/L; and kitchen graywater, 676 mg/L. Total hardness gives a concentration range of 327.8–413 mg/L, in which the highest concentrations were found in the shower graywater (374–455.9 mg/L) (Table 12.3). Thus, graywater produced from all sources is classified as hard water.

The major cations, sodium and magnesium, and nitrate, chloride, and sulfate concentrations are presented in Table 12.3. The data show a low concentration in all wastewater sources compared with the Omani standards for wastewater reuse and discharge.

Other cations, such as potassium and calcium, total alkalinity, and phosphate amount to 11.7, 68.6, 158, and 8.16 mg/L in the CGW, respectively, for which no Omani standards are available.

The sodium absorption ratio (SAR) is an important parameter to consider when wastewater is reused for irrigation. According to Ayers and Westcot (1976), graywater containing a high SAR (13 or more) used for irrigation may result in a soil with reduced permeability and aeration, leading to the degradation of the soil structure. This is because sodium has the ability to disperse soil when it is present above a certain threshold value relative to the concentration of the total dissolved salts. According to the US Department of Agriculture, irrigation water can be classified according to its EC and SAR. Salinity levels (C) as determined by the EC and sodium (S) associated with SAR are classified into four groups: low, medium, high, and very high, varying from excellent (C1 S1) to very poor (C4 S4), indicating the salinity or sodicity hazard encountered when using this wastewater for irrigation.

The results obtained from all sources (Table 12.3) show that the laundry and kitchen graywater have the highest average SAR values (5.37 and 4.79, respectively), whereas the lowest average SAR values were obtained from bathroom graywater (hand basin 2.94 and shower 3.12). A higher detergent content corresponds with high SAR values, indicating that detergents and soaps contain sodium that increases the SAR of the water.

Graywater from the shower, hand basin, and kitchen and the CGW are classified as high salinity–low sodium (C3–S1), whereas laundry wastewater is classified as very high salinity–low sodium (C4–S1). The classification of graywater for irrigation purposes largely depends on the characteristics of the soil (textural, structure, and pore size) as well as the likely treatment (i.e., leaching) that may be necessary for future use, to avoid salt and sodium buildup with continuous use.

12.3.2.2 Site Two

The results of the analysis of the physical, chemical, and biological quality parameters for the different sources of graywater generated from site two (Al Mawaleh) are shown in Table 12.4.

12.3.2.2.1 Physical Properties

The physical and the chemical differences are evident when we compare the quality of the graywater in site one and site two. This is because the source of the water used in site two is municipal (desalinated) water, whereas in site one, the water used is groundwater.

The tap water turbidity, the EC, and the TDS measurements were 0.51 NTU, 131 μS/cm, and 72 mg/L, respectively (Table 12.4). The highest values obtained for turbidity, EC, and TDS were from the laundry graywater (221 NTU, 2728 μS/cm, and 1522 mg/L, respectively). The lowest values obtained for turbidity and TDS were from the hand basin graywater (34.2 NTU and 211 mg/L), whereas the lowest EC was from the shower graywater (307 μS/cm).

12.3.2.2.2 Chemical Properties

The graywater originating from all sources was of acidic nature (average pH ranges from 5.73 to 6.84), except for the laundry graywater, which was alkaline (pH average ranges from 9.36 to 9.90). The freshwater pH was 7.58, indicating that the high pH values found in the laundry graywater were more likely to have originated from the source as well as the addition of the chemical products used in washing soaps.

The lowest DO concentrations were obtained in site two from the kitchen and hand basin graywater, at 1.23 and 1.47 mg/L, respectively (Table 12.4). The bathroom graywater from the hand basin and shower exhibited the lowest BOD_5 values of 124.4 and 77.7 mg/L, respectively. The highest concentration of BOD_5 was found in the laundry graywater, which ranged from 282 to 339 mg/L with an average of 306 mg/L, and in the kitchen graywater, which ranged from 152 to 327 mg/L with an average of 252 mg/L. The CGW contains an average of 356 mg/L. Generally, the BOD concentrations obtained from all sources were still high compared with the Omani standards.

Total hardness (H_T) gives a concentration range of 194–427 mg/L (Table 12.4), which means that the graywater produced from all sources is hard water,

TABLE 12.4

Graywater Sample Analysis for Chemical, Physical, and Biological Parameters at Site Two

	Graywater Sources						
Parameters	Tap Water	HB	SHB	WM	K	CGW	Omani Standard for Reuse
pH	7. 58	6.61	6.51	9.61	6.01	7.24	6.5–8.5
EC (μS/cm)	131	376	307	2728	770	888	—
Turbidity (NTU)	0.51	34.2	146	221	133	171	—
TDS	72	211	172	1522	435	497	1500–2000
TSS	140	0.840	0.710	23.5	8.9	11.11	—
DO	6.1	1.47	2.43	1.77	1.23	0.93	—
BOD5	1.4	124.6	77.7	306	252	356	15–20
Sulfate (SO4)		16.9	9.0	310	29.6	59.6	400
Phosphorus (PO_4)		0.75	0.84	71	2.5	62.5	—
Total alkalinity, as $CaCO_3$		80	87	847	67	215	—
Total hardness (H_T) as $CaCO_3$		408	412	385	327	385	
Calcium (Ca)	20	23.1	19.9	22	19.8	18.7	—
Magnesium (Mg)	0.45	22.6	45.1	8.0	13.4	7.7	150
Sodium (Na)	6.39	37.6	97	186	105	72.7	200–300
Potassium (K)	1.06	19.4	18.6	3.6	11.5	205	—
SAR	0.04	1.72	4.37	8.10	4.46	1.78	10
Chloride (Cl)		35.51	92.5	170.6	120	36	650
Nitrate (NO_3)		<0.15	<0.15	1.5	0.35	0.58	50

Note: Values are in milligrams per liter unless otherwise specified.

according to the USEPA classification system. The highest concentration of total hardness was found in the shower wastewater (374–455 mg/L), whereas the lowest concentration was in the kitchen graywater (194–427 mg/L). The major cations (sodium and magnesium), anions (nitrate, chloride and sulfate), and SAR showed lower concentrations in all sources compared with the Omani standards for wastewater reuse and discharge. The average concentrations for the other parameters, such as potassium, calcium, total alkalinity, chloride, nitrate, and phosphate in the CGW, were 18.7, 20.5, 215, 59.6, 36, <0.15, and 1.78 mg/L, respectively. For these, no Omani standards are specified.

TABLE 12.5

Microbiological Quality for Site One and Site Two from Different Graywater Sources Measured in MPN/100 mL

Site	Parameter	Tap Water	Hand Basin	Shower	Laundry	Kitchen	Combined Graywater
Site One	Coliform	69.7	>200.5	>200.5	>200.5	>200.5	>200.5
	E. coli	0.0	>200.5	>200.5	>200.5	>200.5	>200.5
Site Two	Coliform	0.0	>200.5	0.0	>200.5	>200.5	>200.5
	E. coli	0.0	>200.5	0.0	22.2	>200.5	>200.5

The results for SAR obtained from all sources show that the laundry graywater and the kitchen graywater also have the highest average SAR values at 8.10 and 4.46, respectively, whereas the lowest average values were obtained from bathroom graywater (hand basin, 1.72, and shower, 4.37).

12.3.3 Microbiological Quality

The *E. coli* and the coliform bacteria concentrations were found to be high in most graywater sources (Table 12.5). The results shown are consistent with those reported by Jamrah et al. (2004).

The published results for the graywater quality analysis have shown that graywater from the bathroom and laundry can contain a significant population of organisms represented by a relatively high total bacteria count (Dixon et al. 1999). For the practical application of the graywater reuse system, the major concern appears to be the turbidity, the microbial concentration, and the potential presence of pathogens. The presence of coliforms in the graywater suggests that there is pollution in the water; however, it does not offer information on whether pathogens are present. *E. coli* and other enteric organisms present in water indicate fecal contamination and the possible presence of intestinal pathogens, such as *Salmonella* or enteric viruses (Rose et al. 1991). Coliform bacteria were detected in all graywater samples collected from both sites, except the tap water and the shower water in site two. The laundry water from site two contains a low number of *E. coli* compared with the other sources (22.2 MPN), whereas the shower water from site two contains no coliform or *E. coli*. The wide range of microbial profiles present in the graywater indicates that graywater treatment is required prior to reuse.

12.3.4 Graywater Treatment

In this study, graywater was subjected to filtration, storage, and disinfection treatments, as shown in Table 12.6. The results of the graywater treatments for the site one experiments (filtration and storage) are shown in Figures 12.1 through 12.4.

TABLE 12.6
Various Graywater Treatments Used in the Study

Treatment	Condition	Code Used
Filtration	Filtered graywater	Filt.
	Unfiltered graywater (raw)	Unfilt.
Storage	3-day aerated sample	3DA
	3-day nonaerated sample	3D-NA
	7-day aerated sample	7DA
	7-day nonaerated sample	7D-NA

12.3.4.1 Filtration

Graywater contains solid particles, microorganisms, and nutrients. As such, a filter should be installed to remove these particles, which may otherwise cause the land application system to block. The type of filter required for a graywater system depends on the volume, the flow rate, the quality of the inflowing water, and the desired degree of water purification. The multimedia filter used in this study is well described in Little (2004).

The data show that the use of filtration seemed to be effective in reducing the Ca, Mg, and K concentrations compared with the storage treatments. This is most likely to be an error in the operation of the filtration unit. The data show that the average concentration of Ca in the unfiltered graywater was 80.8 mg/L; however, after filtration, the average concentration of Ca was only 47.4 mg/L. Apparently, a reduction of more than 50% was achieved, indicating the high efficiency of the treatment process.

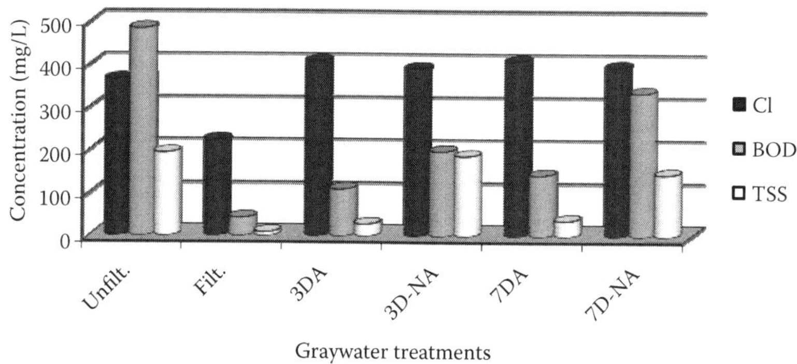

FIGURE 12.1
The effect of the different graywater treatment systems on the concentrations of chloride, BOD, and total suspended solids (mg/L). TDS in mg/L, EC in μS/cm, and turbidity in NTU.

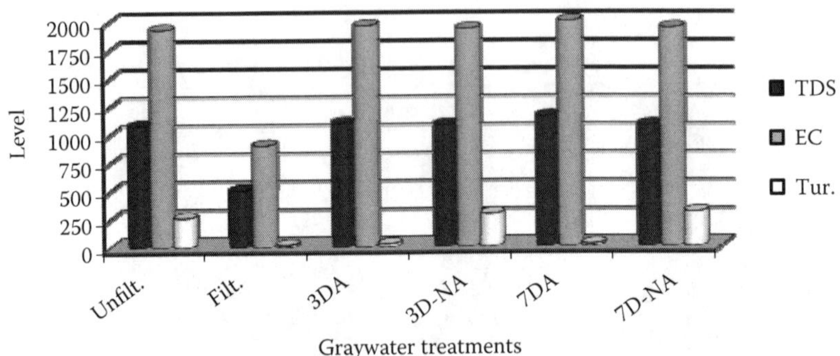

FIGURE 12.2
The effect of the different treatment systems on the level of TDS in mg/L, EC in μS/cm, turbidity in NTV, and DO in mg/L.

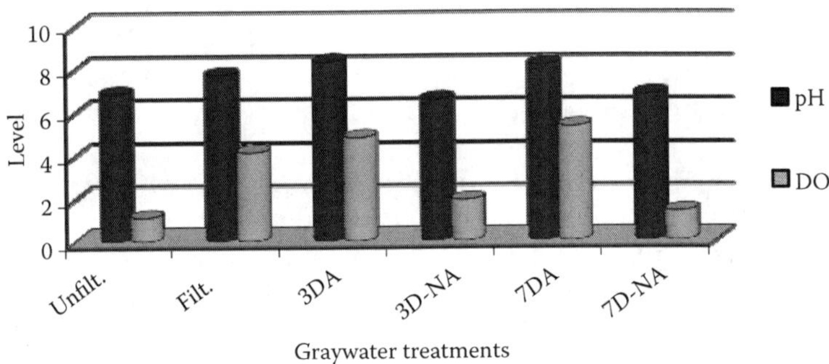

FIGURE 12.3
The effect of the different treatment systems on the level of DO in mg/L and pH.

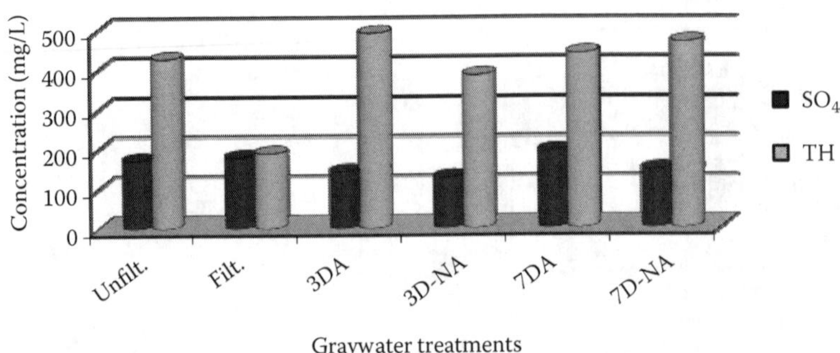

FIGURE 12.4
The effect of the different graywater treatment systems on the concentration of sulfate and total hardness (mg/L).

The average concentration of DO from the filtration treatment was high (4.13 mg/L) compared with the CGW (1.09 mg/L). This means that the filtered graywater has a better DO quality.

The average concentration of the TSS is quite high in the treatment processes 3D-NA and 7D-NA (186 and 144 mg/L) compared with 3DA and 7DA (78 and 36.7 mg/L), respectively. The filtration treatment substantially reduces the TSS concentration of graywater, which gives an average value of 9.0 mg/L compared with the unfiltered CGW concentration of 193 mg/L.

The highest BOD_5 average concentrations were obtained in the treatments 7D-NA and 7DA (314 and 186 mg/L, respectively), which indicates that as the graywater is stored for a longer time it becomes more polluted with microorganisms, resulting in rapid DO depletion. The turbidity average concentrations were found to be higher in the stored nonaerated graywater 3D-NA and 7D-NA (356 and 329 mg/L, respectively) treatments. The aerated graywater shows a noticeable improvement in turbidity from 260 mg/L (CGW) to 28.6 and 23.2 mg/L (for 3DA and 7DA, respectively).

12.3.4.2 Storage

The characteristics of the fresh graywater and the stored graywater can differ substantially. Studies have been conducted to look at the impact that storage has on graywater. It has been observed that storage for 24 h improves the quality of the graywater, but storage for more than 48 h could be a serious problem as the DO is depleted (Eriksson et al. 2002).

The storage of graywater promotes the generation of offensive odors and the growth of microorganisms, such as fecal coliform. However, the direct reuse of graywater without storage minimizes the microorganism growth, thereby reducing the offensive odor and the health risks with contact. However, the storage of graywater is an acceptable practice of treatment if it can improve the quality of the water that can be used for watering lawns and gardens. A higher degree of treatment than simple screening and contact disinfection is necessary to achieve a pathogen-free graywater to use for toilet flushing (Jeppesen 1996).

The DO average concentration shows a substantial reduction for the storage treatments 3D-NA and 7D-NA (1.9 and 1.34 mg/L) compared with the DO concentration of 3DA and 7DA (4.8 and 5.3 mg/L), respectively. There was a major reduction in the graywater TSS concentration with the 3DA and 7DA storage treatments compared with the 3D-NA and 7D-NA treatments.

In the case of treatment processes 3DA and 3D-NA and 7DA and 7D-NA, the results show no change in the concentrations of the cations, anions, EC, and TDS, which means that the process is not effective at removing such parameters from graywater. But there is a slight change in the pH toward acidity (8.5–6.7) and a slight increase in the concentration of the total alkalinity. The BOD value is quite low on the first 3 days (197 mg/L for 3D-NA) of the treatments, but the value increases with time, reaching more than

300 mg/L after 7 days (7D-NA). This indicates that the number of microorganisms in the graywater is increasing with time, requiring more oxygen for the organisms. The BOD levels are high for all treatments showing very high values, which exceed the Omani standard. Further treatment after storage (i.e., chlorination) to reduce the number of microorganisms present in the graywater is essential.

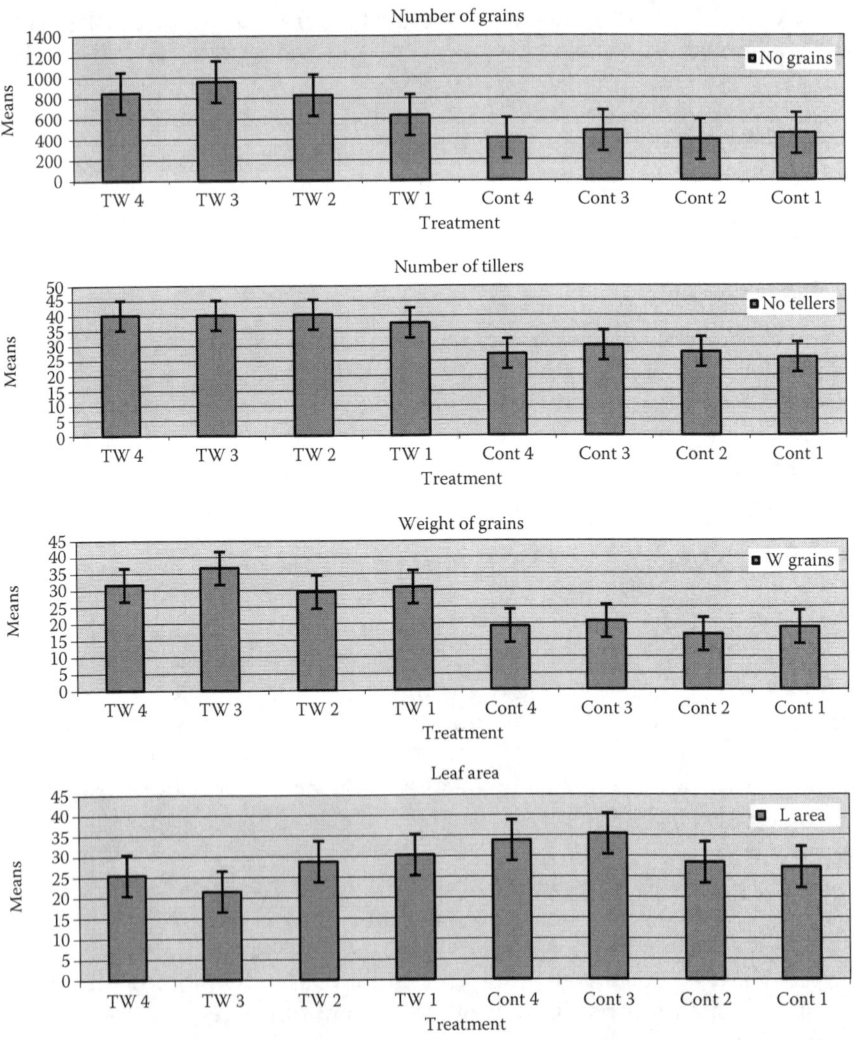

FIGURE 12.5
The effect on barley irrigated with treated graywater (TW) in comparison with tap water (Cont) in terms of the number of grains, the number of tillers, the weight of grains, and the leaf area.

12.3.4.3 Disinfection

Commercially available chlorine tablets were used to disinfect the graywater. The available chlorine in the tablet results in 3000 mg/L of chlorine concentration when 1 tablet is mixed in 1 L of water. The tablets are marketed by Guest Medical Ltd. (UK) and are widely available in Oman. It was observed that a chlorine dose of 1.5 mg/L resulted in no coliform or *E. coli* in the water samples measured 6 h after application.

12.3.8 Use of Treated Graywater for Irrigation

In order to assess the suitability of the treated graywater for irrigation use, pot trials in greenhouses with barley (*Hordeum vulgare* L.) were conducted using treated graywater and tap water. Using Duncan's multiple range test, four parameters—the leaf area, the number of grain heads, the grain weight, and the number of tillers—were found to be statistically different. The results show that there is no likelihood of adverse crop growth if treated (filtered) graywater is used for irrigation. In fact, three of the four parameters showed a positive effect when irrigated with treated graywater (TW) in comparison with tap water (Cont), as shown in Figure 12.5.

12.4 Conclusions

The wastewater that can be collected as graywater from the kitchen and bathroom for site one and site two is 55.5% and 42.8%, respectively, of the total water used. Graywater from the bathrooms (hand basin and shower) is less polluted compared with the laundry and kitchen wastewater, thus it is easier to reuse. The laundry graywater contains a high pollutant level indicated by the high values of the parameters, such as the EC, TDS, pH, turbidity, Cl, and PO_4, and the lowest DO concentration. The kitchen graywater comes second in terms of the level of pollutants. The quantity and the quality of the graywater produced will, in part, determine how it can be reused. Irrigation and toilet flushing are two common uses, but nearly any noncontact use is a possibility. The simple graywater treatments of storage (aeration and nonaeration), filtration, and disinfection have shown a substantial improvement in the graywater quality. The reuse of graywater is a useful component of a water conservation strategy; however, there are technical, social, and economic constraints that need to be thoroughly investigated before reuse. The data imply that there may be some risk associated with the reuse of graywater because of pathogenic bacteria or possibly viruses. Therefore, it is essential that the practice of reusing graywater adequately protects public health. Currently, graywater reuse technology may not be viable in purely economic terms, but its introduction needs to be seen in terms of its contribution to sustainable development and resource conservation without affecting public health or the quality of the environment.

References

APHA. 1995. Standard Methods for the Examination of Water and Wastewater. American Public Health Association. 19th Edition. American Public Health Association. Washington, D.C.

Ayers, R.S. and D.W. Westcot. 1976. Water quality for agriculture. FAO irrigation and drainage paper 29. Rome, Italy.

Dixon, A., D. Butler, and A. Fewkes. 1999. Water saving potential of domestic water reuse systems using greywater and rainwater in combination. *Water Sci. Technol.* 5: 25–32.

Duttle, M. 2001. Use of household greywater. *Guide M – 106.* http://www.cahe.nmsu.edu/pubs/_m/m-106.html.

Eriksson, E., K. Auffarth, M. Henze, and A. Ledin. 2002. Characteristics of grey wastewater. *Urban Water* 4: 85–104.

Ibnouf, M.A. and I.M. Abdel-Magid. 1994. Oman water resources: Management, problems and policy alternatives. In *Proceeding of the 2nd Gulf Water Conference Water in the Gulf Region Toward Integrated Management*, Water Sciences and Technology Association, Bahrain, November 5–9, 1994.

Jamrah, A., A. Futaisi, S. Prathapar, M. Ahmed, and A. Al-Harrasi. 2004. Evaluating greywater reuse potential for sustainable water resources management in the Sultanate of Oman. In *Conference on International Water Demand Management*, Jordan, May 30–June 3, 2004.

Jeppesen, B. 1996. Domestic greywater reuse – Australia's challenge for the future. *Desalination* 106: 311–315.

Little, V.L. 2004. Greywater guideline. The Water Conservation Alliance of Southern Arizona. http://www.watercasa.org.

Rose, J.B., G. Shing Sun, and C.P. Gerba. 1991. Microbial quality and persistence of enteric pathogens in greywater from various household sources. *Water Res.* 25: 37–42.

13

Anaerobic Fixed Bed Reactor for Treatment of Industrial Wastewater

Joseph V. Thanikal

CONTENTS

13.1 Introduction

In many parts of the world, economic, social, and political problems have arisen following rapid industrial development and urbanization, resulting in adverse effects on the quality of life. Urbanization, in general, initially places pressure on and overstrains public amenities. However, long-term and wider issues would eventually be encountered as industrialization and urbanization exert pressure on the larger resource base that supports the community. This larger resource base includes forestry, freshwater, and marine resources, as well as space suitable for further development. The difficulties associated with environmental degradation often originate from industrial development. They are amplified by rapid urbanization, which is responsible for the growth of many major cities. In Asia, urbanization is exacerbated by large rural–urban migrations. These migrations emerge in response to perceived opportunities for a better livelihood in industrialized, economically booming urban areas. Rapid industrialization and its concentration in or near urban centers have placed very high pressures on the carrying capacity of the environment at specific locations. At these locations, water bodies, such as rivers, lakes, and coastal waters, have typically been severely affected. Freshwater is a vital natural resource that will continue to be renewable as long as it is well managed. Preventing pollution from domestic, industrial, and agro-industrial activities is important to ensure the sustainability of the locale's development. Undoubtedly, the water pollution control efforts that have been underway in many countries have already achieved some success. Nevertheless, the problems that are confronted grow in complexity and intensity. The pollution of freshwater bodies with the consequent deterioration in the water quality can only worsen the situation. Such pollution has been brought about by the discharge of inadequately treated sewage and industrial wastewaters. This chapter has attempted to highlight the importance of treating industrial wastewater and the treatment methods. The chapter concludes with an experimental study carried out to treat industrial wastewater with a high pollution load.

Industrial wastewaters (including agro-industrial wastewaters) are effluents that result from human activities associated with raw material processing and manufacturing. These wastewater streams arise from washing/cleaning, cooling, heating, extraction, reaction by-products, separation, and quality control resulting in product rejection. Water pollution occurs when potential pollutants in these streams mix with a receiving water source. In addition to the industrial wastewaters from such processing or manufacturing sites, wastewaters from some domestic sources may also be present; however, they may not constitute a major component. Domestic sewage may be present because of the washrooms and hostels provided for workers at the processing or manufacturing facility. Examples of industrial wastewaters include those arising from chemical, pharmaceutical, electrochemical, electronics, petrochemical, and food processing industries. Examples of agro-industrial wastewaters include those arising from industrial-scale animal husbandry, slaughterhouses, fisheries, fruits and vegetable processing, and oil seed processing. Pollutant concentrations are not usually high in agro-industrial wastewaters, except in the animal processing units.

The impact of the industrial wastewater discharges on the environment and on the human population can be tragic at times. One of the most classic examples is Minamata disease, which spread, some 50 years ago, among the residents in the Yatsushiro Sea and the Agano River basin areas in Japan, because of methyl mercury in the industrial wastewater. However, tragedies as dramatic as the Minamata episode have not occurred frequently. Nevertheless, instances of pollution with potentially adverse effects in the longer term have continued to occur. Examples of these, their recognition, and the efforts made to remedy the situations in the 1980s include the protection of the Malaysian coastal waters from refinery wastewater; the Tamsui River in Taiwan, where pesticides and heavy metals were discovered in the sludge; the Nam Pong River in Thailand, which was polluted by the pulp and paper industry; and the Buriganga River in Bangladesh, which had been polluted by tanneries and other industries. Similar reports in the 1990s include the Kelani River in Sri Lanka, the Laguna de Bay in the Philippines, and the Koayu River, which had occurrences of *Cryptosporidium* oocysts and *Giardia* cysts after receiving inadequately treated piggery wastewater. Such reports are still frequent in the 2000s and cause concern in Vietnam and Korea. Toward the end of 2004, the Huai River in China was reported to have been so seriously polluted by paper-making, tanning, and chemical fertilizer factories that the farmers in Shenqiu County had fallen very ill after using the river water. However, there has been progress, and an example of this is the successful 10-year river pollution cleanup program in Singapore.

Agro-industrial wastewaters, as a subclass of industrial wastewaters, can have a considerable impact on the environment in view of their pollutant concentration and the release of large volumes of such wastes. Citing some examples from ASEAN countries in Asia, in 1981, the Malaysian palm oil and rubber industries contributed 63% (1460 t/day) and 7% (208 t/day), respectively, of the biochemical oxygen demand (BOD) load generated per day, compared with 715 t/day of the BOD from domestic sewage. In the Philippines, pulp and paper mills generated 90 t/day of the BOD load.

13.1.1 Industrial Wastewater Treatment

Industrial (including agro-industrial) wastewaters widely vary in compositions, depending on the type of industry and the materials processed. Some of these wastewaters can be very strong organically, easily biodegradable, largely inorganic, or potentially inhibitory. This means that the total soluble solids (TSS), the BOD_5, and the chemical oxygen demand (COD) values may be as high as 10,000 mg/L. Because of these very high organic concentrations, industrial wastewaters may also be severely nutrient-deficient. Unlike sewage, the pH values are in the range of 6–9 with high concentrations of dissolved metal salts. Water clarity is affected by turbidity, which may be caused by inorganic (fixed suspended solids, FSS) and/or organic particulates suspended in the water (volatile suspended solid, VSS). Turbidity reduces light penetration, thereby reducing photosynthesis. Settleable particulates may accumulate on the plant foliage and on the bed of the water body, forming sludge layers that would eventually smother benthic organisms. As the sludge layers accumulate, they may eventually become sludge banks, and if the material in these is organic, then its decomposition would give rise to mal odors. In contrast to the settleable material, particulates lighter than water eventually float over the surface and form a scum layer. Limits on the wastewater or the treated wastewater discharges typically have a TSS value of 30 or 50 mg/L. Many industrial wastewaters contain oil and grease (O&G). They cause interference at the air–water interface and inhibit the transfer of oxygen, which in turn would reduce the solubility of the oxygen, causing a decline in the dissolved oxygen (DO) levels. The depletion of the free oxygen would affect the survival of the aerobic organisms. A decline to 3–4 mg/L is sufficient to adversely affect the higher organisms, for example, some species of fish. Because of the impact of the DO levels on aquatic life, much importance has been placed on determining the BOD value of a discharge. The typical BOD_5 limits set are values of 20 and 50 mg/L.

13.1.1.1 Inhibition or Toxicity and Persistence

Some organic and inorganic substances, such as pesticides and heavy metals, in wastewater may bias the aquatic population toward those that are more tolerant to these toxicants, while eliminating those that are less tolerant, resulting in a loss of biodiversity. For similar reasons, an awareness of the impact of such substances on the biological systems is not only relevant in terms of protecting the environment, but also in terms of their impact on the biological systems used to treat industrial wastewaters. Even the successful treatment of such wastewater may not necessarily mean that the potability of the water in a receiving water body would not be affected. For example, small quantities of residual phenol in water can react with chlorine during the potable water treatment process, giving rise to chlorophenols that can cause objectionable taste and odor in the treated water. Apart from the organic pollutants, which are potentially inhibitory or toxic, there are those that are resistant to biological degradation.

13.1.1.2 Eutrophication

The discharge of nitrogenous and phosphorous compounds into the receiving water bodies may alter their fertility. Enhanced fertility can lead to excessive plant growth. The latter may include algal growth. The subsequent impact of such growth on a water body can include increased turbidity, oxygen depletion, and toxicity issues. Algal growth in unpolluted water bodies is usually limited because the water is nutrient-limiting. While nutrients would include macronutrients such as nitrogen, phosphorous, and carbon and micronutrients such as cobalt, manganese, calcium, potassium, magnesium, copper, and iron, which are required only in very small quantities, the focus over eutrophication would be on phosphorous and nitrogen. The quantities of the other nutrients in the natural environment are often inherently adequate. In freshwaters, the limiting nutrient is usually phosphorous, while in estuarine and marine waters, it would be nitrogen. The treatment of industrial wastewater (or domestic sewage for that matter) can then target the removal of either phosphorous or nitrogen, depending on the receiving water body, to ensure that the nutrient-limiting condition is maintained. Given the littoral nature of many nations in Asia, the removal of nitrogen would probably be necessary if the wastewater contained excessive quantities of it. When the nutrient-limiting condition is no longer present in the water body, and when other conditions such as the ambient temperature are appropriate, excessive algal growth or algal blooms (e.g., the red tide) may occur. Apart from the aesthetic issues, such algal blooms may affect the productivity of the fisheries in the locale. It should be noted that not all industrial wastewaters contain excessive quantities of macronutrients and micronutrients. This deficiency, if present, results in process instability and/or proliferation of inappropriate microbial species during the biological treatment of the wastewaters. Bulking sludge is a manifestation of such an occurrence. To address this deficiency, nutrient supplementation is required. The quantities used should be carefully regulated so that a condition of excessive nutrients is not inadvertently created, and these excess nutrients are subsequently discharged with the treated effluent. In terms of BOD:N:P, the optimal ratio for biological treatment is often taken as 100:5:1, while the minimum acceptable condition can be 150:5:1.

13.1.1.3 Pathogenic Effects

Pathogens are disease-causing organisms, and an infection occurs when these organisms gain entry into a host (e.g., man or animal) and multiply therein. These pathogens include bacteria, viruses, protozoa, and helminthes. While domestic-related and medical-related wastewaters may typically be linked to such microorganisms (especially, bacteria and viruses), industrial wastewaters are not typically associated with this category of effects. The exception to this is the wastewater associated with the sectors in the agro-industry dealing with animals. The concern here would be the presence of such organisms in the wastewater that is discharged into a receiving water body. With the above effects in view, industrial wastewater treatment would typically be required to address at least the following parameters:

- Suspended solids (SS)
- Temperature
- Oil and grease
- Organic content in terms of the BOD or COD
- pH
- Specific metals and/or specific organic compounds
- Nitrogen and/or phosphorus
- Indicator microorganisms (e.g., *Escherichia coli*) or specific microorganisms

13.2 Reactors Used for the Treatment of Wastewater

Wastewater treatment involving physical operations, chemical unit processes, and biochemical processes are carried out in vessels or tanks commonly known as "reactors." The principal types of reactors used for the treatment of wastewater are (1) batch reactors, (2) complete mix reactors, (3) plug-flow reactors, (4) mix reactors in series, (5) packed reactors (fixed bed), and (6) fluidized bed reactors. A detailed description of these reactors is available in any standard textbook on chemical reaction engineering (Levenspiel 1999). In this chapter, we discuss the anaerobic fixed bed reactor for the anaerobic digestion of industrial wastewater from a vinery processing agro-industry. We start our discussion with a description of anaerobic digestion, in general, and then proceed to discuss vinery waste treatment.

13.3 Anaerobic Digestion

Some waste streams are treated by conventional means, such as aeration, which is both energy intensive and expensive and generates a significant quantity of biological sludge that must be discarded. The generation and disposal of large quantities of biodegradable waste without adequate treatment result in significant environmental pollution. In addition to the health-related problems for the population near the sites where waste is dumped, further degradation of the waste in the environment can lead to the release of greenhouse gases (GHGs), such as methane and carbon dioxide. In the absence of any waste treatment, as is normally the case, the environmental damage caused to the society works out to be more than the financial costs to the industry. In this context, anaerobic digestion offers potential energy saving and is a more stable process for medium- and high-strength organic effluents. Apart from treating the wastewater, the methane produced from the anaerobic system can be recovered. In addition to reducing the amount of GHGs by the controlled use of methane from waste, substituting oil and coal with bioenergy will result in saving the global environment by reducing the use of fossil fuels. The potential of anaerobic treatment is evident from the large number of recent research publications on this process. Up to the late 1960s, aerobic processes were very popular for the biological treatment of waste. The energy crisis in the early 1970s, coupled with increasingly stringent pollution control regulations, brought about a significant change in the methodology of waste treatment. Energy conservation in industrial processes became a major concern, and anaerobic processes have become an acceptable alternative. This led to the development of a range of reactor designs suitable for the treatment of low-, medium-, and high-strength wastewaters. The anaerobic process has several advantages over the other available methods of waste treatment. Most significantly, it is able to accommodate relatively high rates of organic loading. With the increasing use of anaerobic technology for treating various process streams, it is expected that industries would become more economically competitive because of their more judicious use of natural resources. Therefore, anaerobic digestion technology is almost certainly assured of increased use of natural resources in the future.

13.3.1 Development of Anaerobic Treatment Systems

Anaerobic digesters produce conditions that encourage the natural breakdown of organic matter by bacteria in the absence of air. The digestion process takes place in a warmed, sealed, and airless container (the digester), which creates the ideal conditions for the bacteria to ferment the organic material in oxygen-free conditions. The digestion tank needs to be warmed and mixed thoroughly to create the ideal conditions for the bacteria to convert organic matter into a biogas (a mixture of carbon dioxide, methane, and small amounts of other gases). There are two types of anaerobic digestion, namely, mesophilic and thermophilic. The anaerobic digestion of biodegradable wastes involves a large spectrum of bacteria of which three main groups are distinguishable.

The first group comprises fermenting bacteria that perform hydrolysis and acidogenesis. This involves the action of exoenzymes to hydrolyze polymeric matter such as proteins, fats, and carbohydrates into smaller units, which can then enter the cells and undergo an oxidation–reduction process, resulting in the formation of volatile fatty acids (VFAs) and some carbon dioxide and hydrogen. The fermenting bacteria are usually designated as acidifying or acidogenic population because they produce VFA.

Acetogenic bacteria constitute the second group and are responsible for breaking down the products of the acidification step to form acetate. In addition, hydrogen and carbon dioxide (in the case of odd-numbered carbon compounds) are also produced during acetogenesis.

The third group involves methanogenic bacteria, which convert acetate or carbon dioxide and hydrogen into methane. Other possible methanogenic substrates, such as formate, methanol, carbon monoxide, and methylamines, are of minor importance in most anaerobic digestion processes. In addition to these three main groups, hydrogen-consuming acetogenic bacteria are always present in small numbers in an anaerobic digester. They produce acetate from carbon dioxide and hydrogen and, therefore, compete for hydrogen with the methanogenic bacteria. Also, the synthesis of propionate from acetate and the production of longer-chain VFA occur to a limited extent in anaerobic digestion. Competition for hydrogen can also be expected from the sulfate-reducing bacteria in the case of sulfate-containing wastes. It was a long-accepted belief that anaerobic digestion was feasible only for the treatment of concentrated wastes, such as manure and sewage sludge, with long retention times. Around 1950, the anaerobic treatment of wastewater was attempted and the concept of high-rate systems gained importance with the use of mixing devices. The latter helped to break the scum in the digester and increase the contact between the organisms and the substrate. Special reactor types for wastewater treatment, such as the anaerobic contact processes, were also developed.

Fixed bed or fixed biofilm anaerobic reactors have been widely used for the treatment of high-strength wastewaters. In fixed film anaerobic reactors, a large amount of biomass remains in the filter to secure solid retention despite a short hydraulic retention time (HRT). These reactors have several advantages over the aerobic and anaerobic reactors, such as higher organic loadings, lower HRTs, and smaller reactor volumes. Lower sludge and SS quantities can also be achieved in these reactors. In addition, these reactors can tolerate sudden organic shock loads at constant hydraulic loading and recover normal performance within a few days if the alkalinity is high enough to maintain the pH. The reactors can process different waste streams with little compromise in capacity and can adapt readily to changes in temperature. Two kinds of support can be used in this type of reactor: well-ordered and loose supports. Many different materials have been tested for biomass retention in the anaerobic systems, and the performance of these materials appears to be directly related to the ease with which bacteria can become entrapped in or attached to the supports.

13.3.2 Anaerobic Reactors for Wastewater Treatment

Conventional digesters, such as sludge and anaerobic continuous stirred tank reactors (CSTR), have been used for many decades in sewage treatment plants to stabilize the activated sludge and sewage solids. The area is well researched, and sufficient information and operating experience are, therefore, available on the subject. In recent times, the emphasis has shifted to high-rate biomethanation systems, which are based on the concept of sludge immobilization techniques (UASB, fixed films, etc.).

13.3.2.1 Fixed Film Reactor

In stationary fixed film reactors (Figure 13.1), cells are deliberately attached to a large-sized solid support. The reactor has a biofilm support structure (media) for biomass immobilization, a wastewater distribution system for uniform distribution of the wastewater above/below the media, and effluent draw-off and recycling facilities (if required). The fixed film reactors offer distinct advantages, such as simplicity of the construction, elimination of mechanical mixing, better stability at higher loading rates, and the capability to withstand large toxic shock loads (van den Berg et al. 1985). In addition, these reactors can tolerate sudden organic shock loads at constant hydraulic

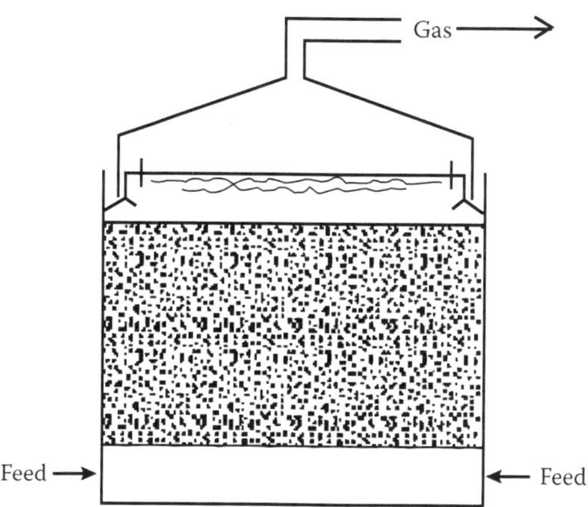

FIGURE 13.1
A schematic diagram of a stationary fixed film bed reactor.

loading and recover normal performance within a few days if the alkalinity is high enough to maintain the pH above 6.2. The reactors can process different waste streams with little compromise in capacity and can readily adapt to changes in temperature. This is important for installations where the wastewater characteristics change rapidly. The reactor start-up can be very quick after a period of starvation (1 or 2 days to reach maximum capacity after 3 weeks of starvation).

The main limitation of this design is that the reactor volume is relatively high compared with other high-rate processes due to the volume of the media. Another common problem associated with stationary fixed film reactors is clogging due to the nonuniform growth of the biofilm thickness and/or a high SSs concentration in the wastewater. The nonuniform growth and the consequent clogging occur especially at the influent entry. Some measures to combat this problem include the recirculation of the effluent and gas to develop a relatively thin film and sloughing of the biomass; the provision for a relatively thin layer of media near the load-entering area to accumulate the excess biofilm; and an improvement in the flow distribution system to avoid very low liquid velocity. The various types of film support that have been tried are activated carbon, polyvinyl chloride (PVC) supports, hard rock particles, and ceramic rings.

13.3.2.2 Effect of Surface Area of Inert Material

A number of inert carrier packaging materials are used for increasing the surface area in the bioreactors (Figure 13.2). Any surface submerged in water is quickly covered by a layer of microorganisms, forming a biofilm. In this way, biofilms grow spontaneously both in fresh and salty aqueous

FIGURE 13.2 **(See color insert)**
Different types of inert carriers.

environments, as well as in water conductors, such as pipes and channels. Apart from these natural biofilms, biological reactors have been developed where the formation of a biofilm is promoted on different materials, in order to treat wastewaters and reach satisfactory purification levels. The initial phase in the biofilm development involves the adsorption of organic compounds over the material, which will be colonized. This initial organic layer is a prerequisite for the later microbial attachment. The biofilm development begins after that phase (Figure 13.3). The biofilm is visible a few hours or minutes after the start-up of the reactor. The adherence, which is strongly influenced by the surface charge, takes place immediately on positively charged surfaces, but can be delayed by several hours if this charge is negative. The duration of this adherence phase will depend on several factors: the nature of the support, the surface charge, the nature and the concentration of the feed, etc. The initial surface colonization occurs at the cavities in the inert material, which has a surface roughness favorable for this development.

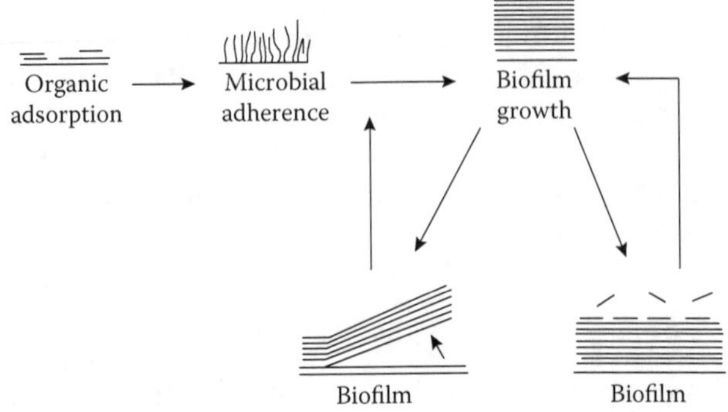

FIGURE 13.3
Biofilm development.

The time taken for the colonization to occur is shorter than the total time necessary for biofilm formation. The growth phase is the sum of the cellular reproduction and the extracellular polymer production. During this phase, a quick biofilm development occurs due to the growth of microcolonies and the adherence of new bacteria, so that at the end of this phase, the surface is totally covered by the biofilm, with a complex structure of microbial cell clusters. This growth phase can be divided into two steps, the first step is a logarithmic biofilm growth and the second step is a constant accumulation rate, which continues until its partial detachment and the steady-state biofilm thickness is reached. Although these phases in the biofilm formation are well defined, the influences that different materials (Figure 13.2) have on them have not been sufficiently studied.

13.3.2.3 Start-Up of Anaerobic Fixed Bed Reactors

The aim of the start-up is to develop an active biofilm on the carrier and to reach the nominal organic loading rate (OLR) with a satisfactory treatment performance. In many cases, the start-up of an anaerobic reactor takes 4 months or more than a year for thermophilic processes before a steady state is reached with respect to removal efficiency. Shortening the start-up time is key to increasing the economic competitiveness of the anaerobic processes. The following are the steps during the start-up (Kennedy and Droste 1985):

1. The inoculation period during which the carrier is put in close contact with an inoculating sludge to initiate biofilm attachment.
2. The progressive increase of the OLR to stimulate the microbial growth of the biofilm.

13.3.2.3.1 Inoculation

In most cases, anaerobic reactors are inoculated as a batch. During inoculation, the carrier material and the active inoculation sludge are brought into contact inside the reactor. The length of the contact time is chosen empirically and can vary from a few days up to more than 1 month. It is generally believed that a long contact time between a concentrated inoculum and the carrier is necessary and will favor biofilm growth in batch conditions. The initial adhesion of bacteria is found from an anaerobic sludge on the mineral particles in an inverse turbulent bed reactor. It requires a minimum of 12 h of contact time for the microorganism to attach to the carrier material, and usually the biofilm will be close to the inoculum. Compared with the traditional inoculation protocol, only a very short period is necessary to obtain adhesion of the microorganisms on the support media and to initiate biofilm formation. Consequently, it is possible to considerably shorten the duration of the inoculation period. The physicochemical properties of the carrier have a significant influence on the early adhesion of the bacteria and Archaea, both quantitatively and qualitatively. The Archaea/bacteria ratio of the adhered microbial communities, as determined by qPCR, was strongly dependent on the nature of the support material.

13.3.2.3.2 Increase in Organic Loading Rate

After the inoculation period, the OLR is normally increased, progressively and continuously. Anaerobic digestion is the result of synergistically interacting microbes, with the limiting step being methanogenesis. The increase in the OLR must be carefully monitored to avoid overloading of the system, which could lead to an inhibition of the methanogens and, consequently, to the failure of the start-up process. The main parameters to tune during this period are the HRT and the hydrodynamic conditions in the reactor.

A conventional way to operate with an increase in the OLR is to feed the reactor at a progressively increasing influent flow rate while keeping the influent COD concentration constant. The flow rate is increased stepwise when a minimum performance (e.g., 80% COD removal) is reached. This conservative strategy is often successful, but needs several months to reach steady state with respect to performance. Such a strategy enhances the competition between the suspended and the biofilm biomasses for the organic substrate. The biofilm accumulation in the reactor results from a balance between growth and detachment, mainly due to shear. The biofilm detachment occurs when the local shear forces exceed the cohesiveness of the biofilm. At steady state, the balance between growth and detachment determines the physical structure of the biofilm and thus the settling and fluidization characteristics in the case of particulate biofilms. Nevertheless, high shear forces lead to the formation of a thin, dense, and active biofilm, but they are suspected to slow down biofilm formation. It is advised to start up a bioreactor by applying minimal shear forces in order to enhance the biofilm growth during the early phase of the biofilm development. Then, the hydrodynamic shear forces can be increased after a sufficient amount of well-adapted biomass has accumulated on the carrier.

13.4 Experimental Case Study

An experimental case study is explained to establish the efficiency of the fixed bed reactors and also the use of a new carrier media (support) made of polyethylene, which was used to treat a highly concentrated vinasse from a wine distillery. A laboratory reactor of 23 L (Figure 13.4), the working volume used in the study, was fabricated out of PVC material, and it consisted of a tubular section of 190 mm internal diameter and 1150 mm total height with

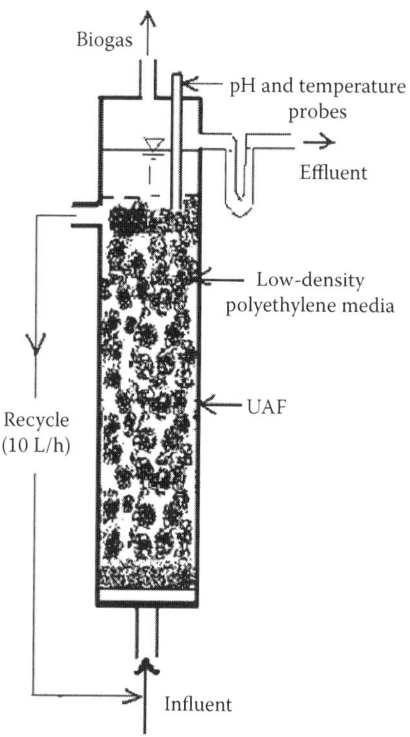

FIGURE 13.4
Experimental setup.

a conical bottom. The system was equipped with a water jacket to keep the temperature of the reactor at 35°C. The reactor was equipped with a substrate feed inlet at the bottom of the reactor and an overflow arrangement was provided such that the effective height of the liquid inside the reactor was maintained at 810 mm. A sampling port was fixed at the bottom of the reactor. A submerged pump (flow rate 480 L/h) was fixed inside the reactor, at the bottom, to facilitate fluidization of the supports.

The reactor was filled with a polyethylene support (Bioflow 30, manufactured by Rauschert) for 60% of the volume of the reactor. This trapezoidal support was 29 mm in height and measured 35 mm at the bottom and 30 mm at the top. It had a density of 930 kg/m^3 and a specific surface area of 320 m^2/m^3. The reactor was fed with a distillery vinasse (wine residue after distillation) in which the total COD varied between 10 and 24 g/L and the soluble COD varied between 10 and 19 g/L. The pH of the feed, which was at 4–5.5, was adjusted to 7–7.5. The reactor was inoculated with anaerobic sludge collected from an anaerobic reactor treating the distillery vinasse and was concentrated to 45 g/L by settling. The volume of the sludge was 10% of the volume of the reactor. The substrate was fed into the reactor through the inlet at the bottom of the reactor, using a peristaltic pump. The inlet substrate was fed at equal intervals of time, sequentially, as per the designed daily volume. The operation of the pump for fluidization was programmed at every 15 min over 3 h. The soluble COD, the VFAs, and the SSs were determined daily through off-line analysis. The COD was measured by a colorimetric method (Jirka and Carter 1975). The VFAs were measured using a gas chromatograph with a flame ionization detector (GC 8000, Fisons Instruments) and an automatic sampler (AS 800, Fisons Instruments). The total and volatile solids inside the reactor and at the outlet of the reactor were measured by standard methods (APHA 1992). The biomass attached to the support was measured by weighing the oven-dried support material (dried at 100°C for 24 h).

At the beginning of the experiment, the reactor was operated with a high HRT and a low OLR. Subsequently, the HRT applied to the reactor was regularly decreased and the OLR increased by increasing the volume of the vinasse treated. The reactor was operated for 180 days, and the total operation period can be divided into three stages, as shown in Figure 13.5:

1. During the first stage, the first 81 days, the increase in the OLR was slow, the HRT was always more than 3.6 days, and the OLR was always less than 5.6 g/L day.
2. During the second stage, day 82–101, the HRT had to be maintained constant at a high value (7.7 days) due to a temporary insufficient availability of vinasse. The OLR was low, that is, between 1.6 and 2.6 g/L day.

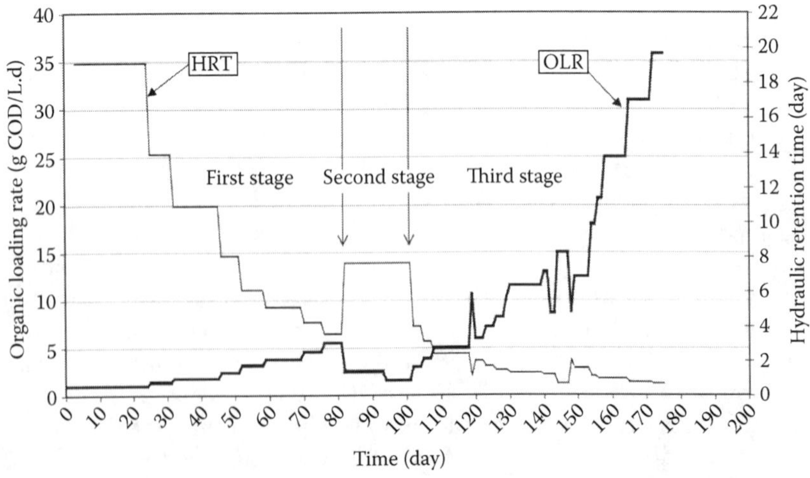

FIGURE 13.5
The evolution of the hydraulic retention time and of the organic loading rate with time.

3. During the third stage, days 102–180, the HRT was rapidly decreased from 7.7 days to a minimum of 0.7 day and the OLR increased from a value of 1.6 up to 36 g/L day. During the first 81 days, when the reactor was fed with a low and slowly increasing OLR, the soluble COD of the treated effluent remained low with values of less than 3.1 g/L (Figure 13.6). The VFA concentration always represented less than 1.6 g COD/L. The slight increase in the soluble COD at the end of period 1 (days 60–81) was linked to the use of a new vinasse in which the nonbiodegradable fraction (1.5 g/L) was higher than that of the previous one (0.55 g/L). During this period, the COD removal efficiency was always more than 85%. During the second stage, the OLR remained low, and at the end of this period, the soluble COD was very low with 0.85 g/L and the VFA concentration was nil. At the third stage, during the rapid increase of the OLR from 1.3 to 36 g COD/L day in 78 days, the global COD removal efficiency was always good with an average value of 85% and the soluble COD at the outlet was always less than 5.5 g/L. Up to an OLR of 12.5 g COD/L day (days 102–153), the average values were 1.4 g/L for the soluble COD and 0.3 g/L for the VFA concentration. The purification efficiency was very good with 89% COD removal on average. For a higher OLR and up to 31 g COD/L day, the purification efficiency decreased slightly but was still more than 80% with an average value of 83%. The soluble COD was always less than 3.5 g/L and the VFA concentration was 1.15 g COD/L on average.

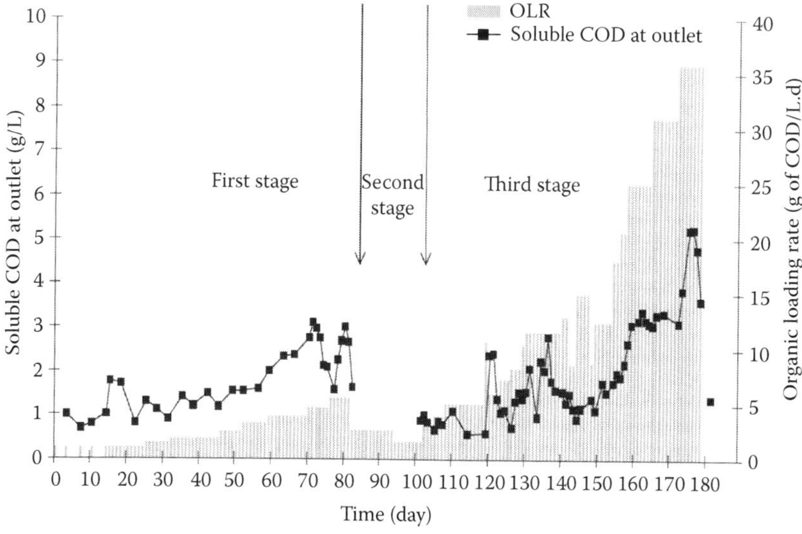

FIGURE 13.6
The evolution of the soluble COD at the outlet and of the organic loading rate with time.

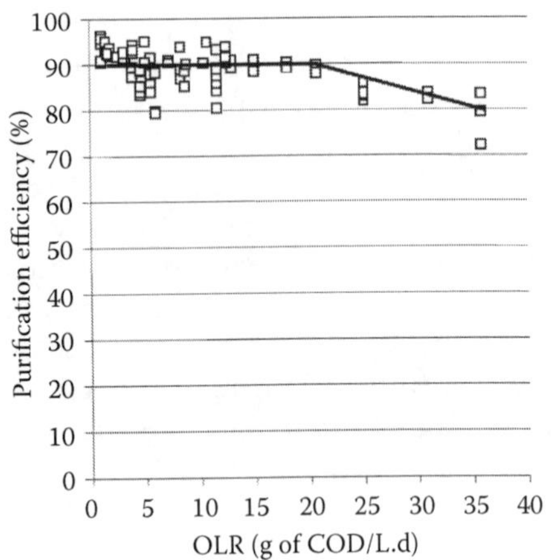

FIGURE 13.7
Purification efficiency as a function of OLR.

The most important data obtained during this experiment are summarized in Figures 13.5 and 13.6, which represent the evolution of the purification efficiency with the OLR (Figure 13.7) or with the HRT (Figure 13.8). These figures clearly show that, when treating a concentrated effluent such as a distillery vinasse, an anaerobic fixed bed with Bioflow 30 can be operated at high OLRs of more than 30 g COD/L day and at a low HRT of less than 1 day with a purification efficiency of more than 80%. It is important to emphasize that the maximum loading rate obtained in this study (>30 g COD/L day) is quite high for a fixed bed reactor treating distillery vinasse,

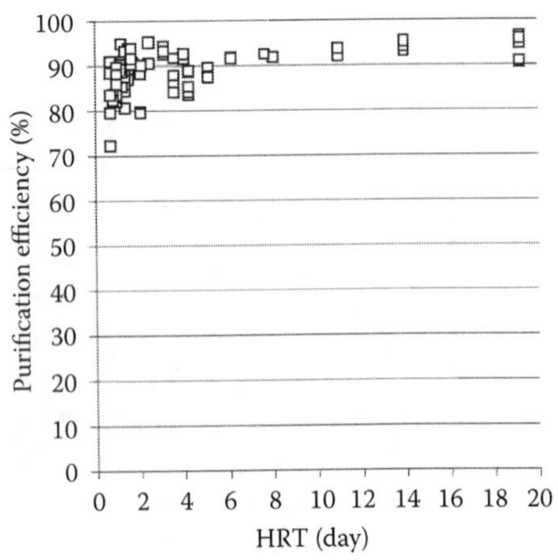

FIGURE 13.8
Purification efficiency as a function of HRT.

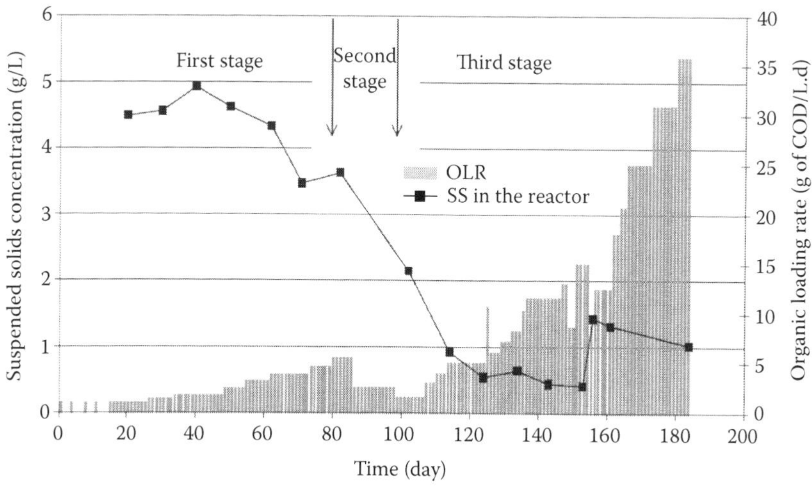

FIGURE 13.9
The evolution of the suspended solids concentration at the outlet of the reactor and of the organic loading rate with time.

which showed that Bioflow 30 was an excellent support that could be used in anaerobic digestion. Indeed, an anaerobic fixed bed containing cloisonyle, which is a well-ordered medium made up of PVC tubes of 102.5 mm in diameter divided into 14 canals with a specific area of 180 m^2/m^3, and treating a distillery vinasse could only reach an OLR of around 14 g COD/L day (Ouichanpagdee et al. 2004). Furthermore, Malina and Pohland (1992) reported that full-scale fixed bed processes have been generally designed for OLRs of up to 16 g COD/L day.

The SS concentration in the reactor was regularly measured to follow the evolution of the biomass in suspension in the reactor (Figure 13.9). During the first stage of the experiment (the first 81 days), the SS concentration remained high with values between 3.5 and 5 g/L. The SSs started to decrease toward the end of the first stage, indicating the washout of the free biomass. During the third part of the experiment, the SS concentration stabilized at low concentrations with values between 0.4 and 1.5 g/L.

After 2 months of operation, the floating supports were taken from the top of the reactor to the first 10 cm below the liquid surface. The first sampling of the support on day 66 showed that the quantity of solids on the supports was around 2.5 g of solids/support. Between day 66 and day 156, the fixed biomass increased by 30% with 3.2 g of solids/support on day 156. However, it was clear that the quantity of floating supports was decreasing with time and that the supports were sinking to the bottom of the reactor. On day 156, the samples were taken close to the surface and as deep as possible inside the reactor, which was of the order of 60–70 cm from the surface.

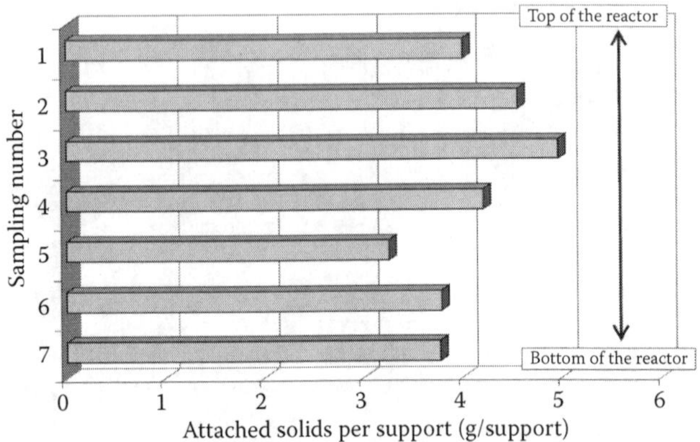

FIGURE 13.10
The evolution of the attached solids per support according to the height of the reactor.

The quantity of solids was 3.2 g on the floating supports and 4.1 g on the supports from deep inside the reactor, which represented a difference close to 30%. The distribution of the supports in the reactor did not seem to be homogeneous, and the quantity of the attached biomass was not constant from one support to another. Thus, it was not possible to make an accurate estimation of the quantity of fixed biomass just by weighing a few supports. However, the quantity of solids on the floating supports after 66 days of operation was quite high, suggesting a good aptitude of the biomass to attach onto the support. At the end of the experiment, after 180 days of operation, the total quantity of fixed biomass was quantified by weighing all the supports. The supports were taken out of the reactor from top to bottom, in batches of 40 supports for the first 5 samplings and of 50 and 67 supports, respectively, for the last 2 samplings (Figure 13.10). The average biomass attached to the supports was not constant and varied between 3.2 and 5 g of solids/support. For the deeper supports in the reactor, the attached biomass was the lowest. The decrease in the attached biomass on the supports close to the bottom of the reactor could be attributed to the detachment of the biofilm because of the high liquid velocity generated near the vicinity of the pump. The total quantity of attached biomass in the reactor was 1300 g. The concentration of attached biomass was then 57 g/L, and the biomass in the suspension concentration was only 1 g/L. When emptying the reactor, it was clear that the supports at the bottom of the reactor were adhered together and that it could no longer be fluidized because of the small diameter of the reactor and the low flow rate of the mixing pump. In these conditions, the bottom of the reactor behaved like an anaerobic filter with a stationary support. A visual observation of the media showed a biofilm formation on the surface of the support, but the biomass was also entrapped inside the support, filling most of the voids (Figure 13.11). Similar results were reported by Young and Dahab (1983) for an anaerobic fixed bed filled with

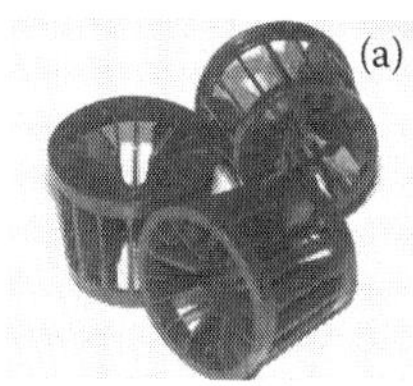

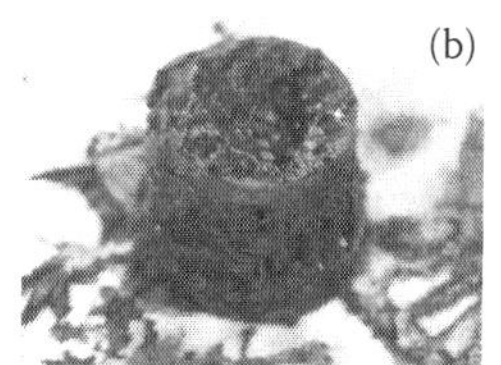

FIGURE 13.11 (See color insert)
Photographs of biological solids attached to the anaerobic fixed bed media. (a) Noncolonized support, (b) colonized support, and (c) support after heat-drying.

cylindrical Pall rings that are 90 mm long with 90 mm diameter, but with much lower loading rates.

From the OLR applied at the end of the experiment and the measurement of the total quantity of the biomass attached to the supports, it was possible to estimate the specific activity of the fixed biomass. This activity was 0.54 g COD/g of dried solids. This value is similar to the specific activity measured by Ruiz (2002) for the biomass in suspension treating sugarcane vinasses (0.52 g COD/g of dried solids) or molasses vinasses (0.48 g COD/g of dried solids/day). With cloisonyle, Ouichanpagdee et al. (2004) found a much lower activity (0.18 g COD/g of dried solids/day) due to the accumulation of mineral solids in the biofilm attached to the surface of the PVC support. Lastly, the specific activity measured in this work is significantly higher than the specific activity reported by Switzenbaum (1983) for an anaerobic fixed bed (0.4 g COD/g day), but lower than the specific activity of an expanded bed (0.8 g COD/g day) and in the lower range of the specific activity of granular sludge (Henze and Harremoes 1983). The results obtained show that the activity of the biomass attached on the support remains good and has a value quite close to that of a suspended biomass. This suggests that the entrapped biomass may play an important role in the global behavior of the reactor and that the support serves not only to create a biofilm on its surface, but also to entrap the biomass in its void space, thereby preventing it from being washed out of the reactor.

The operation of a fixed bed reactor containing Bioflow 30, a polyethylene support with a density lower than 1000 kg/m^3 and a specific area of 320 m^2/m^3, demonstrated that Bioflow 30 is a promising support for application in anaerobic digestion. Indeed, after 6 months of operation, a loading rate of more than 30 g COD/L day could be applied, while maintaining a COD removal efficiency of more than 80%.

The study of the attached biomass showed that it was possible to fix a high quantity of solids on the support. Indeed, the quantity of the biomass in the reactor was increased around five to six times compared with a reactor with a suspended biomass. The activity of the fixed solids on the supports was good with a value close to that of the SSs. It was then possible to operate the reactor with a very high loading rate (more than 30 g COD/L day) as a result of the increase in the quantity of the solids in the reactor with high specific activity. The visual observation of the supports and the specific activity of the attached solids suggested that due to their configuration, the supports entrapped a lot of solids, which played an important role in the overall performance of the reactor. This experimental work illustrated and opened vistas for different inert materials that could hold more biomass for treating high organic loading. It also showed the efficiency of a fixed bed reactor in the treatment of concentrated industrial effluents.

References

Clesceri, L.S., A.E. Greenberd, and R.R. Trussell, eds. 1992. *Standard Method for the Examination of Water and Wastewater*, 18th edn. Washington, DC: APHA (American Public Health Association), American Water Works Association, Water Pollution Control Federation.

Henze, M. and P. Harremoes. 1983. Anaerobic treatment of wastewater in fixed film reactors – A literature review. *Wat. Sci. Tech.* 15: 1–101.

Jirka, A.M. and M.J. Carter. 1975. Micro semiautomated analysis of surface and wastewaters for chemical oxygen demand. *Anal. Chem.* 47(8): 1397–1402.

Kennedy, J.L. and R.L. Droste. 1985. Startup of anaerobic down flow stationary fixed film (DSFF) reactors. *Biotechnol. Bioeng.* 27: 1152–1165.

Levenspiel, O. 1999. *Chemical Reaction Engineering*. New York: Wiley.

Malina, J.F. and F.G. Pohland. 1992. *Design of Anaerobic Processes for the Treatment of Industrial and Municipal Wastes, Water Quality Management Library*, vol. 7. Lancaster, PA: Technomic.

Ouichanpagdee, P., M. Torrijos, P. Sousbie, E. Zumstein, J.J. Godon, and R. Moletta. 2004. Anaerobic fixed bed with Cloisonyle: Study of the colonisation of the support and microbial monitoring by molecular tools. In *Proceedings of the 10th World Congress on Anaerobic Digestion*, August 29–September 2, Montréal, Canada, pp. 1899–1902.

Ruiz, C. 2002. Aplicación de digestores anaerobios discontinuos en el tratamiento de aguas residuales industriales. Escuela universitaria polítechnica. University of Sevilla, Spain.

Switzenbaum, M.S. 1983. A comparison of the anaerobic filter and the anaerobicexpanded/fluidized bed processes. *Wat. Sci. Tech.* 15: 345–358.

van den Berg, L., K.J. Kennedy, and R. Samson. 1985. Anaerobic down flow stationary fixed film reactor: Performance under steady-state and non-steady conditions. *Wat. Sci. Tech.* 17(1): 89–102.

Young, J.C. and M.F. Dahab. 1983. Effect of media design on the performance of fixed-bed anaerobic reactors. *Wat. Sci. Tech.* 15: 369–383.

Index

G

H

I

M

N

O